ADVANCES IN

GEOPHYSICS

VOLUME 7

Advances in GEOPHYSICS

Edited by

H. E. LANDSBERG

U.S. Weather Bureau
Washington D.C.

J. VAN MIEGHEM

Royal Belgian Meteorological Institute
Uccle, Belgium

VOLUME 7

1961

ACADEMIC PRESS · NEW YORK · LONDON

ACADEMIC PRESS INC.
111 Fifth Avenue, New York 3, New York

U.K. edition published by
ACADEMIC PRESS INC. (LONDON) LIMITED
17 Old Queen Street, London, S.W.1.

Library of Congress Catalog Card Number: 52–12266.

PRINTED IN GREAT BRITAIN BY
ABERDEEN UNIVERSITY PRESS LIMITED, ABERDEEN

LIST OF CONTRIBUTORS

A. D. BELMONT, *General Mills, Minneapolis, Minnesota*

DAVE FULTZ, *Department of Meteorology, University of Chicago, Chicago, Illinois*

GEORGE C. KENNEDY, *Institute of Geophysics, University of California, Los Angeles, California*

H. A. PANOFSKY, *The Pennsylvania State University, University Park, Pennsylvania*

MANFRED SIEBERT, *Geophysikalisches Institut der Universität Göttingen, Germany*

J. VAN ISACKER, *Royal Meteorological Institute, Uccle, Belgium*

FOREWORD

After an excursion into the publication of a symposium in Volume VI we return to our previous practice of presenting a volume of individual articles. The selection has been prompted, as always, by the fact that research in a given field has reached a certain plateau which would make a general review useful. For progress in science it is essential that the broader aspects of a field are occasionally surveyed. The solid progress has to be separated from the ephemeral. Various phases of a field have to be related to each other. This consolidation of knowledge is also valuable in overcoming the difficulties inherent in the wide scattering of research papers and their soaring number.

The time lags introduced by the writing, editing, printing, and distribution processes occasionally make it difficult to include the very latest citations. It has been the endeavor of editors and authors to minimize this difficulty. The gratifying frequency with which articles in our earlier volumes have been cited in the subsequent literature is a measure of the success achieved. Some of these articles are serving as precursors to books.

As this volume rolls off the presses another is already in preparation. In it we hope to cover problems dealing with the heat balance of the atmosphere, balloon exploration of the atmosphere, ozone, solar terrestrial relations, ionization in space, paleomagnetism, and other topics now in the limelight of geophysical interest.

The editors are again grateful for the advice received from members of the editorial committee and from interested colleagues in the field.

H. E. LANDSBERG
J. VAN MIEGHEM

CONTENTS

Temperature and Wind in the Lower Stratosphere

H. A. PANOFSKY

Arctic Meteorology (A Ten-Year Review)

A. D. BELMONT

Phase Relations of Some Rocks and Minerals at High Temperatures and High Pressures

GEORGE C. KENNEDY

DEVELOPMENTS IN CONTROLLED EXPERIMENTS ON LARGER SCALE GEOPHYSICAL PROBLEMS

Dave Fultz

Hydrodynamics Laboratory,* Department of Meteorology,
The University of Chicago, Chicago, Illinois

1. Introduction

Meteorology, oceanography, geophysics, astrophysics, and other similar subjects deal with objects of very large masses, energies, and spatial extents or with very long periods of time. There has, in consequence, been a strong historical tendency in such studies to rely almost solely on combinations of theoretical and observational results without direct aid from experimental work. The experimental bases of these sciences have been mainly those of general physics and chemistry that are susceptible of laboratory investigation. For example, the physical properties of materials that are assumed in geophysical discussions are based on laboratory measurements that, however, must often be extrapolated theoretically to pressures and temperatures that are not attainable in the laboratory. Many spectacular advances of the last half-century have resulted simply from extensions of the attainable experimental range for such measurements and from improved ability, since the advent of quantum mechanics, to extrapolate beyond it.

*Part of the preparation of this paper was carried out during tenure in 1957–8 of a National Science Foundation senior postdoctoral fellowship at the Cavendish Laboratory, Cambridge, and at Institutt för Teoretisk Meteorologi, Oslo. I wish to thank the Foundation for the opportunity thus afforded me.

However, many of the most central questions in sciences such as those mentioned have to do with phenomena of the largest possible scales. The broad scale occurrences more or less set the stage occupied by more restricted problems and strongly condition their formulation and interpretation. The general circulations of the atmosphere and oceans are typical examples. The space and time variations and transfers of properties associated with these circulations dominate most more local occurrences in a way that usually makes it difficult to understand local phenomena solely in themselves and in isolation. In relation to such large-scale questions, the role of direct experimentation has for a long time been very tenuous, though not quite nonexistent, in the development of scientific understanding. In the past couple of decades, however, a number of more or less independent advances relevant to this type of problem have taken place on the experimental side in several areas of study. These turn out to have a number of broad connections one with another. A sufficient number of successful applications to the natural phenomena has been established that one can see the outline of further developments that are likely to broaden the study of these geophysical problems in a very fundamental way; namely, to bring about much more of the sort of interplay between experiment, theory, and observation that continually revitalizes the growth of such sciences as physics and chemistry.

1.1 Scope and Point of View

The present review will be devoted to a survey of experimental developments mainly in meteorology and oceanography but with some attention to similar work in geology and certain other fields. An earlier discussion of the meteorological experiments is given in Fultz (1951a) and recently von Arx (1957) has given a more detailed review of the oceanographic work than we shall attempt. Other reviews will be touched on later. Our particular special interest will be to discuss the experiments that have begun to be quantitatively successful (in which, consequently, due regard has been paid to similarity requirements) in connection with medium- and large-scale phenomena as these adjectives might be understood by the ordinary observer. There are large areas of successful experimentation that, though they will be seen to be closely related to some of our topics, will be deliberately excluded. An example is the extensive body of hydraulic model work on flow, waves, tides, etc., in rivers, estuaries, and bays that has been on an engineering basis since the time of Reynolds and Froude. The scale of such experimental models relative to the prototypes is rarely much smaller than the order of 1/10,000 and, while some of the experiments to be considered are at larger scale than this, some of the most interesting and significant have working scales as small as $1/10^7$ and less.

A thread common to almost all the topics and experiments to be touched on is that the point of view is that of macroscopic continuum mechanics. The tacit hypothesis is that some form of the equations of hydrodynamics, thermodynamics, or elasticity governs the phenomena in question. The difficulties of verifying or making effective use of this basic assumption are of two principal kinds. First, there are generally uncertainties as to how variables entering the equations are to be defined in terms of observables and what form the terms representing certain physical effects should have. For example, what stress *vs.* strain or rate-of-strain relation is appropriate? Secondly, and more seriously, the mathematical complexities of the governing equations for realistic physical situations and boundary conditions are so great in these fields that the growth of a body of relevant analytic solutions has been very slow. The consequences of a lack of theoretical solutions have been threefold: the physical effects in the natural phenomena have been difficult to identify quantitatively; experiments for their own sakes have lagged because of the same lack of firm theoretical prediction and control; and experiments regarded as models of some phenomena have often had insufficient quantitative similarity to be of any real scientific value. At least part of the reason for the increasing success of the experimental work we will discuss lies in broad advances beyond the subject matter of classical hydrodynamics and elasticity in the last fifty years and in, what will be of even greater significance in the future, the recent widespread and systematic use of high-speed computers for theoretical work that still remains beyond the capacity of normal analysis.

In connection with the large-scale flow fields of the atmosphere and ocean that are of the principal present interest, the physical and theoretical difficulties are associated mainly with the features that distinguish the "physical hydrodynamics" of V. Bjerknes from the classical hydrodynamics of perfect, homogeneous fluids and with the boundary layer viscous effects associated with the name of L. Prandtl. The essential feature of the resulting flow fields is that they are rotational and not irrotational or potential flows. Several major factors contribute strongly to this characteristic. One is the dominant rotation of the earth operating partly indirectly through friction at the earth's surface. Another is the direct viscous friction that leads to regions of concentrated variation of the boundary layer type. A third is the spatial and temporal density variations that, in association with the gravity field, both directly make possible various sorts of rotational internal wave motions and through heat or mass exchanges drive or modify the circulations (free or forced convection situations). Finally, the consequence of a wide variety of types of instability of the circulations is that there is always more or less fine-scale turbulent variation relative to the actually observable fields and the effects of this fine-scale rotational motion (which is effectively unobservable at most positions and times) are wide and various even though often neglectable for some purposes.

2. Similarity Parameters

Because of the extreme difference in scales of the natural phenomena and of practicable laboratory experiments, the possibility of doing significant experimental work on such effects depends upon being able to establish essential dynamical similarity of the experimental systems and the prototypes; that is, on being able to produce identity of the field equations and boundary conditions in the two cases by linear transformations of the mass, geometrical, time, electromagnetic and thermal variables. This is generally not possible with strict completeness so that considerable care is necessary to insure correctness at least of a few dominant effects. The techniques of dimensional analysis and systematic use of dimensionless variables are the appropriate ones. For reasons of space these cannot be fully described and it will be assumed that the reader is aware of the general principles as discussed, for example, by Bridgman (1931), Birkhoff (1950), Langhaar (1951), and Focken (1953). Aside from individual papers noted later, fairly comprehensive discussions of the application to geophysical problems are given in Fultz (1951a), von Arx (1957), Raethjen (1958), Faller and von Arx (1959) and Fultz *et al.* (1959).

Much of the similarity problem is connected with the dimensionless parameters which appear in the field equations or boundary conditions when they are put in some suitable nondimensional form. We will discuss some of the more important such parameters connected with the previously mentioned physical effects in meteorology and oceanography. For definiteness, imagine a system consisting of a thin layer of fluid such as is involved in the atmospheric or oceanic problem. One of the most crucial parameters for these two subjects is the Rossby number introduced in Fultz (1951b; Dryden *et al.*, 1932, p. 95) that is associated with the rotation of the layer as a whole with the earth. A kinematic form of this parameter can be taken as simply $R_0^* \equiv V/L\Omega$ where V is a representative velocity relative to a rotating coordinate frame, L a horizontal length scale, and Ω the basic rotation. The distinguishing feature of large-scale motion problems for the natural fluid layers is that they are slow in relative coordinates, the kinematic Rossby number being of order 10^{-2} to 10^{-1} in meteorology and of order 10^{-3} in oceanography. Among several possible interpretations, R_0^* can be considered an estimate of the ratio of relative inertia force to Coriolis force and the above small values are associated with the characteristic quasigeostrophic properties of the current systems. In many of the examples to be considered, a suitable Rossby number appears from the experimental evidence to be the most important similarity parameter and fortunately one whose natural range is easily attained experimentally.

The effects of gravity in thin layers of the character of the atmospheric and oceanic ones are ubiquitous but for our purposes are mainly important for making tidal, internal, and external wave motions possible and, in thermal or mass convective velocity fields, providing the quasihydrostatic pressure field and resulting "buoyancy" forces that drive the motion. The nondimensional parameters characterizing the gravitational effects may be taken in a variety of forms. In the convectional situations (mainly for liquids) we shall discuss, two convenient forms are the thermal Rossby number

$$R_{0T}^* \equiv g \frac{E_r^* \delta}{f r_0 \Omega (\Delta r)} \tag{2.1}$$

and a stability number

$$S_z^* \equiv g \frac{E_z^* \delta}{f^2 (\Delta r)^2} \tag{2.2}$$

Here δ is depth of the layer, f the Coriolis parameter (2Ω for a rotating disk), r_0 a reference radius, Δr a reference horizontal (or radial) width, and E_r^* and E_z^* are, respectively, representative total fractional expansions across the horizontal (radial) width and the total depth of the layer. Thus, for example, E_z^* is a suitable volume expansion coefficient times a representative vertical temperature difference.

Here R_{0T}^* may be interpreted as the total meteorological thermal wind difference top-to-bottom, u_T, in units of $r_0\Omega$ and is essentially a measure of the total vertical variation of the quasi-hydrostatic horizontal pressure gradients due to the horizontal density variation. Similarly, S_z^* is a measure of the vertical gravitational stability and can, for example, be interpreted from the point of view of Archimedes' principle as a representative total vertical buoyancy acceleration (for a parcel displaced quasi-statically top to bottom) in units of $f^2 \Delta r^2/\delta$. The specific forms of these two parameters arise in theoretical stability analyses of Kuo (1954, 1955, 1956a,b,c, 1957) which we cannot summarize in detail.

The Richardson number which in some types of problems is the most convenient characteristic measure of relative stability is proportional to $S_z^*/(R_{0T}^*)^2$ (Batchelor, 1954). In internal gravity wave problems, where there is a suitable characteristic current speed V, an internal Froude number

$$F_i^* \equiv V/(gE_z^*\delta)^{1/2} \tag{2.3}$$

(Long, 1953c) is an appropriate parameter since $(gE_z^*\delta)^{1/2}$ is an estimate of internal wave speeds. All of these parameters, and others we will see later in the review of bubble problems, are essentially Froude-type parameters and characterize various aspects of the gravitationally imposed quasi-hydrostatic pressure fields. It is further an exceptionally important circumstance that, for all the experimental work we are reviewing, the natural range of any of the Froude-type parameters has turned out to be easily attainable in the laboratory. This is undoubtedly an important element in the degree of success in strict quantitative modeling that has been attained.

Perhaps the largest group of parameters is that connected with the various diffusion effects: diffusion of momentum by viscosity, of heat by thermal conduction, and finally of mass. The most important historically are the various forms of Reynolds number

$$R_e^* \equiv VL/\nu \tag{2.4}$$

where ν is the kinematic viscosity, but those connected with heat transfer are equally essential in the principal group of convection problems that we consider. We should emphasize at the outset that it is in this group of parameters that the present position with respect to model similarity is least satisfactory. For example, in meteorological experiments, direct Reynolds number equality fails to be achieved for practical reasons by factors around 10^6. Similar discrepancies are met in the Peclet number

$$P_e^* \equiv VL/\kappa \tag{2.5}$$

where κ is the thermometric conductivity, and in other similar parameters. Particularly in the case of the Reynolds number, the practical difficulties in meteorological and oceanographic model experiments were recognized very early (even with Helmholtz, 1873) and it is very probable that this was a major factor in the long failure to pay serious attention to them. Partly in view of this inconclusive state of present knowledge, it is worthwhile to collect a few comments on the parameters of this type, using the Reynolds number as an example, before discussing the individual types of experiments.

It will be sufficient to consider the dimensionless equations of motion of a homogeneous incompressible fluid layer which, with r_0 as length unit and $r_0\Omega$ as velocity unit, take the form

$$\frac{\partial \mathbf{V}}{\partial t} + \nabla . (\mathbf{VV}) + 2\hat{\mathbf{\Omega}} \times \mathbf{V} = \frac{1}{R^*}\nabla^2\mathbf{V} + \text{(pressure and gravity terms)} \tag{2.6}$$

where $(\mathbf{VV})$ is the relative momentum transport tensor, $\hat{\mathbf{\Omega}}$ is a unit vector in the direction of the system absolute rotation, and $R^* \equiv \Omega r_0^2/\nu$ is a rotation Reynolds number appropriate to the other units. All variables are dimensionless with the time t in units of Ω^{-1}, a choice which makes the time for one revolution correspond to one day.

In most practical situations the Reynolds numbers are sufficiently high that the strong viscous influences are confined to boundary layer regions especially over the bottom surface (meteorology) and the air-water interface (oceanography). By an analysis exactly comparable to that for aerodynamic boundary layers (Goldstein, 1938; Schlichting, 1955) the depth δ_v of the layer of viscous influence (in r_0 units) is

$$\delta_v \sim R^{*-1/2} \tag{2.7}$$

Even more suitably (Fultz *et al.*, 1959) the ratio of δ_v to total depth δ

$$\delta_v/\delta \sim \delta^{-1}R^{-1/2} \sim R_\delta{}^{*-1/2} \tag{2.8}$$

where $R_\delta{}^* \equiv \Omega\delta^2/\nu$ is another form of Reynolds number which is appropriate when the principal frictional effects arise from velocity variations in the vertical (z) direction.

This estimate of $\delta_v/\delta \sim R_0{}^{*-1/2}$ is precisely the same (to within a numerical factor) as that given by the Ekman layer solution (Brunt, 1939, p. 252). Now taking molecular values of ν, values of $R_\delta{}^*$ for the atmosphere (troposphere) are about 4×10^8, and for the ocean are about 8×10^8 using an average depth of 4 km. Somewhat smaller values would apply if a local component along the vertical is used for Ω. Typical experimental values that cannot be greatly exceeded are, say 10^3 to 10^4. Analogous relations, expressions, and values occur for the bottom thermal boundary layers with κ replacing ν.

In spite of the value discrepancies in these and other comparable diffusion parameters, a variety of experiments have already led to quite detailed quantitative comparisons with atmospheric and oceanographic prototype phenomena. Considering first the Reynolds number, there appear to be two principal alternatives (not necessarily exclusive): the first is that the values, while very different in model and prototype, are sufficiently high to have reached an asymptotic state in which the dependences of the phenomena on Reynolds number are very weak or absent. An important variety of evidence for this interpretation is the following: in several types of experiments, regular or irregular motions varying in space occur which are convenient to analyze by the techniques common in turbulence theory; separation of the field variables into means and deviations, etc. When averaging is over space intervals of the order of r_0 (e.g., around latitude circles) and the velocities are measured in $r_0\Omega$ units, it has been found that both of the momentum transport terms, $\nabla \cdot (\overline{\mathbf{V}}\overline{\mathbf{V}})$ of the averaged flow and $\nabla \cdot (\overline{\mathbf{V}'\mathbf{V}'})$ of the fluctuations that arise in equation (2.6), in certain meteorological experiments correspond quite quantitatively to atmospheric values (Starr and Long, 1953; Corn and Fultz, 1955; Fultz *et al.* 1959). In eddying motions, one can usually expect contributions from $\nabla \cdot (\overline{\mathbf{V}'\mathbf{V}'})$ far to exceed those from $1/R^*\nabla^2\overline{\mathbf{V}}$ and one thus feels safer in the hypothesis initially mentioned of assuming the phenomena independent of R^* in the range of the model to prototype values.

The second alternative, for which there is also some evidence both in meteorological and oceanographic experiments, is that Reynolds number similarity is attained by having molecular laminar friction in the models play the role of medium- and small-scale eddy friction in the prototypes. Thus, for the atmospheric case, with an eddy viscosity of 5×10^4 cm^2/sec, $R_\delta{}^*$ goes to 2000, and for the oceanic case, with a vertical eddy viscosity of 500 cm^2/sec, $R_\delta{}^*$ also goes to 2000. In fact, a qualitative inspection of a range of meteorological experiments suggests that the best values are around $R_\delta{}^* \sim 1000$ and of the oceanographic ones (von Arx, 1957) around $R_\delta{}^* \sim 1600$. The fact that these values mean in each case that the layer of molecular or eddy friction is, respectively, 1/30 and 1/40 as a fraction of the total depth makes the whole idea considerably more plausible though the evidence is still not decisive and any final appraisal must await a good deal of further work. The same alternative is described by Long (1957) in connection with model experiments on mountain waves.

Especially in the free convection experiments of a meteorological nature, parameters of the same general type as the Nusselt, Grashof, and Rayleigh numbers of technical convection problems occur. For example,

$$N_u{}^* \equiv \frac{H}{kA\Delta_r T(\Delta r)^{-1}} \tag{2.9}$$

where H is the heat transfer per unit time across area A, k is the thermal conduction coefficient, $\Delta_r T$ is a radial temperature difference (in the direction of transfer), and Δr is the radial distance corresponding to $\Delta_r T$. Here, $N_u{}^*$ is a measure of actual (dominantly convective) heat transfer in units of the pure conductive transfer corresponding to a similar temperature field,

$$G_r{}^* \equiv \frac{gE_r{}^*\delta^3}{\nu^2} \tag{2.10}$$

and

$$R_a{}^* \equiv \frac{-gE_z{}^*\delta^3}{\nu\kappa} = \frac{-gE_z{}^*\delta^3}{\nu^2} \cdot P^* \tag{2.11}$$

where $P^* \equiv \nu/\kappa$ is the Prandtl number and other quantities have been defined previously. (The Rayleigh number is usually taken positive for vertically unstable stratification.) The $G_r{}^*$ and $R_a{}^*$ are parameters that essentially characterize the balance of buoyancy against viscous and thermal conductive dissipative effects. Parameters of this type can often be interpreted as a kind of Reynolds number in which the velocity scale is determined by the buoyancies (Batchelor, 1954). Just as with Reynolds numbers, the attainable experimental values lie in general far below those of typical natural phenomena as shown in Table I. The evidence here is so far less definite, but there are

TABLE I. Typical characteristic values and nondimensional parameters

	Cylinder convection experiment	Atmosphere troposphere circulation		Oceanic general and surface circulation	
	(160156)	Molecular coefficients	Eddy coefficients	Molecular coefficients	Eddy coefficients
Ω	$0.40\ sec^{-1}$	$7.3 \times 10^{-5}\ sec^{-1}$		$7.3 \times 10^{-5}\ sec^{-1}$	
r_0	19.5 cm	6.4×10^{8} cm		6.4×10^{8} cm	
$r_0\Omega$	7.8 cm/sec	4.65×10^{4} cm/sec		4.65×10^{4} cm/sec	
δ	4.2 cm	11.7 km		4 km	
ν	$7.1 \times 10^{-2}\ cm^2/sec$	$0.23\ cm^2/sec$	$5 \times 10^{4}\ cm^2/sec$	$1.5 \times 10^{-2}\ cm^2/sec$	$5 \times 10^{2}\ cm^2/sec$
P^*	4.8	0.76	1.5		
R_0^*	0.11	0.14		0.003	
R_{0T}^*	0.03_7	0.04		5×10^{-5}	
S_z^*	0.03_2	0.04		8×10^{-5}	
R_i^*	94	70		1×10^{5}	
F_i^*	0.09	0.19		0.03	
R_e^*	500	3×10^{10}	8×10^{4}	4×10^{9}	1×10^{5}
R_δ^*	990	4×10^{8}	2000	8×10^{8}	2000
δ_v/δ	1/31	1/20,000	1/44	1/28,000	1/44
N_u^*	(4200)	1×10^{11}	6×10^{5}		
G_r^*	1.57×10^{6}	2×10^{21}	4×10^{10}		
R_a^*	-1.32×10^{7}	-4×10^{21}	-2×10^{11}		

indications again of some validity in successful experiments for a pair of alternatives very similar to those outlined in the case of Reynolds number similarity. For example, in partitions into mean and eddying flow the contributions locally in the heat equation due to heat transfer by convection occur in terms like $\nabla \cdot (\bar{T}\bar{\mathbf{V}})$ and $\nabla \cdot (\overline{T'\mathbf{V}'})$ very much resembling the momentum transfer situation. In some meteorological convection experiments there is evidence (Riehl and Fultz, 1957, 1958) for relative contributions of these convective heat flow terms to the local thermal balance that have quite similar distributions in space to those estimated for typical atmospheric motions in spite of the fact that Nusselt numbers are much lower in the experiments and the convective transfer consequently is much less strongly dominant over conductive contributions. [In the particular case mentioned $N_u{}^*$ for horizontal (radial) transfer was only about 60.]

The foregoing does not exhaust the list of parameters that may be important in even purely hydrodynamic problems. Further examples will be touched on in connection with specific types of experiments; in particular, those that also involve electromagnetic phenomena.

3. Large-Scale Phenomena

In discussing the experimental work on large-scale natural phenomena we will take up first the meteorological and oceanographic work in two parts: one involving properties of homogeneous (barotropic) fluids and one involving density differences and convection (baroclinic fluids); second, geological work on the elastic and plastic behavior of the earth; and third, work connected with cosmical electromagnetic phenomena such as the aurora phenomena.

3.1. Rotational Influences, Homogeneous Fluids

In their application to large-scale meteorology and oceanography, the experiments that come first in point of principle, though it happens not in point of time, are some dealing with ideal properties of a layer of uniform incompressible (or autobarotropic) fluid. They have close connections with corresponding theoretical calculations that, of course, are much easier to obtain for such a simple medium. All those in question, whether for frictionless or viscous fluids, involve strong effects of the absolute rotation of the fluid layer and, in meteorological terms, more or less highly geostrophic conditions. This means in essence that the term $2\hat{\Omega} \times \mathbf{V}$ in equation (2.6) dominates on the left and a principal condition for this is that suitable Rossby numbers be small. If velocities are measured in $r_0\Omega$ units (i.e., $\mathbf{V}/r_0\Omega$) as in (2.6), in general, values below 0.1 are sufficiently small. The relative inertia terms $\mathbf{V} \cdot \nabla\mathbf{V}$ will then usually be of order 10^{-2} or less.

The effects of this quasi-geostrophic property are very far-reaching. In the instances to be noted they can be described quite simply as arising from the stabilizing effects of the rotation and the properties of the vortex tubes of the absolute motion. The Helmholtz vortex theorems (Lamb, 1932, p. 202 ff.) in the variables used in equation (1) follow from

$$\frac{d\boldsymbol{\zeta}_a}{dt} = \boldsymbol{\zeta}_a \cdot (\nabla \mathbf{V}_a) \tag{3.1}$$

where

$$\boldsymbol{\zeta}_a \equiv 2\hat{\boldsymbol{\Omega}} + \nabla \times \mathbf{V}$$
$$\mathbf{V}_a \equiv \hat{\boldsymbol{\Omega}} \times \mathbf{r} + \mathbf{V}$$

$\mathbf{r}$ is the position vector and $\hat{\boldsymbol{\Omega}}$ is a unit vector in the direction of the rotation axis. The theorems state, in the absence of friction and density differences, that the vortex tubes (Fig. 1) move with the fluid and have constant strength $\bar{\zeta}_a \cdot A$ both along an arbitrary tube and in time. (Here, A is the cross-sectional area of the tube and $\bar{\zeta}_a \cdot A = \int_{\sigma_1} \boldsymbol{\zeta}_a \cdot d\boldsymbol{\sigma}$ where σ_1 is any simple surface spanning the tube and $d\boldsymbol{\sigma}$ is the normal vector area element of σ_1.)

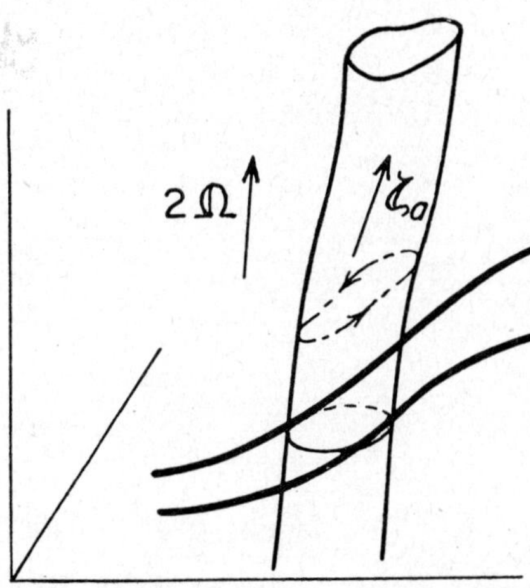

FIG. 1. Absolute vortex tube (light lines), relative streamlines (heavy curved arrows).

If the strength $\bar{\zeta}_a \cdot A$ is constant along a tube and in time while, at the same time, $\mathbf{V}$ always and everywhere remains small (say 0.1) in the coordinate system rotating at 2Ω, then some reflection shows that, on the whole over appreciable volumes, the vortex tubes cannot tilt appreciably from the $\hat{\boldsymbol{\Omega}}$ direction or undergo substantial fractional changes of cross-section area (Taylor, 1917, 1921). If either type of evolution were to occur, the vorticity of the relative motion would have at least one component with values not small compared to 2 and over appreciable volumes this would imply $\mathbf{V}$'s in $r_0\Omega$ units comparable to, rather than much smaller than, 1. The consequence is, that for slow relative motions of the above kind (quasi-geostrophic), the rotational stability expresses itself as a strong tendency for motions to occur, if it is at all possible, which merely translate the vortex tubes parallel to themselves. They behave in many respects elastically, showing distinct resistance to bending and stretching deformations. The experimental instances of this tendency are extremely striking and have been studied in a number of theoretical and experimental papers since the initial work of G. I. Taylor (1917, 1921, 1922, 1923) and Proudman (1916). (A recent review of

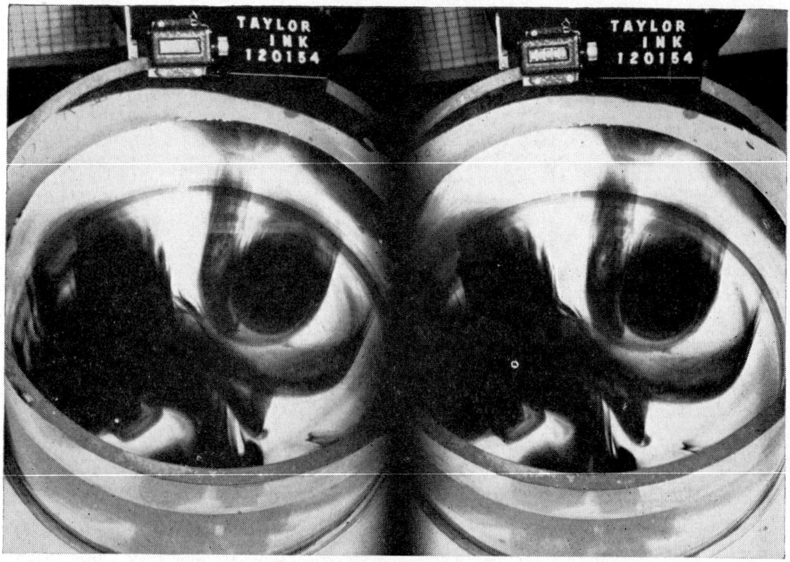

FIG. 2. Oblique stereoscopic photograph pair of Taylor ink walls in a deep rotating cylinder of water (120154–6). (*Note*: The stereoscopic effect may be obtained by placing a piece of stiff paper 10 to 15 in. long edgewise along the division between the left and right photos. Be careful to orient the eyes parallel to the horizontal and in such a position that the paper prevents the left eye from seeing the right photo and vice versa. The eyes should be trained as though observing a distant object.) Note the thinness and accurately vertical arrangement of many of the ink filament walls. The detailed pattern depends on the accidents of initial pouring of the inky water. The walls develop rapidly as the relative currents become slow. Conditions: liquid: water, outer radius 14 cm, depth 13 cm, rotation 3.0_8 sec^{-1}, kinematic $R_0{}^*$ probably around 1/2%, $R_\delta{}^* = 5.3 \times 10^4$ $\delta_v/\delta = 1/230$.

this work has been given by Squire (1956). (See also Ekman (1923)). Perhaps the most striking of Taylor's experiments was one in which he showed that if a small sphere was towed slowly across a rotating rectangular tank it carried with it a cylinder of fluid while the vortex tubes outside underwent a two-dimensional motion as though a solid cylinder were the moving object

(Taylor, 1923). A second very impressive phenomenon occurs if one has a rotating disk of liquid of roughly uniform depth[1] and disturbs it in an arbitrary manner by inserting a small quantity of ink; the ink becomes distributed rapidly in the axial direction (almost a squeezing between the vortex tubes). These walls parallel to the rotation axis remain so during the subsequent slow motions (Taylor, 1921; Long, 1954a; Fultz, 1956a). Figure 2 is a stereophotograph pair of such an experiment in which the ink was initially poured in and

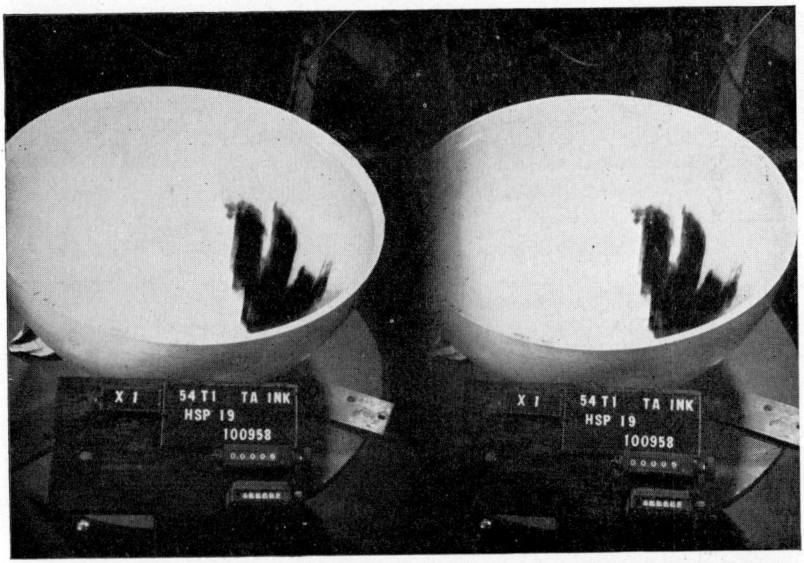

FIG. 3. Oblique stereoscopic photograph pair (see note with Fig. 2) of Taylor ink walls in a rotating hemisphere filled to the equator with water (100958–1–5). Initial stages are qualitatively similar to those of the case in Fig. 2. Ultimately, the ink columns and walls tend to flatten and line up along zonal circles as a result of resistance of the vortex tubes to lengthening in nonzonal motions. Conditions: liquid: water, radius 29 cm, rotation 2.3_0 sec^{-1}, kinematic R_0* similar to Fig. 2.

is now undergoing a very slow two-dimensional, highly geostrophic motion. In meteorological terminology, this type of parallel translation of the vortex tubes occurs because the thermal wind is zero.

Still a further ramification of this stability occurs when the initially undisturbed fluid is not of uniform depth but, for example, is contained in a complete hemisphere so that the individual absolute vortex tubes vary in length from r_0 on the axis to zero near the equator. Then, if ink is poured in rather vigorously, a situation very like Fig. 2 develops first but very soon the individual walls become distorted and tend to line up in arcs along the latitude circles, as appears in Fig. 3. The motions become rapidly very nearly zonal

[1] This experiment is very easy to arrange on a $33\frac{1}{3}$ rpm phonograph turntable.

because of the resistance of the vortex tubes to the changes of length that would be required by displacements in the radial direction.

In passing we may comment that, for motions in which the rotation effects are strong, this vortex tube stability makes it possible to get around another of the old objections to meteorological and oceanographic model experiments. This is the problem that geometrically similar fluid layers are so thin that the practicable Reynolds number values are hopelessly low. It is necessary to exaggerate the model in the vertical and this may be expected to throw vertical motions out of scale, etc. What now has been found to occur experimentally is that, if Rossby numbers are suitably adjusted, the vortex tube stability suppresses vertical motions relative to the horizontal and couples upper to lower layers of the fluid in about the correct degree. Quantitative correspondences in vertical velocities can be established (Riehl and Fultz, 1957; Fultz *et al.*, 1959) and depths can be varied through a wide range, as may be most convenient for other purposes, without changing the essential nature of the phenomena.

3.1.1. The Rossby β-Parameter and Long Waves. The most important experimental developments in this area of homogeneous fluids are some in which quantitative relations have been established between the experimental results and full theoretical solutions. In both the meteorological and oceanographic cases, it happens that these have had to do with theories that depend on the horizontal (latitude) variation of the Coriolis parameter $f \equiv 2\Omega \sin \phi$ where ϕ is latitude, and incidentally the interpretation of these in terms of the vortex tubes has helped to illuminate the still active problem of geometrical distortion in these models from spherical to other shapes.

The meteorological case is that of the famous Rossby long waves (Rossby, 1939) which are a two-dimensional perfect fluid wave motion depending on $\beta \equiv \partial f / r_0 \partial \phi$, the rate of variation of f. In spite of the simplicity of these motions, they have been shown to be closely related to some of the major atmospheric perturbations, and their study has formed the starting point of some of the major theoretical and practical meteorological developments of the last two decades. They essentially arise from the absolute rotational stability and from the dynamical law, $(f + \zeta) =$ individual constant, can be interpreted as propagating relative vorticity patterns ζ in accordance with meridional motions that vary f (ζ is the normal component of the relative vorticity $\nabla \times \mathbf{V}$). This solution was extended by Haurwitz (1940) to the case where two-dimensional motion on a spherical surface is involved, the stream function of the motion being given by tesseral spherical harmonic functions. It was noticed by Long (1952) that the quantitative vorticity changes expected in this theory are precisely those which are calculated in middle and high latitudes for a spherical shell of a small depth δ and radius r_0 when one assumes that the vortex tubes are displaced in an axiparallel way. They consequently are forced to change length nearly proportionally to $\delta/\sin \phi$. The important preceding experimental discovery by Long was that,

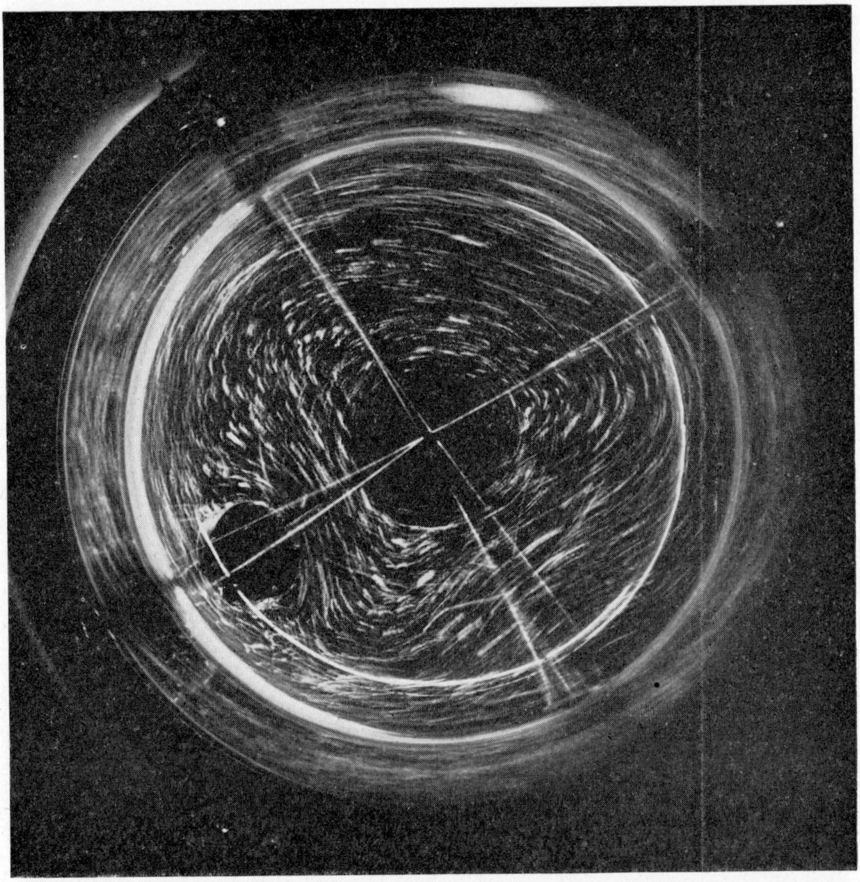

FIG. 4. Time-exposure (streak) photograph (260351–3) of a Rossby-Haurwitz barotropic wave motion with wave-number three in a rotating hemispherical shell of liquid. The exciting obstacle is a small circular cylinder of radius 10° latitude centered at about 45° latitude. The motion is seen in a coordinate system rotating with the obstacle and looking along the rotation axis toward the equatorial plane. Relative direction of flow of the liquid and of the absolute rotation is counterclockwise. The wave motion is stationary with respect to the circular obstacle. Conditions: liquid: water, inner radius 9.2 cm, outer radius 10.8 cm; absolute rotation of obstacle 7.0 sec^{-1}, relative zonal current $\omega_r/\Omega \leqslant 0.11$ (from Fultz and Long, 1951).

in a rotating spherical shell of moderate depth, one can generate Rossby waves by rotating a small object at an independent, somewhat slower, absolute rate. The experimental evidence suggests that the motion tends to preserve orientation of the absolute vortex tubes and the above comment establishes the connection with the two-dimensional theories of Rossby and Haurwitz. In addition, since the density is uniform, a consideration of the governing equations shows that the hydrostatic part of the pressure can be subtracted out completely and this, eliminating also the gravity potential, makes the experiment as a model independent of the difference of the gravity field from the earth prototype (Fultz, 1951b). Figure 4 shows an example of a stationary three-wave pattern in a westerly current relative to a circular obstacle of 20° latitude diameter. Several descriptions have been

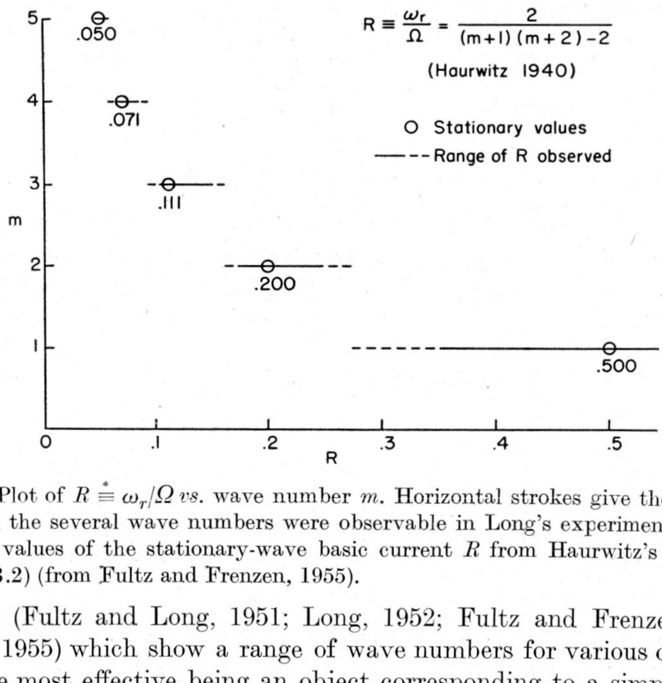

FIG. 5. Plot of $R \doteq \omega_r/\Omega$ *vs.* wave number m. Horizontal strokes give the R ranges over which the several wave numbers were observable in Long's experiments. Circled points are values of the stationary-wave basic current R from Haurwitz's frequency equation (3.2) (from Fultz and Frenzen, 1955).

published (Fultz and Long, 1951; Long, 1952; Fultz and Frenzen, 1955; Frenzen, 1955) which show a range of wave numbers for various obstacles, one of the most effective being an object corresponding to a simple north-south mountain range.[1] The principal quantitative evidence for the identification of these experimental waves with Rossby-Haurwitz waves is the agreement of the ranges of relative current speed, ω_r/Ω, in which certain wave numbers were observed in Long's experiments, with Haurwitz' frequency equation for stationary waves with an equatorial node only:

$$(\omega_r/\Omega)_s = \frac{2}{(m+1)(m+2)-2} \tag{3.2}$$

where m is the integer wave number. Figure 5 from Fultz and Frenzen (1955) shows these results which appear quite decisive at the higher wave numbers.

[1] Such an obstacle shows easily recognizable local mountain effects, especially a ridge in the motion near the object and a strong lee trough.

3.1.2. The Ocean Gulf Stream Problem. The oceanographic developments in this class of rotating, homogeneous fluid problems began actively in connection with the Munk–Stommel (Stommel, 1948; Munk, 1950) theory of the general wind-driven ocean circulations which again depends essentially on the β-parameter whose importance was first emphasized for the ocean problem by Ekman (1923). Treating either a homogeneous layer (Stommel) or vertically integrated quantities (Munk), these authors showed by explicit solutions that the general features of the anticyclonic ocean circulations with intense currents (e.g., the Gulf Stream in the North Atlantic) on the western sides of the oceans could be approximated by a purely wind-driven circulation. The mechanism was that, in the vertical component vorticity equation without inertia terms, the curl of the wind stress must be balanced by the curl of the eddy frictional force and the vorticity changes due to β for north-south currents. On the eastern sides of the ocean, the β and wind terms roughly balance with broad slow currents while on the western side they are of the same sign and can only balance with intensified friction, i.e., intensified currents and vorticity. Von Arx (1952) set out to test experimentally the broad features of this theory and to see whether clues to finer details of the ocean circulation could be so obtained. He constructed a generalized coastal pattern of the Northern Hemisphere in a 2-meter-diameter paraboloid of 0.5-meter focal length (equilibrium rotation rate 3.13 rad/sec) so as to rotate a uniform density layer of water and then apply zonal wind stresses having negative curls comparable to those of the climatological wind stresses. The initial result, with the paraboloid rotating at the equilibrium rate, was negative; only a slow, broad clockwise flow developed in response to a suitable distribution of westerly and easterly winds in the air above the layer. Von Arx then found that, if the paraboloid was rotated at greater than the equilibrium rate, westward intensification occurs and a narrow Gulf Stream is present in the Atlantic basin on the west side. This effect was explained by C.-G. Rossby as being due to the variation of normal depth from pole to rim (since the free surface is an exactly similar paraboloid shifted parallel to the axis). The relative vorticity changes expected from the variation of the Coriolis parameter are exactly compensated by the variations of normal depth. In terms of the absolute vortex tubes, which are all of equal length parallel to the axis, a quasi-geostrophic motion can take place without any changes of relative vorticity except those directly produced by the curl of the wind stress. When the rate of rotation is increased, the free surface rises at the rim so that the absolute vortex tubes are forced to stretch when approaching the rim and thus to produce positive relative vorticities in the manner of the β-effect. On the other hand, in an even more striking confirmation of the Munk–Stommel idea, if the paraboloid is rotated at a subequilibrium rate with the same wind torque distribution, the rapid narrow current occurs in the flow from the north on the eastern side of the basin.

It is a fairly immediate step (made in discussion at the Geophysical Models Symposium, von Arx, 1955) from this use of properties of the vortex tubes to

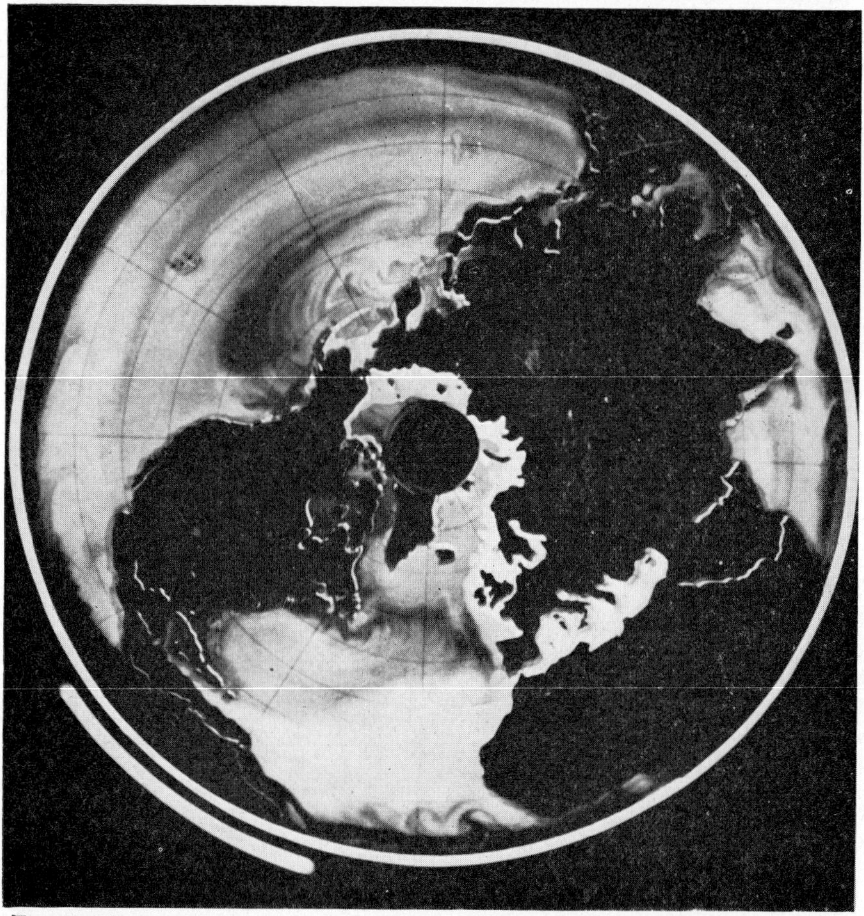

FIG. 6. Photograph of a plane-disk ocean circulation model of the Northern Hemisphere. Mean depth 4 cm, equator to pole depth ratio 2.72, Rossby number of the Gulf Stream about 3×10^{-3}. While black and white gives only an impression of the motion developing the ink tracer patterns, the Sargasso Sea coincides with the light gray area near Bermuda. The concentrated Gulf Stream current flows in the middle gray zone seaward of the coastal dark gray and the hooks near Newfoundland result from some of the transient eddies. The Pacific currents and the Kuroshio current also correspond well (photograph by W. von Arx).

taking an arbitrary geometry of the fluid layer (in particular, a plane disk which has many practical advantages) and arranging depth variations to produce the equivalent of the β-effects. It is not possible to make a completely quantitative transformation in all respects from a sphere to a paraboloid or a plane disk (for example, any correspondence adopted to give nondimensional numerical equality between integrated vorticity effects for northerly currents over a finite displacement will not do the same for southerly currents) but Faller and von Arx (von Arx, 1957) have computed a transformation of latitude to radius on the disk which, utilizing the depth variation of the paraboloidal free surface, makes the vorticity changes due to β alone on the earth correspond to those due to the depth variation in each small neighborhood on the disk. This condition calls for the rim depth, corresponding to the equator, to be $e \approx 2.72$ times the depth at the pole.

With a rotation adjusted to give depth variations of this magnitude, when the Rossby number $R_0{}^* = V/r_0\Omega$ is adjusted by changing the wind intensity to values of about 10^{-3} in the western current, the entire pattern of currents on such a disk has been shown by von Arx to be very realistic. Figure 6 gives one of his photographs for a Northern Hemisphere model in which not only is the over-all picture correct, for example, in the North Atlantic but even many of the small transient eddies in the Gulf Stream appear to fit with types of synoptic circulations that have been observed (Fuglister and Worthington, 1951). Many other details of correspondence are discussed by von Arx (1952, 1955, 1956, 1957).

More recently some very striking experimental geostrophic flow systems have been obtained by Stommel *et al.*, (1958) in connection with ideas of Stommel (1957) on the deep sea circulation. [Stommel was led by his argument to predict a south-setting current below the Gulf Stream which was then looked for and found (Swallow and Worthington, 1957).] We cannot describe these ideas in detail except to mention that they also depend on β-effects and have been investigated experimentally using depth variations on a plane disk so that the same vortex tube stabilities are involved. The experiments show that at very low Rossby numbers narrow western boundary currents develop in enclosed basins (e.g., a triangular sector) under much more general circumstances than wind stress generation. For example, a weak source flow at the apex of the sector proceeds along the western radius to the rim and then fills the sector as a broad, slow southerly current. Or a source and an equal sink spaced along the eastern radius cause a flow in which a narrow current passes at constant radius to the western wall, flows north or south to the radius of the sink, and then crosses at constant radius to it. This effect, except for the radial walls which make possible the western current, is clearly the same as that of Fig. 3.

3.1.3. Ekman Layers and Oscillatory Flows. In all of the above-mentioned experiments (and many others) an easily observed feature of the motion is the presence of pronounced Ekman layers (Ekman, 1905) (i.e., frictional boundary layers (Prandtl, 1904)) in which the current direction changes in a characteristic spiral through the layer of appreciable viscous forces until the geostrophic flow region is reached. Many closely related or identical problems have been investigated in the general hydrodynamic literature (rotating disks, secondary flows) and some early experiments were carried out specifically to check Ekman's calculation at least qualitatively (V. Bjerknes in Ekman, 1905; Sandström, 1914). In fact, this is one of the cases where better quantitative evidence is available from experiments than from meteorological or oceanographic observation (e.g., Thiriot, 1940). However, the question of the detailed effects of the frictional layers on some of the more complicated motions to be discussed later is an area ripe for future examination and is, of course, intimately connected with the unresolved Reynolds number similarity problems mentioned earlier.

Before passing on to the next group of variable density (baroclinic) phenomena we may note briefly another class of homogeneous fluid motions where much interesting work is likely to be done in the future. These are various wave and oscillatory motions in rotating fluids in ranges of medium and high frequencies (measured relative to 2Ω) where more or less ageostrophic effects occur. Where gravitation is involved, these may be tidal in type and become closely allied to the hydraulic work on gravity wave motions. Several investigations of another kind are reviewed by Squire (1956) and careful measurements have been carried out by Long (1953a) on waves generated by translating an object along the rotation axis. Some extremely interesting motions generated by a small oscillating body in which the velocity fields exhibit discontinuity surfaces have been investigated by Morgan (1951, 1956), Görtler (1943, 1944, 1957) and Oser (1957, 1958). Several types of theoretical oscillations were systematically discussed by the Norwegian school (Bjerknes *et al.*, 1933) and Fultz (1959) has recently shown that very precise quantitative agreement with theory is obtained experimentally for the class they call "elastoid-inertia" oscillations. Both in meteorological and oceanographic contexts (see also Arons and Stommel, 1956; Miles, 1959), experiments of these and similar types are likely to be of value in studying the properties and interactions both of geostrophic (Taylor-type) and ageostrophic flows.

3.2. Variable Density Flows and Atmospheric Convective Motion

The next major class of experiments on large-scale phenomena we consider is that which depends essentially on density variations in space and time and on the associated gravitationally induced pressure forces. These baroclinic experiments are most highly developed in connection with meteorological problems though there is little reason to doubt the eventual applicability also to other fields. The most strikingly successful are a large group in which the density variations are produced in a more or less natural manner by heating and cooling and in which, as we will see, convective heat transfer occurs in a manner quite directly analogous to its role in the general atmospheric circulation. The history of experiments of this kind is a long and spotty one, much longer than is still generally realized, and is a very interesting example of how attempts from widely different points of view were made at intervals over more than a century without succeeding in penetrating the main stream of scientific development in meteorology until the ground had been prepared by a number of developments in other subjects. It is intriguing, for instance, that in one of the first clear recognitions of the significance of convection for heat transfer, the celebrated Count Rumford in the late 1790's happened to construct a small glass box containing salt solution and powdered amber in which, being placed between the room and outside air, he was greatly surprised to see a series of superposed, nearly horizontal, instead of vertical circulations. In his enthusiasm "it really seemed to me . . . that I now saw the machinery at work by which wind and storms are raised in the atmosphere" though "I am . . . far from being desirous that much stress should be laid on this single experiment . . . (but) the hint given us is too plain not to deserve some attention" (Rumford, 1870, pp. 392–398).

3.2.1. Early Meteorological Experiments. Most of the relevant historical experiments and suggestions differ in arrangement from Rumford's little box and pertain to a rotating circular disk of air or liquid, playing the role of the atmospheric layer, which is heated in some manner at the rim (equator) and cooled at the axis of rotation (pole). Experiments of this type that were far ahead of the time were carried out by F. Vettin in Germany in the 1850's and later (Vettin, 1857, 1884–1885), explicit suggestions (not followed up so far as we can determine) were made by J. Thomson (1892), Abbé (1907a,b), Bigelow (1902a,b), and experiments of a qualitative nature were made by Exner (1923) and Rossby (1926, 1928). In addition, unpublished trials were made by L. F. Richardson in February, 1918 (manuscript notes) and by L. Prandtl at Göttingen in the 1920's (Fultz *et al.*, 1959). About the only evidence of persistent interest in these and other types of meteorological experiment until quite recently was a long survey article by Weickmann (1929).

The meteorological starting point of the recent work was some slightly different convection experiments in a rotating hemispherical shell that were originally suggested by Professors C.-G. Rossby and V. P. Starr (Rossby, 1947) at the University of Chicago. The motions were of a quasi-turbulent character and had a certain disadvantage in that the combined gravitational and centrifugal potential surfaces had serious deviations from the boundary shape. In spite of this, with convection caused by heating at the lower pole, it was first found on January 17, 1947 that average zonal circulations divided into easterlies and westerlies were occurring in the shell and that measured in $r_0\Omega$ units (i.e., what were later called Rossby number units) these were

(a)

(b)

FIG. 7. Photographs of convective motions driven thermally by heating at the lower pole in the same hemispherical shell as that of Fig. 4. Two ink clouds were injected from a rotating hypodermic needle at approximately latitudes 5° and 50°. Conditions: liquid: water, rotation 0.86 $\sec^{-1}$ from right to left as seen, interval between photographs 14.6 sec = 2.0 "days" (from Fultz, 1949).
a. Ink clouds just after injection. b. Ink clouds two "days" later showing zonal drifts that are easterly in low latitudes (to the right) and westerly in high latitudes (to the left). At about 10° latitude, the zonal speed is about 0.01_5 easterly in $r_0\Omega$ units while at about 45° latitude it is near 0.04 westerly.

quantitatively in the range below and around 0.1 that is *typical of the general tropospheric currents* (Univ. of Chicago, 1947; Fultz, 1949) and even showed a not unreasonable variation with latitude. Figure 7 shows a photograph from the side of the displacement during two revolutions of two ink clouds released from a hypodermic needle in about latitudes 10° and 50°. The apparent rotation is right to left and easterlies of about 0.01_5 appear at the lower latitude while westerlies of about 0.03_5 in $r_0\Omega$ units occur at the higher latitude. It was the prospect, implicit in these results, of extended quantitative experiments and quantitative atmospheric comparisons that sparked later work of a number of types and suggested that significant scientific results could be expected in an area where a purely qualitative character had fatally handicapped almost all previous work.

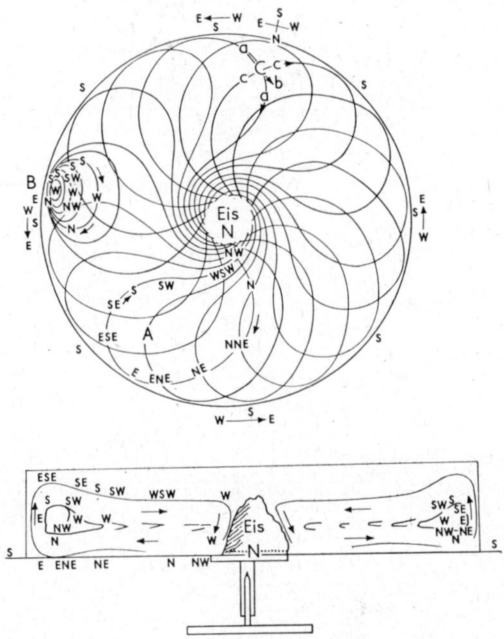

FIG. 8. Diagram from F. Vettin (1884–1885) showing plan and cross-section views of his experimental estimate of the Hadley regime axisymmetric trade-cell motion in a rotating disk of air (2 in. height, 12 in. diam). Heat is applied at the rim and cooling by ice at the center during a slow rotation. Vettin gives little quantitative data but repetitions of his arrangement in a slightly larger disk at 0.1 $\sec^{-1}$ rotation give very vigorous anti-trade zonal motions with kinematic R_0*'s reaching 2 and R_{0T}*'s of 10 to 20 (Fultz *et al.*, 1959).

3.2.2. Rotating Cylinder Convection. Much of the most effective work to the present has been of the rotating disk or cylinder type mentioned above.

Several reviews have been published (Fultz, 1951a, 1956a, Fultz *et al.*, 1959; Hide, 1956b) and only some illustrative highlights can be touched on here. If one rotates a circular disk of fluid at a moderate rate, heats the rim, and cools the pole the easiest expectation is that an axially symmetric convective flow of the type shown in Fig. 8 will ensue. This is suggested by classic ideas of the

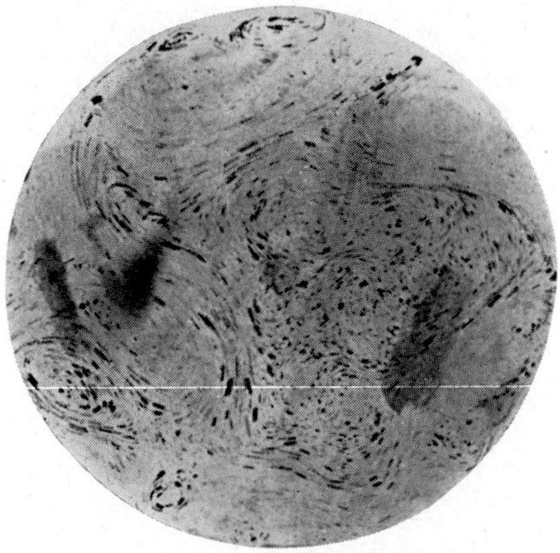

(a)

(b)

FIG. 9. Time-exposure (streak) photograph of rim-heat Rossby regime rotating dish-pan convection experiment with no imposed cold source (the dynamical effect of this as noted by Faller, is the equivalent of a volume-distributed heat sink). The flow pattern is a typical example of the meteorologically realistic types that occur in the Rossby regime quasi-geostrophic range. Such fields of motion are continually changing in time. Conditions: liquid: water, rim radius $r_0 = 15.7$ cm, depth $\delta = 6$ cm, rotation $\Omega = 0.75$ sec^{-1} counterclockwise, heating begun 3.5 min prior to photograph, mean water temperature = 33°C, kinematic $R_0{}^*$ in jets = 0.15, $R_{0T}{}^* \sim 0.02_4$, $S_z{}^* \sim 0.02_8$, $R_i{}^* \sim 190$, $R_e{}^* \sim 1400$, $R_\delta{}^* = 3600$ (from Fultz *et al.*, 1959). a. Top-surface aluminum powder streak photograph in the rotating coordinate system (031150–1–2). b. Streamline analysis of this photograph. Stippled areas are the jet regions of maximum speeds,

atmospheric circulation going back to Hadley (1735) and was shown to be realized experimentally by Vettin (1857, 1884–1885; Skeib, 1953) using air as the working fluid. The initial trials at Chicago with water in 1950, however, showed a completely different type of flow field in which irregular, time-varying currents, often in the form of narrow jets, and a variety of vortical circulations were visible at the top surface. An example is given in Fig. 9 where the currents, at a thermal Rossby number $R_{0T}{}^*$ of 0.02_4, have a general appearance and structure strongly resembling a hemispheric weather map for the upper troposphere. It was found in the next year that a broad separation existed between the types of flow in Figs. 8 and 9, sufficiently pronounced to consider them quite distinct regimes of motion. The axially symmetric convective flows are called the Hadley regime (Fultz, 1956a) and occur in general at high Rossby numbers (1/2 or more) while the irregular, meteorological types of flow are called the Rossby regime. This occurs in general at low Rossby numbers, i.e., under quasi-geostrophic conditions and, in fact, at Rossby number values ($R_0{}^*$, $R_{0T}{}^*$) precisely through the atmospheric range. The transitions between the two can be quite sharp (see further discussion below) and, in going into the Rossby regime, are of the nature of an unstable breakdown of the flow. It is one of the interesting general conclusions (supported by a variety of lines of theoretical evidence) even at this early stage that, since the imposed conditions are axially symmetric in both cases, this spontaneous instability, which is general for the Rossby regime experiments, is a strong indication that a uniform earth without continents and oceans would have essentially the same types of atmospheric disturbance provided only that the mean latitudinal temperature and density fields were left substantially unaltered.

Concerning the fundamental correspondence of the Rossby regime experiments to atmospheric dynamics, a number of interesting points can be selected. At suitable values of experimental parameters (mainly Rossby numbers and stabilities) in the Rossby regime:

1. Details of the jet velocity distributions at the top surface can be compared quantitatively in nondimensional units to atmospheric jets as shown in Fig. 10 (Fultz, 1956a; Fultz *et al.*, 1959).

2. It is possible without great trouble to select runs of 6 or more days of Northern Hemisphere 500-mb maps in which broad features (and sometimes even details) of the velocity fields not only resemble individually a corresponding series of experimental top surface fields but show a corresponding evolution in time (Corn and Fultz, 1955; Fultz *et al.*, 1959). Since this is the fundamental feature of dynamic similarity requirements, it is a clear indication, that, in spite of the difficulties, the over-all dynamics is essentially equivalent to that of the large-scale atmospheric fields.

3. Vertical structure of the density and motion fields, variations of patterns with height, thermal wind effects, cyclones and anticyclones at the bottom, polar front structures, and cyclone families all have been shown to be present in a qualitatively very convincing way to the observer familiar with synoptic maps (Fultz, 1952; Faller, 1956; and see Fig. 16).

4. The eddy statistics of the flow at the top surface, in one case so far in the interior, and, in particular, the values of the momentum transports or

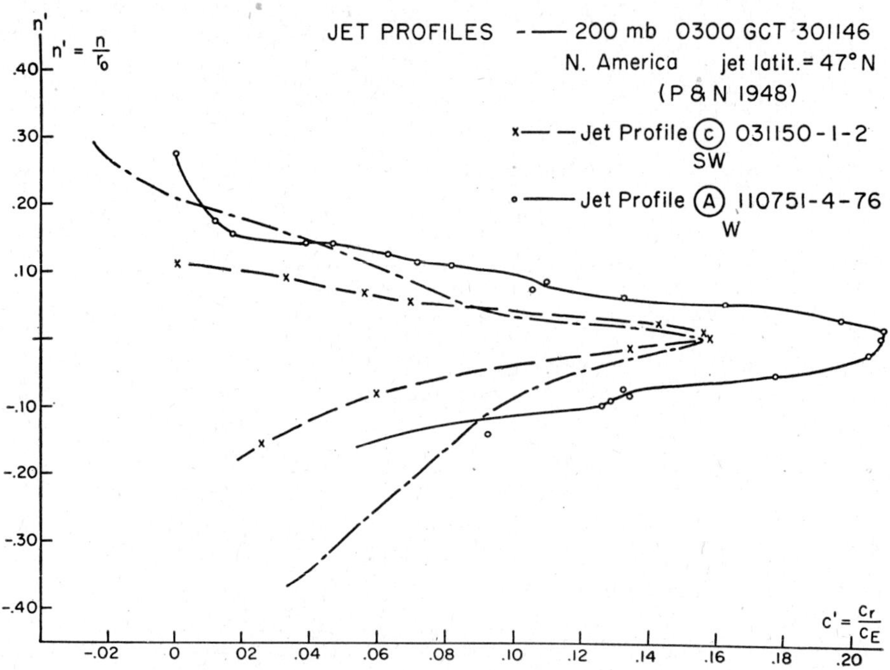

FIG. 10. Jet speed profiles along lines roughly normal to the currents. Velocities are in $r_0\Omega$ units and the distance in r_0 units. The first profile is from analyses by Palmén and Nagler (1948) at 200 mb over North America. The second profile is measured along the line marked "C" in Fig. 9b and pertains to that experiment while the third is from another similar experiment (from Fultz, 1956a).

Reynolds stresses $[\overline{u'v'}]$ have been shown to have both reasonable radial distributions and also nondimensional values in $r_0{}^2\Omega^2$ units which lie precisely at the atmospheric values (see Fig. 22) (Starr and Long, 1953; Corn and Fultz, 1955; Fultz *et al.*, 1959; Riehl and Fultz, 1957). As mentioned earlier, evidence on the eddy transport values for momentum, heat, etc., is extremely important since it is a major aspect of the dynamic similarity problem.

3.2.3. Two-Layer Polar Front Waves. Before proceeding to another important group of thermal convection experiments, we will notice another type of baroclinic experiment in which the density fields are much simpler and, in fact, correspond precisely to a possible idealized version of the Norwegian polar front theory. Two (or more) homogeneous layers of slightly different densities play the role of air masses separated by a frontal discontinuity surface (water solutions and suitable organic liquid mixtures, for example,

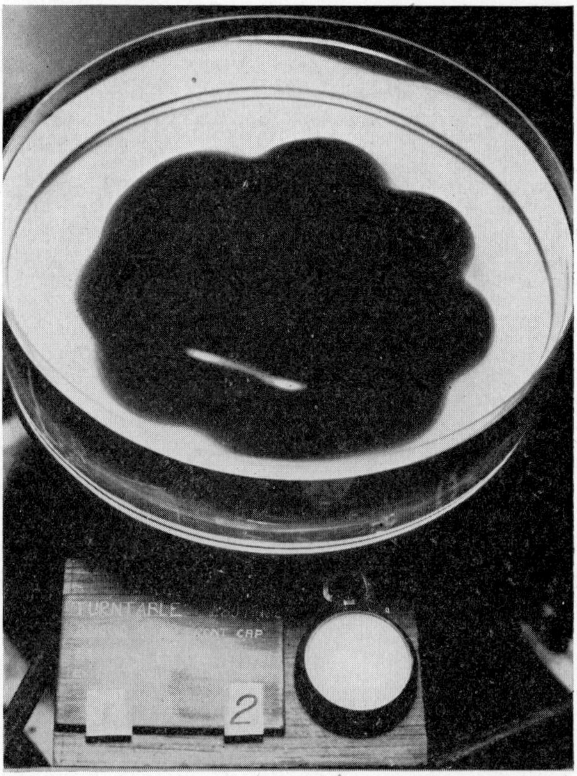

FIG. 11. Oblique photograph (280152–1–2) of a denser dome (the dark mass) underneath a water layer in a glass cylinder. There are eight regular polar-front-type wave disturbances around the periphery of the dome with corresponding wave disturbances in a westerly jet above in the water layer. Conditions: total depth 6 cm, outer radius 12.2 cm; initial rotation 3.80 sec^{-1} counterclockwise, photograph about 2 min 42 sec after start of deceleration to 1.86 sec^{-1}, probable density difference $\sim$0.005 gm/cm^3, parameter analogous to $R_{0T}^* \sim 0.30$ (from Fultz, 1956a).

will work or salt solutions may be used though in this case the surface is not permanent and is eventually destroyed by mixing). We found accidentally, in the course of tests for a quite different purpose on a two-layer system contained in a circular cylinder, that it was possible by systematic acceleration or deceleration of the container rotation to produce axially symmetric relative flow fields in which the interface is either cupped or domed relative to the equilibrium paraboloid. At the extremes, the lower layer forms a ring near the rim or an isolated dome at the axis (Helmholtz, 1888; Fultz, 1952, 1956a; Faller, 1958). The slopes of the interface correspond to the zonal flows by the Margules rules that are familiar in meteorology and involve velocity shears across the internal surface. (These velocity shears, due to the Coriolis effects, are associated with pressure gradients that can support a sloping interface against gravity.) Just as in the thermal convection experiments, it was found that at suitably low Rossby numbers (of the same order as for the other experiments) the axially symmetric flow is unstable and develops usually first regular waves (starting as short wavelength shear ripples but rapidly growing and changing type) and then goes into irregular Rossby regime motions indistinguishable qualitatively from those of the thermal experiments. Figure 11 shows an example at a stage when the waves on a denser dome are very regular. The later evolution of these waves, both so far as the field of motion and the changes of form of the interface are concerned, has many points of identity with that of the Norwegian wave cyclone including, for example, formation of a sweeping cold front, of a low level cyclonic vortex, and in some cases a clear process of occlusion. Figure 12 shows the top-surface flow on the upper layer in the final stages of a Rossby regime motion. Because of the method of generation, these particular experiments are not steady on the average and last only as long as the initial kinetic energy and potential energy of the density distribution allow. They have barely been opened so far as quantitative exploitation is concerned but are extremely promising especially in the possibilities for studying the mechanics of the waves where a number of nonlinear features are evident. In addition to the wave cyclone evolution mentioned above, it is possible at lower rotations (higher Rossby numbers) to get a sequence in which the waves develop to a maximum amplitude and then decay to a Hadley regime motion in which the dome and currents are axially symmetric and have distributions strongly resembling typical trade-cell convection experiments. Both with the domes and even more uniformly with the ring arrangements, certain stages in the wave development have such regular waves that they suggest amplification of a single relatively simple mode. This feature has stimulated a detailed theoretical attempt by Lowell at calculating wave solutions which has attained very promising success (Lowell, 1958). In addition, while nothing quantitative has been done so far partly because of lack of suitable experimental data, valuable insights should be gained from a systematic investigation of the relations between suitable experiments and high-speed computer integrations carried out using the numerous two-level theoretical models developed in the past decade for meteorological forecasting purposes.

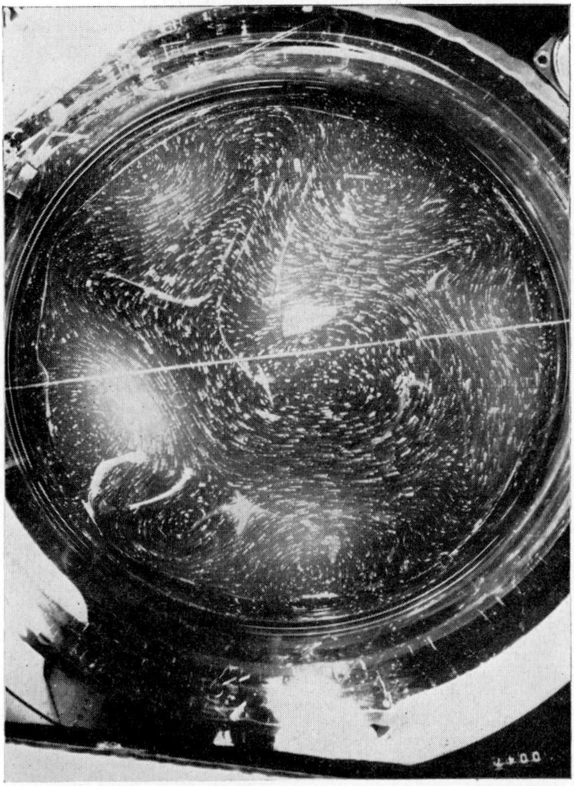

FIG. 12. Streak photograph (200757–2–47) of the top-surface flow in a two-layer Rossby regime experiment after the development of an irregular quasi-geostrophic flow above a lower, denser dome of liquid. There is a close correspondence between the westerly meandering jet current above and regions of maximum slope on the interface. The lower liquid is dark colored, and the lighter patches correspond to depressions of the interface almost down to the bottom solid surface. These occur under the anticyclonic ridge features of the upper flow. Compare Fig. 9. Conditions: liquid: water above chlorobenzene-toluene solution, total depth = 7.0 cm, undisturbed lower layer depth = 2.0 cm, outer radius = 19.0 cm, initial rotation = 3.09 $\sec^{-1}$ counterclockwise, photograph about 5 min 22 sec after start of deceleration to 1.65 $\sec^{-1}$, density difference = 0.004_6 gm/cm^3, upper kinematic $R_0{}^* \sim 0.05$, parameter analogous to $R_{0T}{}^* \sim 0.02$.

3.2.4. Concentric Cylinder (Annulus) Convection. Returning now to thermal convection experiments, a decisive step was taken in 1950 when Hide (1953, 1956a, 1958), in connection with problems of the motions of the earth's core (see below), began a series of experiments using concentric cylindrical containers. The arrangement is very similar to the dishpan experiments mentioned above but with the new feature that the central cold (or heat) source consisted of a concentric cylinder core provided with bath liquid and thus the working liquid was confined in an annular ring. A later version of one of the setups at Chicago which has just about the size and proportions of Hide's is seen in Fig. 13. The same sort of contrast between Hadley and Rossby regime motions occurs as in the open cylinders but with the highly important addition that, in the quasigeostrophic Rossby regime, the wave and jet motions under many conditions become very much more regular in

FIG. 13. Photograph of a version of a tall annular cylinder assembly similar to Hide's (1953). The experiments of Figs. 14, 19, and 23 were carried out in such a container. The largest cylinder contains the outer source liquid, heaters, thermometers, etc. The two smallest cylinders contain the working liquid between them and the inner source liquid on the rotation axis. Inner radius 2.46 cm, outer radius 4.92 cm, liquid depth usually 13.0 cm.

pattern and behavior in time. The central core source has the effect of "purifying" the response of the working liquid and eliminating much of the "noise" and instabilities of various sorts that are almost invariably encountered in the Rossby regime dishpan motions.

Depending on the ratio, η, of the inner radius to outer radius of the ring, on the thermal Rossby number R_{0T}^*, on the vertical stability S_z^*, and to some extent on other dimensionless parameters, one can obtain in the Rossby regime range of values, simple baroclinic wave forms of almost any wave number. When the depth-radius ratio is fairly high, aluminum powder tracer on the top surface of the working liquid spontaneously collects in a ribbon along the upper jet current and thus outlines both the principal current and the waves. Examples are seen in Fig. 14 for a cylinder combination ($\eta = 0.50$) at different positive R_{0T}^* values (rim hot, the normal meteorological analog). The jets at the top surface in each case are westerly (counterclockwise), and the waves propagate toward the east as do the atmospheric upper waves, but with very little or no change in shape (unlike most meteorological situations). It is known from several lines of evidence that the dynamics of these waves must be largely identical with those of the irregular meteorological-type motions; for example, from the fact that, at suitable Rossby numbers, the annulus motions can become irregular and qualitatively indistinguishable from those in open cylinders. As subjects of investigation, the regular waves have a classic suitability and it is difficult to exaggerate either their theoretical or practical importance. Not only do the phenomena present much more clear-cut theoretical questions but also it is possible to make measurements on them of types and densities that are still not feasible in the irregular Rossby regime states (Fultz, 1952, 1956a,b; Fultz, *et al.* 1959).

With some care in adjusting experimental conditions, annulus waves of any wave number can be obtained in a state that is almost completely steady in a coordinate system rotating with the waves. It is then feasible to make complete three-dimensional temperature measurements by placing a single thermocouple junction, fixed relative to the container, at a number of radius and height positions, making a time record of the values as the waves pass, and converting from time variation to the corresponding space variation. One such major experiment has been analyzed so far for wave-number three (Riehl and Fultz, 1957, 1958). Figure 15 shows the top-surface velocity field relative to the pan obtained from a time-exposure photograph in this experiment. A very smooth continuous counterclockwise (westerly) jet current passes around the three sinusoidal wave troughs, the maximum speeds being shown by the greater length of the aluminum powder displacements during the exposure. From the quantitative top-surface velocity field extracted from such photographs, by methods familiar in the analysis of meteorological flow fields, it is possible to calculate approximately the relative pressure field. This field can then be used with the internal temperatures to integrate hydrostatically downwards, determine thus the internal pressure fields, and from these return to an estimate of the internal velocity field (Riehl and Fultz, 1957). Present techniques do not yet allow very accurate direct measurements of such

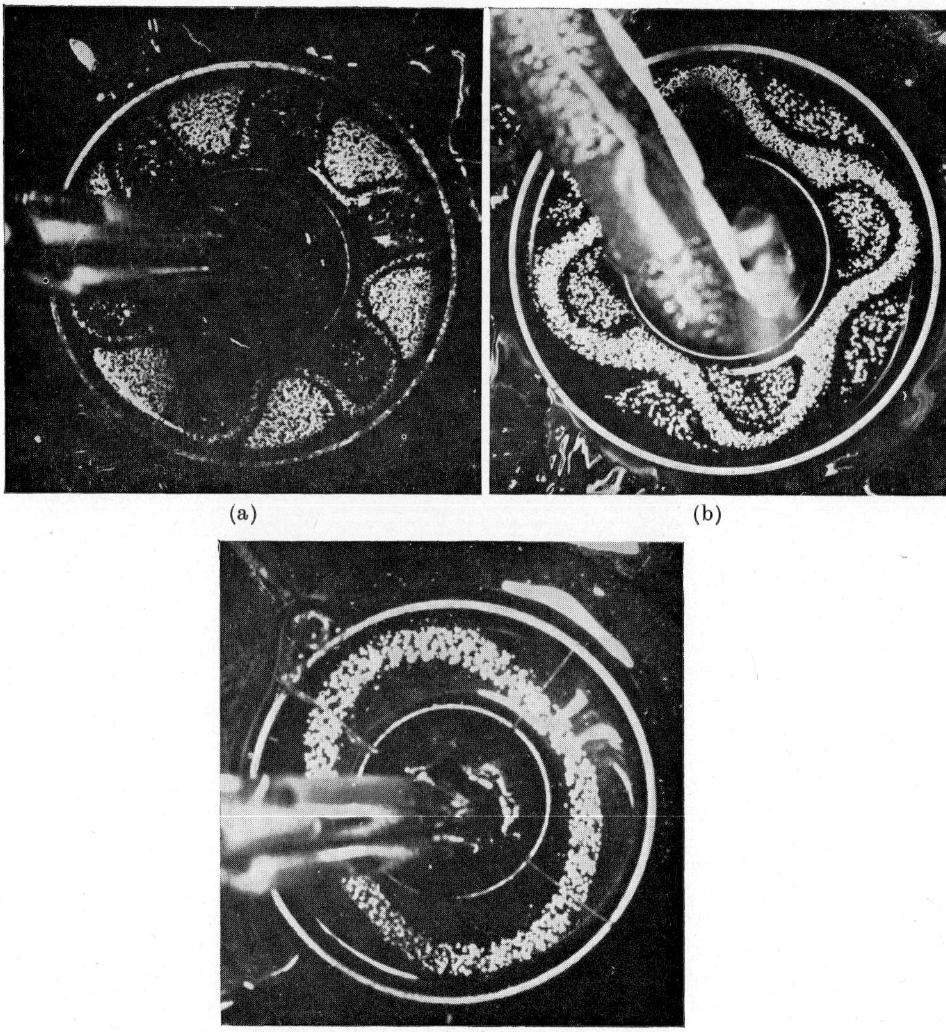

(a) (b)

(c)

FIG. 14. Top-surface flash photographs of alumnium powder patterns during Rossby regime steady baroclinic wave experiments in the annulus of Fig. 13. The outside source is hot, the inside source is cold, and the direction of relative motion at the top is counterclockwise (absolute rotation also counterclockwise). The aluminum powder ribbon in each case marks the westerly jet while the powder patches outside the ribbon mark anticyclonic circulation centers. The liquid is distilled water with a slight concentration of surface-active solute. a. Wave-number six (090457–22): Rotation = 4.0 sec^{-1}, $R_{0T}^* = +0.04_5$, $S_z^* = +0.07_7$, $R_i^* = 38.0$, $N_u^* = 12.0$; $\Delta_r T = 5.7°C$, $\Delta_z T = 6.9°C$, bath temperature difference = 12.1°C. b. Wave-number four (250259–3): Rotation = 2.0 sec^{-1}, $R_{0T}^* = 0.08_5$, $S_z^* = 0.10$, $R_i^* = 14.0$. c. Wave-number two (211156-6): Rotation = 2.0 sec^{-1}, $R_{0T}^* = 0.24$, $S_z^* = 0.37$, $R_i^* = 6.2$, $N_u^* = 6$; $\Delta_r T = 6.8°C$, $\Delta_z T = 7.4°C$, bath temperature difference = 10.9°C.

internal velocities. These velocities were later checked on repeat experiments by direct ink displacement measurements with good results. In addition, measurements were made in the frictional layers near the boundaries where the previously mentioned techniques do not give good results. Figure 16 shows the result of measurements near the bottom. This figure compares with Fig. 15 very much as a surface weather map compares with a high tropospheric upper-air chart. The cold front shown is also confirmed by almost discontinuous drops in some of the temperature trace records for the lowest half-centimeter or so.

The most important result of the internal velocity field analyses by Riehl was that it proved possible to make consistent estimates of the vertical velocity components, again by certain standard meteorological methods. This is the first experimental case where such estimates have been possible. The values turned out, when corrected for the vertical exaggeration relative to atmospheric troposphere depths, to have values (corresponding to 2 or 3 cm/sec at full scale) just of the order of those associated with large scale meteorological disturbances (Riehl and Fultz, 1957; Rossby 1926; Fultz *et al.*, 1959). This is still another, rather sensitive, confirmation of dynamical similarity. The availability of the vertical velocity fields even more importantly, makes it possible to evaluate most of the energy and momentum mechanisms operating in the waves (Riehl and Fultz, 1958). Much observational and theoretical research has been carried out in recent years on these mechanisms in the atmosphere (e.g., Starr and White, 1954; Mintz, 1955; Kuo, 1953; Phillips, 1956; Lorenz, 1955), and even this comparatively regular experimental case turns out to have a surprising number of features in common with the atmospheric results. Most of these studies take Reynolds' (1895) point of view of writing all quantities as the sum of an average and a deviation value. Usually at least two types of averaging are chosen; one in space and one in time. As an example, if the space average is over a latitude circle, the time and zonally averaged northward and vertical velocity components in the classical pictures give a meridional cellular structure (Rossby, 1949) with two direct cells in low and high latitudes and an indirect (Ferrel) cell circulating in the opposite direction in middle latitudes. Such a result has been found by Phillips (1956) in an epochal numerical integration experiment and, for the corresponding quantities, the experimental result is the same, as shown in Fig. 17. This average meridional flow produces fluxes of heat, zonal momentum, etc., and the total fluxes can be separated into an average flux and an eddy flux (Reynolds stress in the case of momentum). Generally, the atmospheric results show that fluxes by the average meridional cells dominate in the low-latitude Hadley cell while eddy fluxes dominate in middle latitudes. The analogous experimental results for heat transports is shown in Fig. 18. Still more striking aspects connected with the energy transformations from eddy

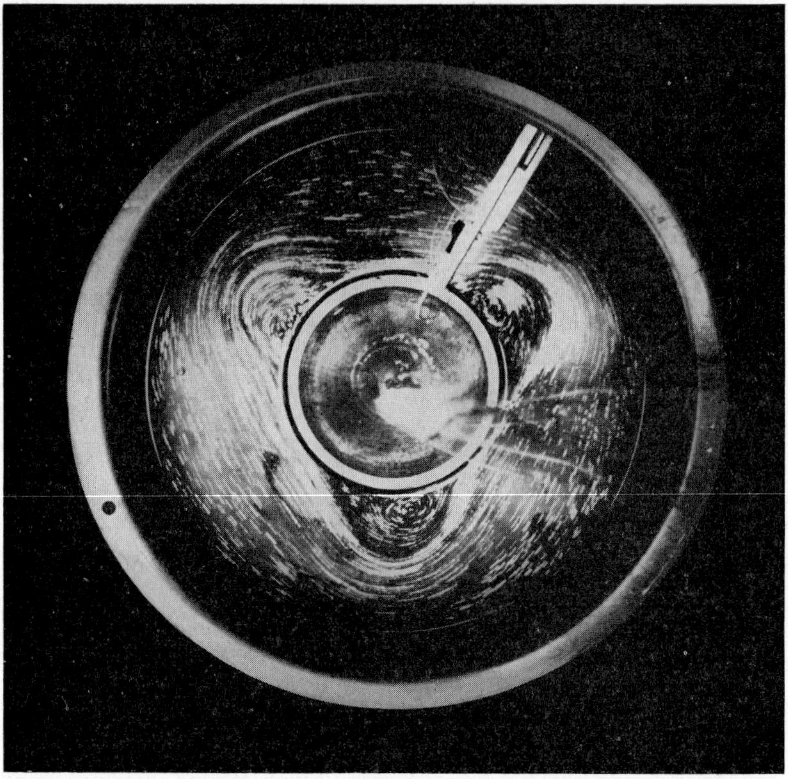

FIG. 15.

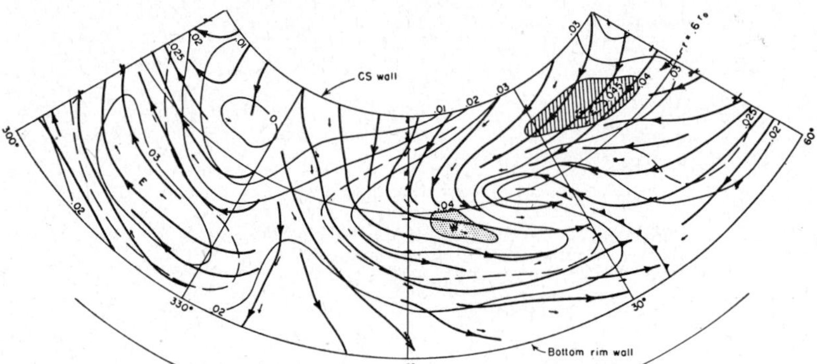

FIG. 16.

to average motions are considered, for example, by Kuo (1953), Lorenz (1955), Phillips (1956), and by Riehl and Fultz (1958) but cannot be discussed here.

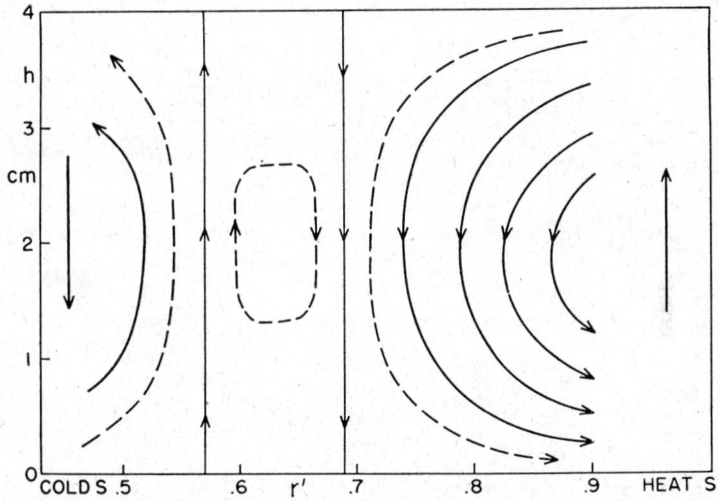

FIG. 17. Streamlines of the zonally averaged flow in a vertical meridional plane for the experiment of Fig. 15. The cold source wall is on the left and the rim on the right. Three cells of average meridional motion are present as in classical pictures (Rossby, 1949; Phillips, 1956). The right-hand circulation corresponds to a Hadley trade-cell and the middle cell to a Ferrel middle-latitude indirect cell (from Riehl and Fultz, 1958).

FIG. 15. Top-surface streak photograph (120854–3B–123) of a steady baroclinic three-wave flow in an annulus. The rim is heated and the central cylinder cooled. Note the smooth sinusoidal shape (though the troughs tilt slightly back towards southwest) of the waves and the long streaks along the length of the continuous westerly (counter-clockwise) jet. These waves propagate toward the east in the pan at about 25° longitude per day. Conditions: liquid: water, rim radius = 15.4 cm, radius of cold source cylinder = 6.5 cm, water depth = 4.4 cm, rotation = 0.30_3 $\sec^{-1}$ counterclockwise, nominal heating intensity = 150 watts, mean water temperature = 19.5°C, kinematic $R_0{}^* = 0.20$, $R_{0T}{}^* = 0.19$, $S_z{}^* = 0.24$, $R_i{}^* = 8.1$, $R_e{}^* = 400$, $R_\delta{}^* = 560$, uncorrected $N_u{}^* = 62$ (from Riehl and Fultz, 1957).

FIG. 16. Analysis of the flow near the bottom in experiments (2202–010357) identical with that of Fig. 15. Direct visual displacement observations were made on small ink clouds released 1 to 2 mm above the bottom. Heavy lines with arrows are streamlines while the light lines (isotachs) give speeds in units of $r_0\Omega$. The maximum speeds are about 0.04_5. The cyclonic eddy to the right of the 0° longitude line (which is placed at the top-surface wave trough line) corresponds to a surface cyclone and is in quite a normal position relative to the upper wave. The spiked line extending rimward from the cyclone center is a wind-shift line associated with a cold front that is also substantiated by the temperature traces in the lowest half to one centimeter (from Fultz *et al.*, 1959).

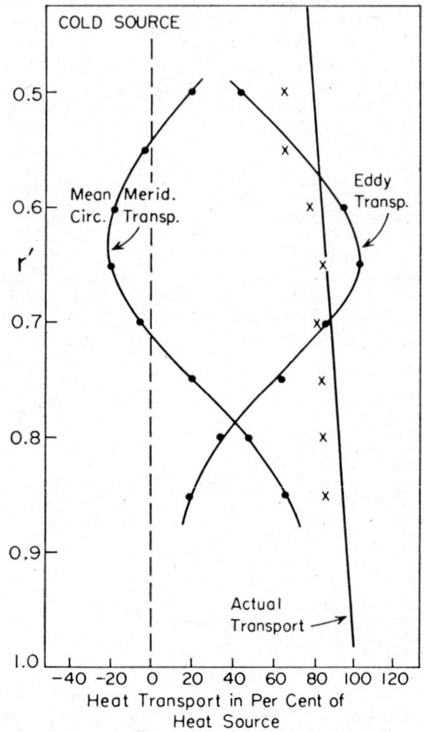

FIG. 18. Contributions to the total heat transport toward the cold source as functions of radius in r_0 units for the experiment of Fig. 15. The transports are in per cent of the estimated receipt per unit time at the rim and are positive for transport toward the cold source. The curve for heat transported by the meridional circulations goes to negative values in the Ferrel cell. The other curve gives the analyzed values of contributions to the transport by eddy terms of the type $[\overline{T'V'}]$ (from Riehl and Fultz, 1958).

3.2.5. Transition Spectra for Annulus Waves. The rather tall cylindrical annuli first used by Hide and used for the experiments in Fig. 14 have the important advantage of making the changes from axisymmetric motions to Rossby regime waves and from one wave number to another occur quite sharply and reproducibly as the experimental conditions are changed. Figure 19 gives two examples of diagrams of transition curves from an extensive series of experiments (Fultz *et al.*, 1959; and forthcoming publication) in which the procedure was as follows: at a fixed rotation rate, water in an annulus for which the radius ratio η was 0.50 was subjected to a variable radial temperature gradient between the inner and outer bath. This gradient was either positive (Fig. 19a, rim hot) or negative (Fig. 19b, rim cold). The bath temperature difference is initially zero so that there is then no relative flow. One bath is very gradually heated and the other very gradually cooled, at rates sufficiently small that the working liquid is never very far from equilibrium; the changes of state are "quasi-static" in the same sense as in thermodynamics. At the lower temperature differences this is found to mean practically that the difference should not change faster than a couple of hundredths degrees Centigrade in five or ten minutes. On the diagrams, at a fixed abscissa value, one proceeds parallel to the ordinate axis toward higher magnitudes of R_{0T}^*. There is, first a very slow purely axisymmetric motion with kinematic Rossby numbers for the zonal flow of order 10^{-3} to 10^{-4}. At the lower curve (which corresponds for positive R_{0T}^* to a nearly constant radial temperature difference of about $\frac{1}{4}$°C) there is a sharp unstable development to waves of the kinds mentioned above. The curve defines a marginal or neutral stability state with respect to amplification or maintenance of the wave disturbances that is just of the type associated with the development of Bénard–Rayleigh cellular convection in a horizontal layer of fluid. As the radial temperature difference and R_{0T}^* continue upward, successive transition curves from wave number to lower wave number are passed until finally the last curve marks a transition again to an axisymmetric Hadley regime motion of the same type as in Fig. 8. While the diagrams for positive and negative R_{0T}^* resemble one another strongly on the upper left, there are drastic differences on the lower right. Similarly, the transition diagrams obtained when the radial temperature gradient is decreased quasi-statically in magnitude are quite different, especially in the lower halves. There are strong "hysteresis" effects in the sense, for example, that the four- to three-wave transition curve lies at considerably higher R_{0T}^* values than the three- to four-wave transition curve. In general, these effects must be results of the nonlinear characteristics of the waves.

The definiteness of these "spectrum" results and of other similar results in the convection experiments has stimulated a considerable theoretical effort at accounting for the main features. This effort in turn has helped guide the experiments. Both approximate solutions for the field variables, taking partial account of convection, and linearized stability analyses for estimating amplification of wave perturbations have been worked out by Davies (1953, 1956), Lorenz (1956), Chandrasekhar (1954a), Lance (1957, Lance and Deland 1958), in a series of papers by Kuo (1954, 1955, 1956a,b,c, 1957), M. H. Rogers (1954), and recently by R. H. Rogers (1959). Especially Kuo has succeeded in obtaining reasonable estimates of most favored wave-number regions and quite accurate theoretical loci for the transition from symmetry curves (Kuo, 1957).

3.2.6. Index-Cycle Convection. The final aspect of these annulus convection experiments that will be discussed is one that involves pronounced time changes of the wave systems and, consequently, is strongly related to ultimate

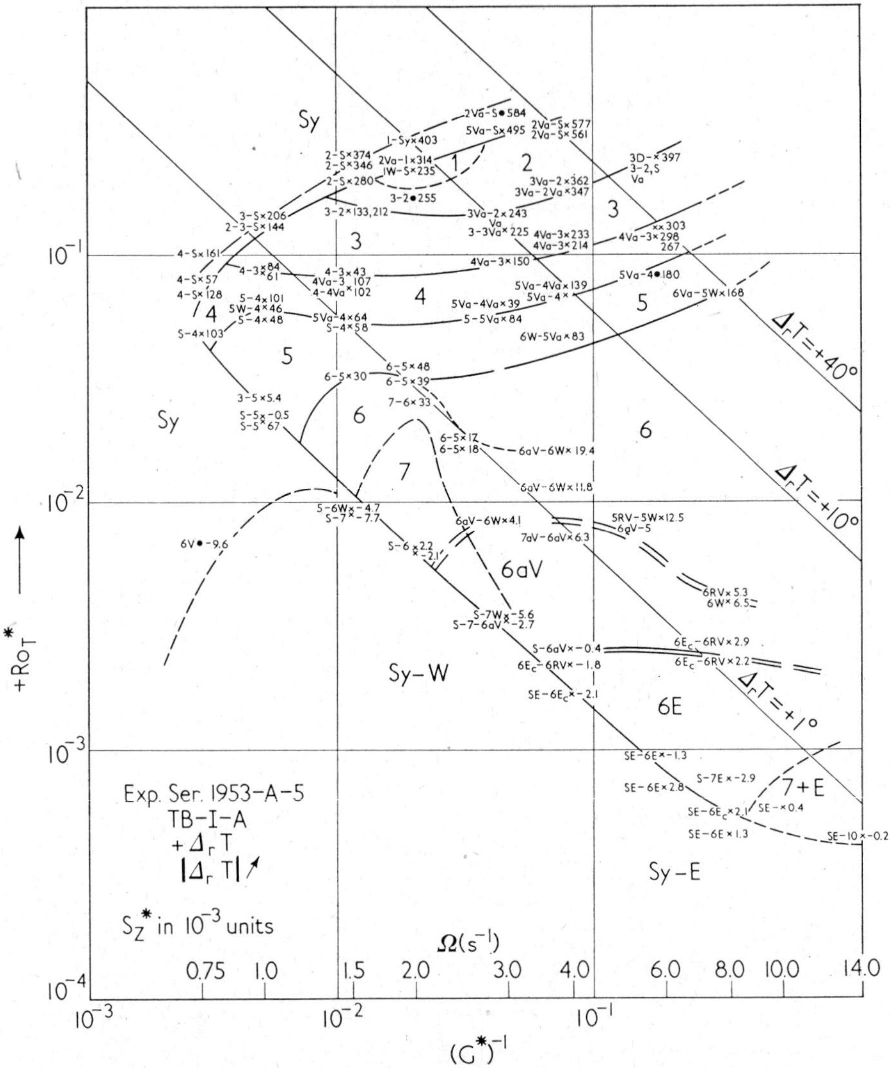

FIG. 19. Two transition-curve or spectrum diagrams for annulus convection in the tall cylinder (brass) assembly of Fig. 13. The abscissa is a rotation parameter ($G^{*-1} \equiv r_0\Omega^2/g$) and the ordinate is the thermal Rossby number $R_{0T}{}^*$. Experiments for these diagrams were conducted at a constant rotation, with water as nearly as possible at a constant mean temperature of 21 to 22°C, and for very slowly increasing (quasi-static) values of the magnitude of the radial temperature difference $\Delta_r T$ (increasing $R_{0T}{}^*$). Here, $\Delta_r T$ was measured internally with thermocouples as close as possible to the rim walls. Large figures give the wave number observed. Axisymmetric motions are observed outside and to the left of the enveloping transition curve. The large-scale meteorological range of $R_{0T}{}^*$ is in the percent logarithmic cycle. a. Diagram for increasing positive $R_{0T}{}^*$ $(+\Delta_r T)$, rim heated. Currents and jets at the top are predominantly westerly.

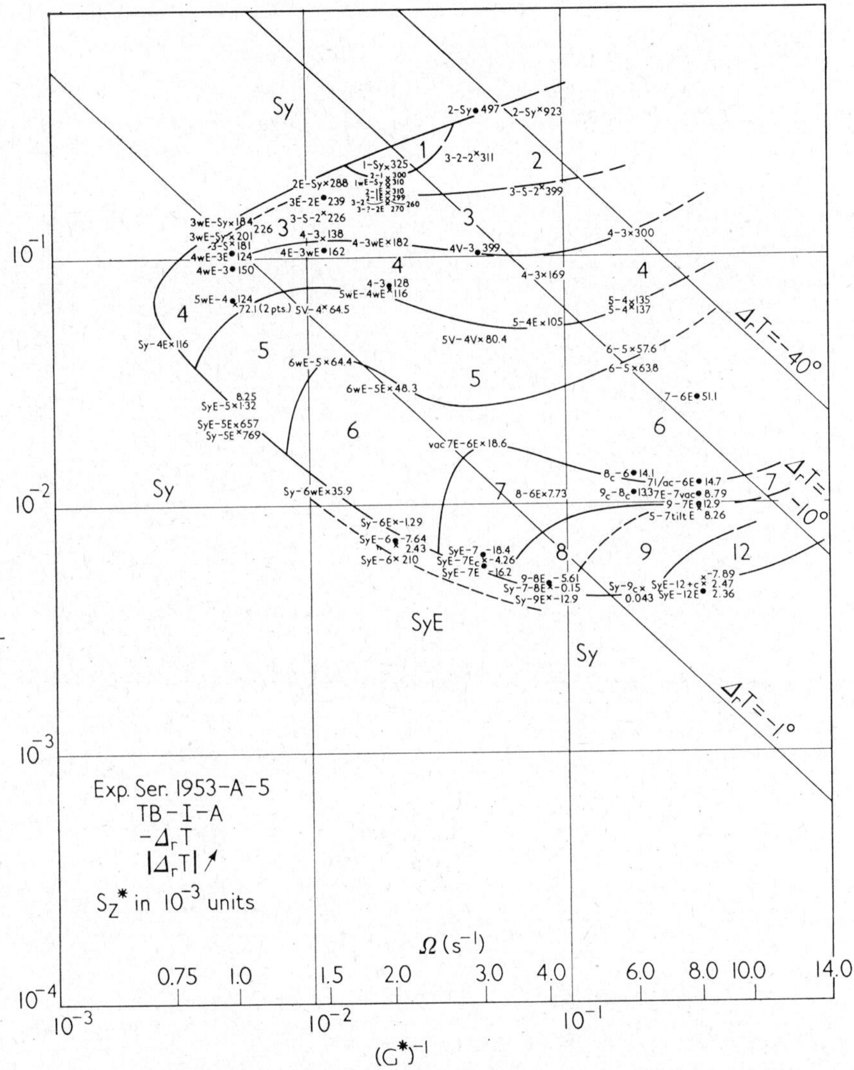

FIG. 19. b. Diagram for increasing (magnitude) negative R_{0T}^* $(-\Delta_r T)$, core heated. Currents and jets at the top are predominantly easterly.

meteorological questions of the mechanisms and causes for both shorter term and climatic fluctuations. Hide, in the earliest experiments in 1951 (Hide, 1953; Runcorn, 1954), discovered one type of very marked fluctuation in time which he called "vacillation." This phenomenon can take various forms but generally consists of a quasi-periodic variation in the wave intensities, shapes, eddy properties, etc. In extreme cases, it can involve periodic changes from

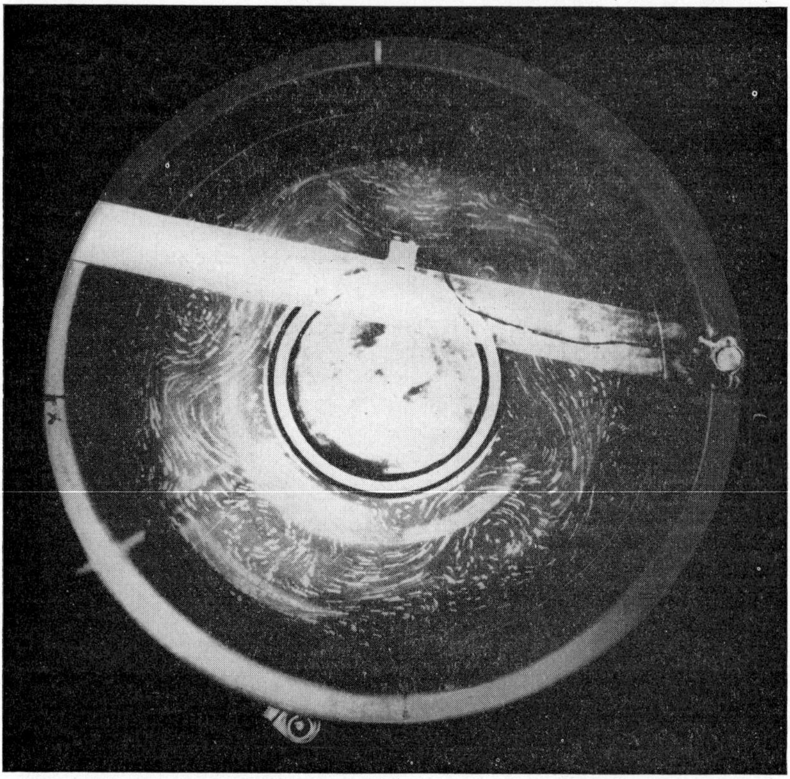

(a)

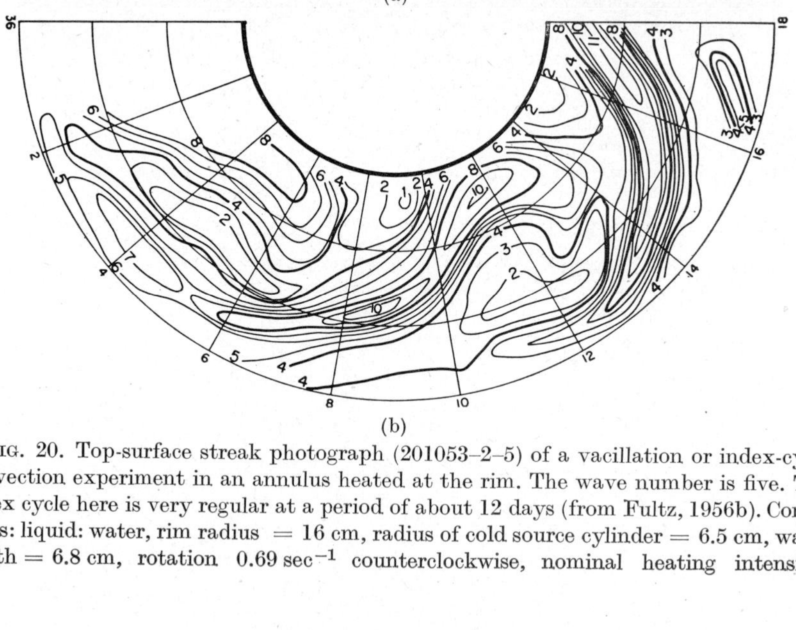

(b)

FIG. 20. Top-surface streak photograph (201053–2–5) of a vacillation or index-cycle convection experiment in an annulus heated at the rim. The wave number is five. The index cycle here is very regular at a period of about 12 days (from Fultz, 1956b). Conditions: liquid: water, rim radius = 16 cm, radius of cold source cylinder = 6.5 cm, water depth = 6.8 cm, rotation 0.69 sec^{-1} counterclockwise, nominal heating intensity.

one wave number to another and back in a regular sequence. The length of period of the cycle is one of the most interesting characteristics involved. It lies, for most of the experimental work so far, in the range from the time for ten up to that for a couple of hundred revolutions of the cylinder. (This, for most satisfactory choices of nondimensional variables, corresponds to the same number of days in the atmospheric situation.) This characteristic fluctuation time at the lower end of the range is just of the order of that of certain meteorological oscillations, the "index cycle" (Namias, 1954; Riehl *et al.*, 1952), which occur irregularly especially in winter and last two weeks to a month. There are associated changes in the mean zonal currents, in the development of upper disturbances, and in the eddy fluxes which are quite analogous in both the experimental and full-scale index cycles.

An example of typical wave disturbance changes in an experimental vacillation or index cycle of moderate intensity is shown in Figs. 20 and 21 (Fultz, 1956b; Fultz *et al.*, 1959). Figure 20 gives a streak photograph and speed (isotach) analysis at a phase when the jet waves have rather open troughs, relatively weak amplitude, and the corresponding mean zonal currents are high. This phase is succeeded by a development in which the troughs change orientation to tilt toward the southeast and deepen rapidly to closed upper cyclones as in Fig. 21. Simultaneously, the mean top-surface zonal currents fall to a minimum ("low index" in meteorological terminology). In this particular case, there is then a slower return to the open wave, "high index" phase; the whole cycle occupying twelve revolutions or days. This sequence is then repeated indefinitely for as long as the imposed experimental conditions are unaltered.

The eddy properties associated with the index-cycle variations show corresponding changes. For example, the convective heat flux to the cold source undergoes a substantial amplitude variation with the twelve-day period. The eddy momentum transport at the top (negative Reynolds stresses), relative to zonal averages, undergo actual reversal in the cycle from -1 to $+3 \times 10^{-4}$ in $r_0{}^2\Omega^2$ units. The time-averaged eddy momentum fluxes in this case, in spite of the regularity of behavior compared to atmospheric time sequences, show surprisingly similar characteristics. This can be

---

250 watts, mean water temperature $= 26°$C, kinematic $R_0{}^* = 0.10$, $R_{0T}{}^* = 0.02_8$, $S_z{}^* = 0.07_6$, $R_i{}^* = 125$, $R_e{}^* = 800$, $R_\delta{}^* = 3{,}700$, uncorrected $N_u{}^*$: maximum 83, minimum 76. a. Streak photograph at a developing high-index stage with open wave troughs on the westerly jet. This photograph occurs about two days after time of maximum heat transfer to the cold source and one to two days before the maximum averaged zonal current at the top is reached. b. Speed (isotach) analysis of photograph in Fig. 20a. Note the narrow long jet streaks. The speed units are $10^{-2}r_0\Omega$ and the maxima thus reach 0.10 to 0.11.

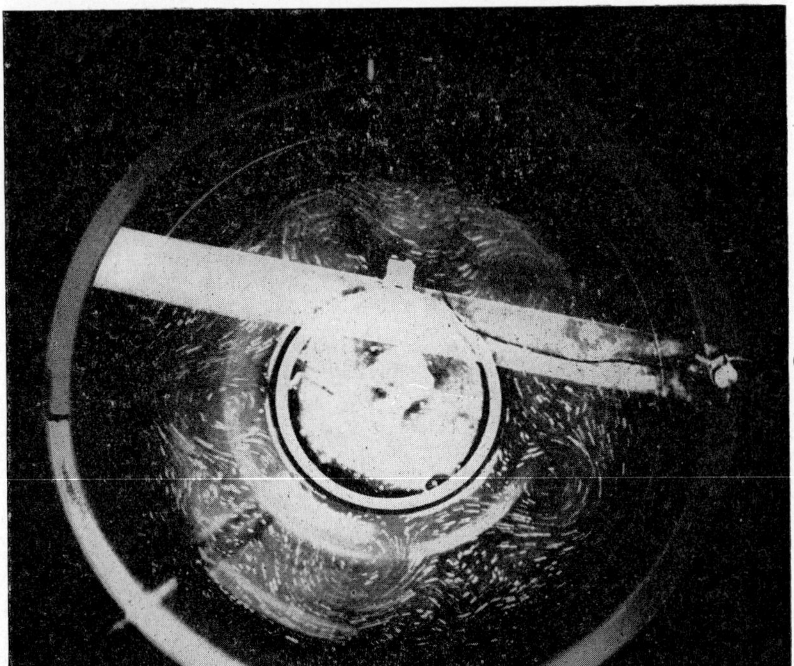

(a)

(b)

FIG. 21. Top-surface streak photograph (201053–2–11) for the same experiment as Fig. 20 after an interval of 6 days (revolutions) (from Fultz, 1956b). a. Streak photograph at a low-index stage in which upper closed cyclones are deepening rapidly in all the five wave troughs (though the phases of development differ slightly). Maximum heat transfer to the cold source occurs about three days later and the minimum averaged zonal current value at about that same time. b. Speed (isotach) analysis of photograph in Fig. 21a. Note the drastic changes in the jet pattern from Fig. 20b. (Units as in Fig. 20b.)

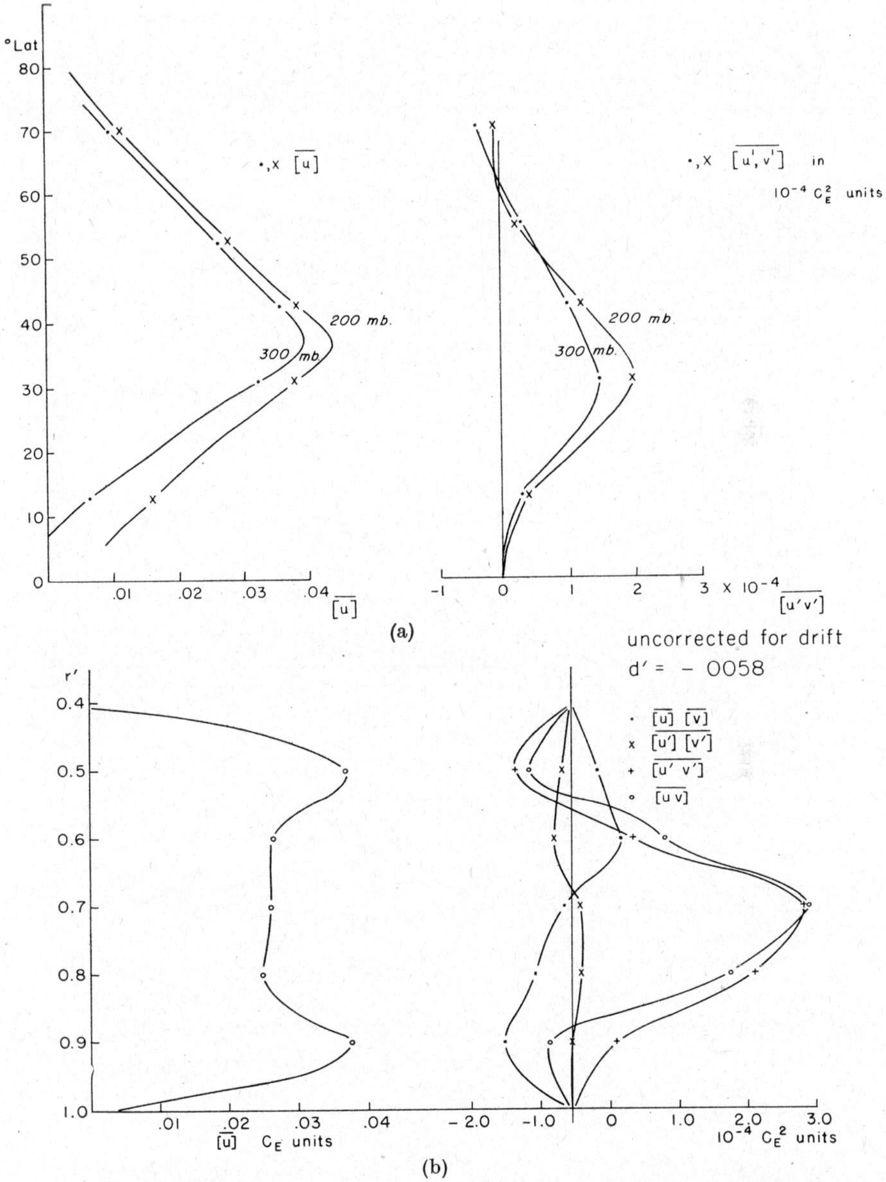

FIG. 22. Comparison profiles of longitude- and time-averaged quantities for the 200- and 300-mb levels in the atmosphere from Starr and White (1954) and for the experiment of Figs. 20 and 21 (from Fultz *et al.*, 1959). a. Averages for the year 1950 for the 200- and 300-mb levels of: on the left, zonal current $[\bar{u}]$ in $r_0\Omega$ units and on the right, relative eddy zonal momentum flux $[\overline{u'v'}]$ in $10^{-4}r_0{}^2\Omega^2$ units. b. Averages over two index cycles at the top surface for the 201053 experiment. On the left, zonal current $[\overline{u}]$ in $r_0\Omega$ units and on the right, eddy momentum flux $[\overline{u'v'}]$ in $10^{-4}r_0{}^2\Omega^2$ units.

seen in Fig. 22 which compares the experimental average zonal flow and momentum fluxes as functions of radius to 200 and 300 mb curves averaged

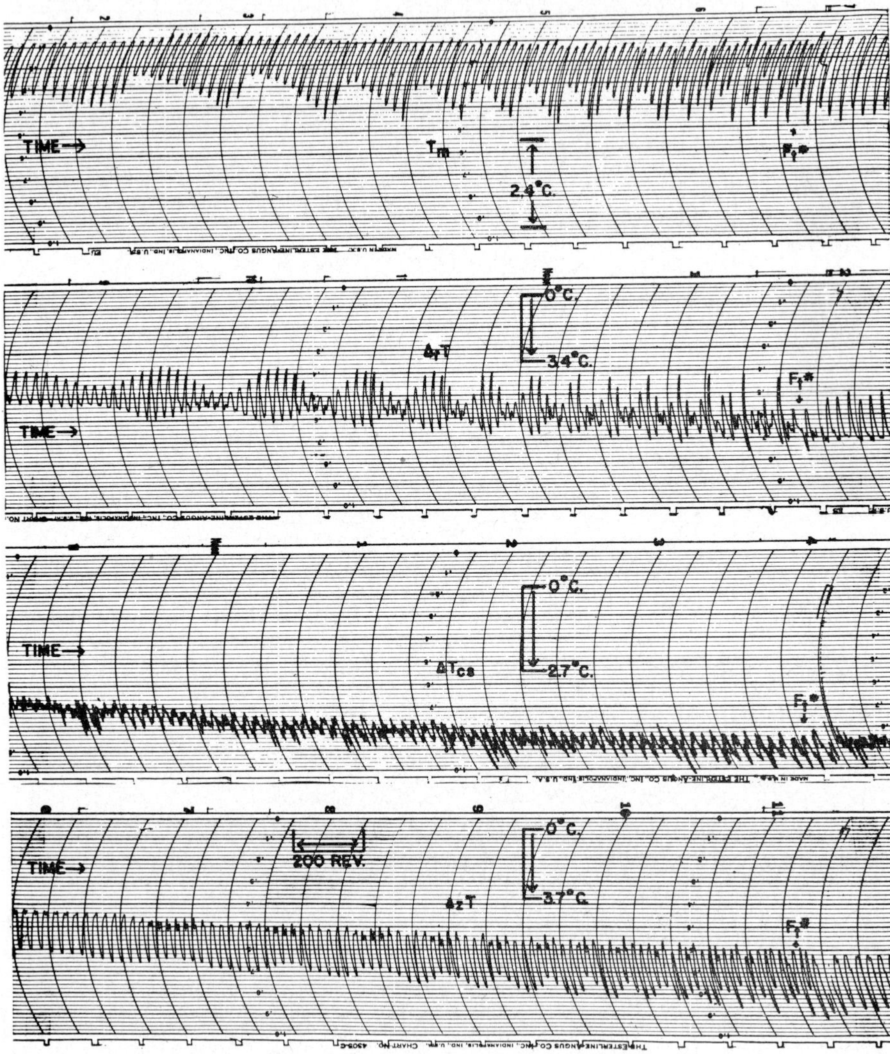

FIG. 23. Traces (190357) of a midpoint temperature (T_m, top) a radial temperature difference ($\Delta_r T$, second), a temperature difference proportional to the heat transport to the cold source (ΔT_{cs}, third), and of a vertical temperature difference ($\Delta_z T$, bottom) during an index-cycle (vacillation) experiment with four waves in a tall annulus (Fig. 13). The cyclic variations of the heat transport (ΔT_{cs} trace) exhibit an index-cycle period of about 30 days. Time increases to the right. Conditions: liquid: water, rotation 3.0 rad/sec, $r_0\Omega = 14.9$ cm/sec, mean water temperature about 24°C, $R_{0T}{}^* = 0.07_4$ increasing to 0.09_7, $S_z{}^* = 0.11_3$ increasing to 0.15_0, bath temperature difference = 13.7°C at end.

for the full year 1950 by Starr and White (1954). This sort of correspondence, as commented on earlier, is extremely important to questions concerning the degree of full dynamic similarity of the experiments to the atmosphere.

Figure 23, finally, shows the impressively regular and long period fluctuations that occur in the presence of an index cycle for quantities measured at fixed locations relative to the containers. Various temperature quantities are being measured in one of the tall annulus experiments during a fairly slow increase of the magnitude of the radial temperature difference (second strip chart down in Fig. 23 with time increasing to the right). In all except the third strip chart, the individual oscillations correspond to the passage of individual waves and the modulation envelope is produced by beating between the index-cycle period and the wave period at the measuring element. Here, the index-cycle period itself is about thirty days while the modulation period is almost four-hundred days at the beginning. Other long-period variations not produced in such a simple way also occur in other experiments; all types eventually should be investigated in great detail. In spite of the fact that theory so far has made only slight progress with these fluctuation phenomena, in contrast to the considerable success with other aspects of the waves, it is already clear that the study of the fluctuations is going to be extremely important in extending understanding of atmospheric variations toward longer time scales and in the question of extraterrestrial causes for such variations. For example, quite aside from questions of how far detailed comparison between experimental index cycles and atmospheric ones can be pushed, it is an important piece of general background in these questions to know that such behavior of a fluid system can be produced and determined solely by internal convection mechanisms. This is undoubtedly true of the experiments at constant control settings, though some aspects may be partly determined by properties of the walls and source baths.

3.3. *Large-Scale Geological Processes*

We turn now to some rather different problems in geophysics and some related areas. In geology, perhaps more than in any other geophysical subject, experiments have played an important direct role in the development of fundamental ideas on large-scale phenomena in the earth and the earth's crust. This is particularly true of ideas on the evolution of the crust and the formation of mountains and other structural details. Two crucial demonstrations that were critically dependent on experiments were carried out by Sir James Hall soon after 1800. He showed that metamorphic rocks must have been produced by high temperatures and pressures due to deep burying of the rock strata (Hall, 1803, 1812). By sealing limestone in gun barrels welded shut with a plug of iron and heated strongly in a founder's furnace (at considerable danger to life and limb, since several exploded before he learned to exclude moisture rigorously), he was able to show that he could produce marble (Hall, 1812).[1] Secondly, he correctly identified the powerful folding and convolution, observed during a trip with Hutton and Playfair, in a series of exposed strata on the Berwickshire coast as produced by mechanical compression of the layers when under a superincumbent load. This idea obtained in the field, he immediately proceeded to test by experiments on layers of linen and wool and clay beds compressed between the plates of a strong screw press (Fig. 24; Hall, 1815).

3.3.1. Mountain Folding, Similarity. Experiments similar to Hall's on folding, faulting, and other rock features have continued ever since. Adams (1918) and Summers (1933) mentioned over fifty investigators up to 1930, mostly concentrated in the 1880's and 1890's and after 1910. All sorts of materials, layer combinations, and stress arrangements were used: wax on a rubber balloon (De Chaucourtois, 1878); glass in torsion (Daubree, 1879a,b); stucco and sand (Cadell, 1890); beeswax and plaster of Paris with heavy overburdens of lead shot (Willis, 1891–1892); sand on carpet felt (Avebury, 1903); mixtures of iron powder, machine oil, and paraffin (Königsberger and Morath, 1913); soft clay (Cloos, 1930); and so on. Now most of this work until quite late, even though many valuable results and analogs to many types of geological features were attained, was almost purely qualitative. Considered as strict models, nearly all had major defects which were not recognized for a century until clear and correct principles were stated by Königsberger and Morath (1913), Hubbert (1937), and others. It is instructive and chastening to consider how such a state of affairs could have lasted so long, with all sorts of scientists concerned with one aspect or another of the very famous questions of earth history involved, when, as we will see in a moment, the critical point was made by Galileo (1638). As Willis (1891–1892) said: ". . . it is the lesson of experience in many directions that it is less difficult to imitate one of nature's processes than to understand either the imitation or, through it, the original."

Following Hubbert (1937, 1945), one can introduce the question we will consider in terms of the apparently contradictory evidence around the turn of the century on the behavior of the earth as a whole. On the one hand there was the evidence, such as that from seismology, for the propagation of elastic waves through at least the crust and mantle over the outer half-radius

[1] He thus became the initiator of an extremely active modern field of high pressure, high temperature mineralogical research.

(a)

(b)

FIG. 24. Two of Sir James Hall's (1815) sketches on the lateral compression folding of strata in the earth's crust. a. Field sketch of folds in 200–300 ft cliffs on the coast of Berwickshire between Fast Castle and Gun's Green. b. Sketch of the screw press and the result of an experiment on a set of laminated clay beds.

of the earth and that from the direct measurements of earth tides by Michelson and Gale (1919). Both of these indicated behavior like that of an elastic solid with a rigidity similar to that of steel. On the other hand, there was the evidence of plastic or essentially fluid behavior such as that from the intense and ubiquitous folding of strata in mountainous regions and from gravity measurements indicating isostatic adjustment of the main mountain masses (Pratt, 1855) and continents, as nearly hydrostatically floating bodies on the substratum materials. Without entering into any of the unsettled detailed questions of stress response of earth materials, the resolution of this contrast between "hard rock" and "soup" schools of thought may be considered to come from two principal ideas. First, there are many reasons to expect that the response of earth materials to short-period stresses, as in seismic waves, is not the same as to long-period sustained stresses (Dobrin, 1939). Second, from the point of view of the folding experiments, Königsberger and Morath (1913), Hubbert (1937, 1945), and others showed from a simple dimensional analysis that materials corresponding in the laboratory to an earth considerably stronger than steel are still extremely weak. To illustrate the argument, consider a block of strata 1000 km on a side and, say, 50 km deep. For long-term geologic processes, inertia effects are negligible and area stresses must be essentially balanced by body forces (gravity). Rocks are certainly imperfectly elastic and the ultimate strengths (compressive or shearing) will be important characteristic quantities. Deep-focus earthquake results indicate fracture strengths comparable to those of surface rocks at least to several hundred kilometers in depth. Further, at least one of the principal stresses will be comparable to the hydrostatic weight of the rock column. Consequently, important nondimensional parameters of model and prototype will be of the form $S/g\rho L$ where S is a strength (dimensions of stress), ρ is density, and L is a characteristic length. Similarity will require $S_M/g\rho_M L_M = S_P/g\rho_P L_P$ where M and P refer to model and prototype, respectively. Since g is constant and little variation between ρ_M and ρ_P is available, $S_M/L_M = S_P/L_P$. If S_P is, say, a typical steel strength of 4×10^9 dynes/cm^2 and the model block has depth 20 cm, its strength should be $S_M = 1.6 \times 10^4$ dynes/cm^2. At a density ρ_M of 1 gM/cm^3, a column of such a material 20-cm high would not sustain its own weight. Hubbert (1937) calculates the example of a sphere of 4-ft diam with even the constant of gravitation altered so as to keep the surface acceleration of gravity the same as the earth. With the strength of steel assigned to the earth, the equivalent 4-ft sphere should be made of soft mud with a strength of 400 dynes/cm^2. These arguments, as Hubbert points out, are just the equivalent of those given by Galileo in *Two New Sciences* (1638) for the impossibility of indefinitely large animals, because total weights increase as the cube of a linear dimension while the supporting power of the skeleton increases only as the square.

This sort of similarity condition was seriously violated in most of the nineteenth century folding experiments. For example, in Willis' (1891–1892) experiments, which were very carefully conducted and compared with many geologic structures, the mixtures of beeswax and plaster of Paris with overburdens of lead shots were so strong that excessive faulting and fracturing occurred unless such heavy loads of lead shot were placed on top, that highly unrealistic depths of burying for the strata were implied for the folding process. Since the twenties there has been a definite movement in the direction of more quantitative work in which results are more consistent and convincing, though often not widely different qualitatively from the older work. For example, work in weak or granular materials on folding (Clarke, 1937) or extensive studies of fracture and faulting patterns in soft muds especially by Cloos (H. Cloos, 1930, 1931; E. Cloos, 1955; Bain and Beebe, 1954) (Fig. 25) in which definite efforts have been made to estimate relations of the deformtions to the local stresses, to time rates of change in the model, etc. (also Hubbert, 1951).

Experiments especially in weak muds and sand slurries are on the whole the most successful and this whole general class of phenomena in the crust can be said to be fairly well understood in a schematic way. However, this does not really represent a very advanced state of understanding and at least the beginnings of the next important developments in this area can be seen in the direction of theories that have sufficient quantitative content to guide and provide verifiable predictions for experimental checking. An example on a small-scale problem is Ramberg's (1955) experimental study of boudinage (formation of sausage chains) in a relatively strong layer sandwiched between comparatively plastic layers, all under compression perpendicular to the stratification. A theoretical example, that is obviously a necessary step in the direction of properly accounting for folding either in experiments or nature, is Biot's (1957) recent calculation of folding as an instability phenomenon in a finite layer bounded by an infinite medium. Biot's analysis leads, for example, to a definite most-unstable wavelength, depending on the properties of the media, that can and should be checked experimentally. That there are not many more such calculations presumably has depended on the need for general theoretical continuum mechanics to advance to a certain level. As more such solutions of the field equations for a variety of stress responses of the media are calculated, there will be a revived need for experiments of the above types carried out with much closer control of the materials and more detailed measurements. It does not seem at all over-optimistic to expect that proper experiments could be used to work backward from theory and, by cross-checking against the geological evidence, to obtain much improved estimates of the stress-deformation properties of the upper crust.

(a)

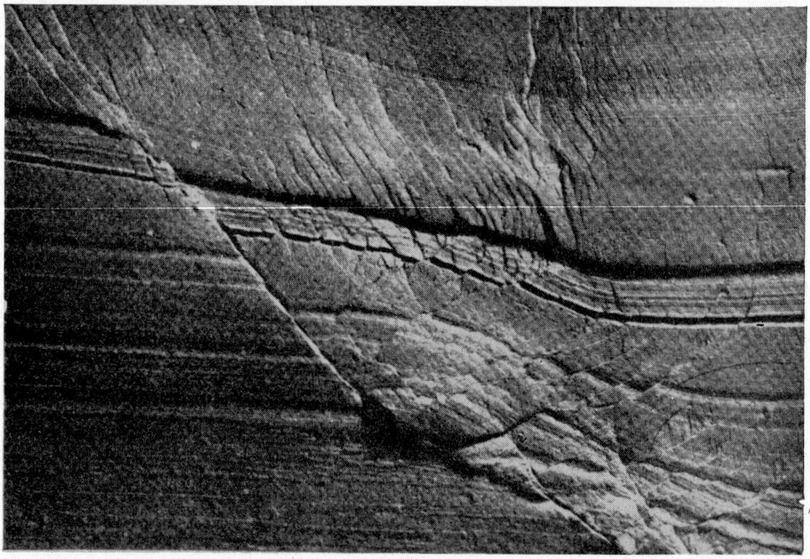

(b)

FIG. 25. Photographs by H. Cloos of details of fracture and fault patterns in soft wet clay layers subjected to tension stresses. a. Right edge of a graben trench produced by allowing the layers to collapse into a gap in the base which lies below and to the left. The markings were initially circles and allow quantitative estimates of the local finite deformations to be made (from H. Cloos, 1930). b. Details of tension and step faults produced in soft clay layers by combined sinking and horizontal stretching (from H. Cloos, 1931).

3.3.2. Mantle Convection. The really large-scale questions in this area are, of course, the tectonic ones of accounting for the existence and global distribution of mountain-building zones, continents, etc. No satisfactory, even qualitative, pictures of these processes and structures have yet been offered though a great variety of theories are vying with each other. We will describe only some experiments connected with two classes of these theories since all results are still quite inconclusive. A number of experiments have been constructed on the contracting earth theory with the idea of determining what, if any, characteristic failure patterns occur on a thin spherical shell subjected to such contraction. De Chaucourtois (1878) did this, for example, by placing a wax layer on a rubber balloon and collapsing the balloon, Rimbach (1913) similarly with a sand layer on the balloon, and Bull (1932) slightly more elaborately with various layers on a rubber sheet. Recently, similar experiments have been carried out by Bucher (1951, 1956) with a casting plastic poured in the space between a wooden core and a thin spherical shell of Plexiglas. The shrinkage of the cast causes some very suggestive fracture patterns in the Plexiglas which Bucher compares with respect to some features to the global orogenic belts.

The contrasting point of view of regarding the crustal materials as behaving on a long time scale like an extremely viscous fluid has had some semiquantitative results in several problems. The salt-dome problem to be discussed later (Section 4.2.3) is one, and another, perhaps the best known, is that of the post-glacial elevation rises in Scandinavia and around the northern Great Lakes. Interpreted as a viscous recovery since removal of the ice loads, the uplifts of the order of a couple of hundred meters in 10^4 years over areas of the order of 10^3 km in span imply viscosities of about 10^{22} poises from Haskell's theoretical solution (Haskell, 1935; Hubbert, 1937). Gedanken experiments on this picture behave correctly in time for viscosities similar to warm asphalt and recovery constants of the order of hours. The ideas that are concerned with really large-scale phenomena of this type are those connected with various versions of convection theories in the mantle, say for the first thousand kilometers of depth. These theories have been advanced in the attempt to provide some motivation for orogenic processes especially, through drawing on the drags applied to the crust by viscous motion driven by density convection (whether thermally produced or otherwise). The general idea is very old (going back at least to Hopkins, 1839), but the recent revivals have drawn fairly heavily on some qualitative experiments and on some theoretical calculations (e.g., Pekeris, 1935; Chandrasekhar, 1952a, 1957b) which indicate reasonable required density gradients and velocities. The most famous of the experiments are probably those of Kuenen (1936) who, in connection with Vening-Meinesz' ideas, derived from a famous set of gravity-deficiency observations in the East Indies, on the nature of the buckling of the crust, carried out a variety of experiments on very soft layers of paraffin floating on water and subjected to mechanical compression. Somewhat different experiments on various powder layers floating on and subjected to differential motions in a fluid layer below were carried out by Terada and Miyabe (1928) to study structures to be expected along the borders of continents. Finally Griggs (1939) suggested that, if a strength limit obtains, the resulting pseudo-plastic convection could be cyclic and consistent with recurrent orogenic periods. He carried out some experiments with oil and glycerine or waterglass layers with motion in the lower layer caused by two mechanically rotated cylinders. These showed very interesting geosynclinal features but in common with the others can be taken only as suggestive. A serious difficulty is that the experimental motion has to be produced mechanically instead of by the actually postulated convective mechanism and the experiments consequently have even more of an *ad hoc* flavor than the usual type of qualitative model. Nothing appears to have been done recently in experiments along these lines although in the next section we will see that some real progress has occurred with respect to a deeper problem: that of the earth's core.

3.4. Large-Scale Electromagnetic and Hydromagnetic Phenomena

Some extraordinarily interesting developments with unexpected interrelations have occurred in connection with studies of some geophysical phenomena of an essentially electromagnetic nature; for example, the problems of the earth's magnetic field, the magnetic fields in sunspots, and of strongly magnetic stars. We will attempt to comment on some of them even though our coverage will have to be highly compressed in view of the extensive recent activity in hydromagnetics in general.

3.4.1. Motions in the Earth's Core. With the establishment from seismic evidence of the existence of a liquid core (probably metallic) extending out to a radius of 3500 km in the earth, the possibility of motions in this liquid medium is immediately plausible and has contributed to an extensive discussion of a dynamo action associated with such motions and simultaneous electric currents as accounting for the magnetic field of the earth. The motion is usually presumed to be a density convection. The principal proponents of this idea have been Bullard and Elsasser and many reviews of the subject have been published in the last few years (Elsasser, 1950, 1955, 1956a,b; Bullard and Gellman, 1954; Runcorn, 1954; Hide, 1956b; Inglis, 1941, 1955). The essential feature of these discussions is that in the large-scale phenomena of geomagnetism and others to be mentioned, the electromagnetic and hydrodynamic effects are coupled in the manner of Alfvén's hydromagnetic waves (Alfvén, 1942, 1950) by the induction term $\mathbf{V} \times \mu\mathbf{H}$ in

Maxwell's equations and the ponderomotive force $\mathbf{j} \times \mu\mathbf{H}$ in the equation of motion. Here μ is the permeability, $\mathbf{H}$ the magnetic field vector, and $\mathbf{j}$ the current. Details are discussed in the above reviews and by Alfvén (1950), Elsasser (1954), and others. If L is a length scale of the motion, V a characteristic velocity, and σ the electrical conductivity,

$$R_m^* \equiv VL/(\mu\sigma)^{-1} \tag{3.3}$$

is a parameter ("magnetic Reynolds number") measuring this coupling in the absence of speeds high enough to entail relativistic effects. If the electrical conductivity σ is sufficiently high, $(R_m^*)^{-1}$ will be small and the consequence is to produce a coupling remarkably like the properties of the vortex tubes in Section 3.1; namely, that the magnetic tubes of $\mathbf{H}$ move with the fluid. The equations for the magnetic induction $\mathbf{B} \equiv \mu\mathbf{H}$, in fact, have the same form in the limiting case of infinite σ as equation (3.1) for $\boldsymbol{\zeta}_a$. For example, an experiment which corresponds precisely to the Taylor two-dimensional properties in a rotating fluid (Section 3.1) has been carried out using a circular cylinder of mercury by Lehnert (1955, Fig. 26). A strong magnetic field parallel to the cylinder axis corresponds to $\boldsymbol{\Omega}$ of the rotating problem. In Fig. 26, the circulating velocities in a narrow ring are caused by a rotating copper disk at the bottom of the mercury. They extend top to bottom and are driven by dragging of the axiparallel fluid magnetic tubes with the ring-shaped exposed section of the copper disk. A similar motion could be produced in the low Rossby number rotating-fluid experiments of Section 3.1 by rotating a ring-shaped paddle wheel at the bottom at a slow rate relative to the container rotation. In this particular experiment and in most conditions attainable in the laboratory, R_m^* is small instead of large and the induction effects do not dominate to anywhere nearly the extent that is common in nature.

Three other dimensionless parameters, among the many that arise in these problems, will be convenient to introduce at this point in connection with the hydromagnetic coupling effects. If $B \equiv \mu H$ is the magnetic induction, a characteristic ratio of hydrodynamic kinetic energy to magnetic energy or of inertia force to magnetic force is

$$K_m^* \equiv V^2(\mu\rho)/B^2 \equiv V^2/V_A^2 \tag{3.4}$$

where $V_A \equiv B/(\mu\rho)^{1/2}$ is the Alfvén velocity of a simple hydromagnetic wave (Alfvén, 1950). A characteristic ratio of magnetic force on the fluid to viscous forces is the parameter Q^* introduced by Chandrasekhar (1952b);

$$Q^* \equiv \sigma B^2 L/\rho\nu \tag{3.5}$$

Finally, since we will be concerned with the rotating core, a parameter of the same form as the Rossby number that measures the ratio of magnetic force to characteristic Coriolis forces is

$$C_\sigma^* \equiv \sigma B^2/2\rho\Omega \tag{3.6}$$

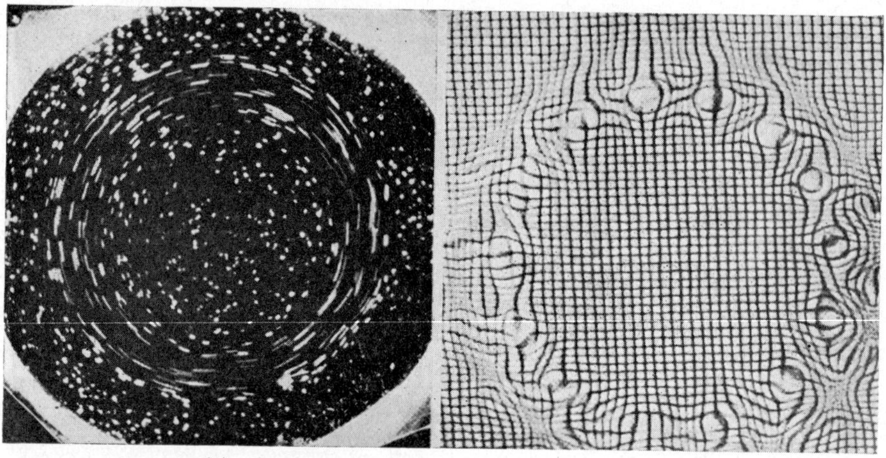

(a) (b)

FIG. 26. Photographs from Lehnert (1955) of the motion in a horizontal circular cylinder of mercury in the presence of a strong, uniform, vertical magnetic field. The base of the cylinder is fixed except for a copper ring which is rotating under the ring of motion seen. Velocity discontinuity surfaces like those generated in some rotating fluid experiments are seen (Fultz and Long, 1951; Long, 1952). Primarily because of the small laboratory sizes, R_m^* is small and the induction effects do not dominate as they do in large-scale problems. Conditions: cylinder radius 7 cm, mean radius of copper ring 3.5 cm, mercury depth above base 0.6 cm, rotation of copper ring 1.26 $\sec^{-1}$, ring velocity at mean radius 4.4 cm/sec, maximum fluid velocities 3.2 cm/sec, $R_m^* \sim 10^{-3}$, $Q^* \sim 4700$, $K_m^* \sim 1.8 \times 10^{-4}$, $C_\sigma^* \sim 5.7$. a. Streak photograph (exposure time 0.2 sec) of sand grains on the mercury surface. b. Photograph of the reflected image in the mercury surface of a rectangular grid.

Now in fact much of the experimental work in this area has not been directed specifically to the geophysical problems, though stimulated by them in greater or less degree, but much more importantly to establishing confidence in purely physical aspects of the theoretical analyses. Thus, Hartmann's, which was almost the first hydromagnetic work (on mercury) (Hartmann and Lazaraus, 1937) was pure experimental physics. Alfvén's group at Stockholm has been extremely active in following up their theoretical analyses with experiments on simple hydromagnetic waves in mercury and liquid sodium (Lundquist, 1949, 1952; Lehnert, 1954) and on other nonoscillatory types of motion in the same liquids (Lehnert, 1955, Lehnert and Little, 1957). Lochte-Holtgreven and his colleagues, in connection with ideas on plasma dynamics, have by ingenious arrangements been able to generate and measure magnetic fields of the order of 10^{-4} to 10^{-5} gauss both in a mechanically rotated ring-vortex in mercury and in a flame in very rapid motion in a vortex tube (Burhorn *et al.*, 1954; Lochte-Holtgreven and Schilling, 1953; Schilling and Lochte-Holtgreven, 1953, 1954). The currents which induce the field arise from electron diffusion affected by the intense velocity variations of the hydrodynamic motions.

The most fully worked-out area of this kind in hydromagnetics, however, is that of the Bénard-Rayleigh type of cellular convection (see Section 4.2.1) in the presence of a uniform external magnetic field. Theoretical discussions of a series of cases have been published especially by Chandrasekhar (1952b, 1954b,c, 1956a,b, and see 1957a for other references) and followed up by a series of experiments by Nakagawa and others. Here, Q^* is the parameter that arises in the theoretical analyses [where B in equation (3.5) must be taken as the component perpendicular to a horizontal layer of fluid and L becomes the depth]. The onset of convection is governed by a critical Rayleigh number $R_a{}^*$ which generally increases with Q^*. Excellent verifications of the predicted properties have been obtained by Nakagawa (1955, 1957a,b), Jirlow (1956), and Lehnert and Little (1957). Similar work reviewed by Chandrasekhar (1957a) has been done on the effects of rotation on Bénard convection and on the situation where both a magnetic field and rotation are present. This latter comes closest to being of direct interest in the geomagnetic problem and it possesses some remarkable complications that one is happy to have well settled and at hand in any consideration of the core convection problem even though the geometry is quite different in the two cases. The theory for a layer both rotating and in a magnetic field parallel to Ω has been given by Chandrasekhar (1954b, 1956a,b), and the corresponding experiments were carried out by Nakagawa (1957a,b, 1959) using an old cyclotron magnet. The three principal dimensionless parameters of the problem are $R_a{}^*$. Q^*, and the Taylor number T^* which is $(2R_\delta{}^*)^2$ in terms of the rotation Reynolds number used earlier. Figure 27 gives calculated and Nakagawa's observed critical numbers for zero rotation and one value of T^* over a range of Q^*. The striking and unprecedented feature is that, with rotation, the cellular convection can set in in two distinct ways as Q^* and the magnetic field increase. At low Q^*'s, it sets in as "overstability"; that is as oscillating cells of definite periods whose predicted values have been confirmed. On the other hand, at high Q^* to the right of the peak on the curve the convection begins as the ordinary type of steady, unidirectional cellular motion. Perhaps even more surprisingly there is an abrupt change in the horizontal size of the individual cells at the same point.

The preceding results are a warning, if any were needed, that a problem involving thermal convection, rotation, and hydromagnetic effects is likely to contain many surprises and to call for great caution in reasoning either qualitatively or from theoretical calculations whose approximations are not known for certain to be suitable. Runcorn and Hide's original work that was discussed in Section 3.2 (Hide, 1953, 1956a,b, 1958; Runcorn, 1954) was intended primarily to give some guidance on the type of convective motions to be expected in the presence of a strong rotation even before the question of magnetic fields and the possibility of self-excited dynamo action as a means of producing and maintaining the field is raised. The concentric cylinder arrangement of Hide's experiments was partly suggested by the supposed presence of a solid inner core though it mainly represented a convenient cold source form.

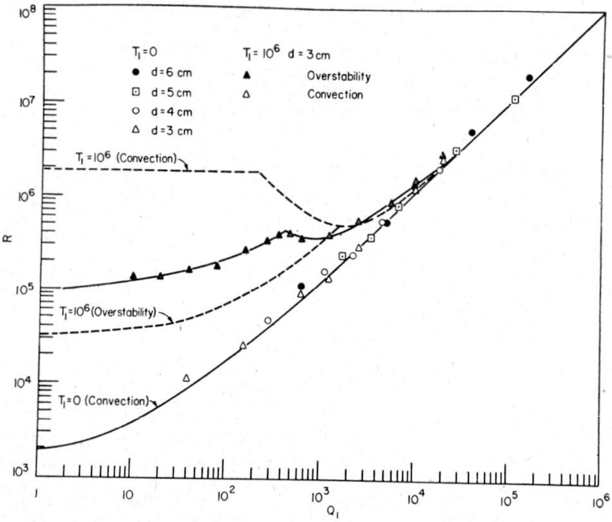

FIG. 27. Critical Rayleigh number curves calculated by Chandrasekhar (1956a) for the onset of convection in a horizontal layer of mercury subject to a magnetic field and to rotation with experimental points from Nakagawa (1957b). The lowest curve and points for zero Taylor number (zero rotation) rise from $R_a{}^* = 1708$ at zero magnetic field to over 10^7. (Maximum field was about 5000 gauss.) The observed upper solid curve for $T^* = 7.8 \times 10^5$ consists of two branches. On the branch left of the peak, convection sets in as oscillating cells (overstability) while on the right it sets in as steadily circulating cells. Theoretical curves are the lower solid curve for $T^* = 0$ and the upper dashed curves for $T^* = 1.0 \times 10^6$.

The especially interesting property from our point of view is that if, as has been assumed in a number of such estimates, the order of magnitude of the core velocities is the same as that of the established westward drift of the geomagnetic secular variations (about 2° or 3° longitude per 10 years), then as Runcorn and Hide particularly emphasized, the Coriolis forces are a major term in the characteristic force balance in the core. Kinematic Rossby numbers on this basis are about 10^{-5} to 10^{-4} which suggests that motion, if any, should be of the irregular sort found in highly geostrophic convection experiments. The magnetic Coriolis number C_σ^* in Hide's estimate is about 10^{-1} which shows that magnetic forces are comparable and while, according to him, kinematic viscosity values for the core are extremely uncertain the extreme values give ordinary Reynolds numbers of 10^2 to 10^{10} and Q^* values of 10^3 to 10^{15}. The core flow consequently is not strongly viscosity-limited as must be the case for any possible convection in the mantle.

In consequence of the very low Rossby number, the core motion must tend to exhibit the Taylor two-dimensional properties as modified, however, both by the meteorological effects and instabilities due to density gradients that we have discussed in Section 3.2 and by the competing tendencies of the fluid to stick to the absolute vortex lines and the magnetic field lines simultaneously. No experiments or theory on this score are as yet satisfactory but Hide's work and the meteorological experiments have been suggestive in the developing ideas on the types of motions that might be admissible. These are discussed in Inglis' most recent review (1955) and in Elsasser's later reviews (1955, 1956a,b), where definite qualitative motions that depend strongly on the Coriolis effects and are capable of dynamo action, are advanced as a result of his and Parker's work (1955). That a dynamo theory of the magnetic field must depend on highly asymmetric motions if it is going to work at all, has, of course, been known since Cowling's proof long ago that no axisymmetric motion is capable of sustaining a dipole field (Cowling, 1933).

3.4.2. Auroral and Cosmic Motions. The other similar major area of now enormously expanding experimental and theoretical work is that of the dynamics of ionized gases in the presence of magnetic and electric fields, particularly at very low pressures. The earth's ionosphere, outer regions of the sun and stars, and interstellar space are all media of this character; in the latter case at much higher vacuum than is attainable in the laboratory. Where a nonmicroscopic continuum point of view is appropriate in treating large-scale problems the particular difficulties in this type of hydromagnetic analysis arise from the presence of at least three collections: electrons, ions, and neutral particles whose interrelations must be correctly approximated in, for example, calculating the currents to appear in Maxwell's equations. We will not go into details of the various theories of magnetic storms, aurorae, etc., but will describe several groups of experiments that are particularly associated with the natural problems and note a couple of the interesting similarity problems where experiments are interpreted as models.

The first and most famous aurora experiments are the "terrella" experiments of Birkeland in 1896 and later (Birkeland, 1908). Birkeland (Fig. 28) applied a high voltage between two electrodes in an evacuated chamber, one electrode being a sphere (the terrella) containing a coil to produce a dipole magnetic field. For auroral and magnetic storm experiments, the terrella was the anode and cathode-ray discharges of many forms were found including the auroral zone pattern of Fig. 29. Birkeland also investigated a number of discharge conditions with the terrella as cathode which he interpreted as analogous to zodiacal light, sunspots, Saturn's rings (Fig. 30), and other phenomena. It would, in fact, be very much worthwhile to investigate with modern resources many features of Birkeland's experiments that are not usually referred to.

FIG. 28. Birkeland with his larger terrella chamber and a discharge in operation (Birkeland, 1908).

Birkeland's work inspired Störmer to his lifetime work of calculating the orbits of charged particles in the field of a magnetic dipole (Störmer, 1955) and has been followed by a number of investigations designed along the lines of later theories. Brüche (1930) used gas-focused electron streams in mercury vapor and Malmfors (1946) and Block (1955, 1956) used arrangements similar to Birkeland's with an electron-gun source in connection with Alfvén's electric field theory of the aurorae. More recently Bennett, in connection with his magnetic self-focusing theory (Bennett and Hulburt, 1953; Bennett, 1955,

1959), has produced a wider series of beautifully coherent discharge streams including motion pictures of the periodic equatorial ring-current orbits that develop for certain orientations of the stream relative to the terrella. Not all

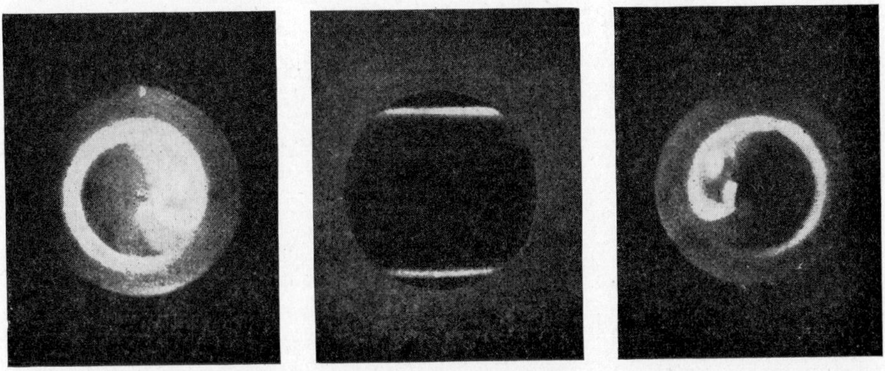

FIG. 29. Photographs of luminescent auroral zones where cathode rays are striking the terrella (Birkeland, 1908).

FIG. 30. Photograph of a double ring-current discharge around the equator of Birkeland's terrella (Birkeland, 1908).

the similarity conditions can be simultaneously satisfied in these experiments, but Block (1956) shows that mean free paths can scale roughly correctly and drift orbits can be correct even though the terrella dipole is too weak and the spiraling of the particles around the magnetic lines is not sufficiently tight to correspond to natural conditions. On a slightly different problem, using the

Stockholm apparatus, Malmfors (1945), Brunberg (1953, 1956), and Brunberg and Dattner (1953) have also been able to produce accurate enough, narrow, cathode-ray orbits to serve for analog computation of the orbits of high-energy cosmic-ray particles in the earth's field.

Finally, in the extensive present activity on plasma physics, mention may be made of details reminiscent of aurorae in electron-beam work by Webster (1957) and of some extremely intriguing phenomena obtained in plasma-gun discharges of small ionized regions into magnetic fields by Bostick (1956; Bostick and Twite, 1957). The evolution of small plasma rings, often several fired simultaneously in various configurations at speeds up to 10^7 cm/sec, has produced some forms remarkably reminiscent of galactic forms, for example, the barred spirals. These results would suggest a fundamental role for the galactic magnetic fields, suggested, e.g., by Chandrasekhar and Fermi (1953) in determining galactic evolution. If borne out by future work, such model interpretations will constitute by all odds the largest scale natural phenomena to which modeling ideas have been applied.

4. Medium-Scale Phenomena

With length scales from kilometers to hundreds of kilometers in mind, so far as terrestrial examples are concerned, the author proposes to survey three groups of investigations in which an active interplay of theoretical with experimental developments has occurred in the last couple or more decades. The first two groups involve almost purely hydrodynamic effects of density fields under the influence of gravity, respectively, for primarily vertically stable density distributions and for essentially unstable arrangements (positively acting buoyancy forces). The third concerns rather different problems of elastic wave propagation relevant to seismic waves, either artificial or natural.

4.1. Stable Density Stratification

The first group comprises studies of a number of interesting phenomena of the atmosphere and hydrosphere that are mainly of the nature of internal gravity waves (relatively uninfluenced by the earth's rotation). These waves are made possible by the restoring buoyancy forces associated with disturbances of a vertically stable density stratification. From a gross point of view, the atmospheric troposphere, many lakes, and the upper parts of the oceans often have such stable stratifications, so far as many short-term types of disturbances are concerned. In the case of the latter two, the arrangement is often of an upper warm, light layer divided from a lower denser layer by such a narrow, nearly horizontal zone of transition (the thermocline) that a fairly easy idealization to two homogeneous layers of different density separated by a (zero-order) discontinuity surface is often effective in understanding important quantitative aspects of the natural occurrences.

4.1.1. Orographic Waves. In the flow of relatively uniform atmospheric currents across topographic features such as mountain ranges and even small hills, a variety of wavelike motions, often quasi-stationary, are excited. Many recurrent mountain-induced (lenticular) clouds are well known—the Moazagotl of the Riesengebirge, the cap clouds of Mt. Fuji, and many others —that are produced by the vertical motions associated with wavelike flows like the stationary waves induced in a running stream by bottom irregularities. Many observational studies have been carried out on these phenomena; some of the most detailed were stimulated by gliding activities in Europe since the 1930's. These activities early showed that great altitudes could be attained in unpowered aircraft by making use of the regions of upward motions in the waves that were found to extend to heights far exceeding that of the mountains (see Fig. 33). An extensive theoretical development has taken place along lines foreshadowed by classical work of Kelvin (Thomson, 1886) on stationary waves in streams and of Lamb (1916) on waves in stratified incompressible currents. The initial recent theoretical investigations were by Lyra (1940, 1943) of Prandtl's school and these were followed by many investigations of which Queney (1947), Scorer (1949), and Palm (1953) may be mentioned. Reviews have been given by Görtler (1941), Scorer (1951), and Corby (1954).

The above investigations have firmly established that the atmospheric motions in question are primarily gravity-wave motions of an internal type and depend on, in addition to the geometry of the topography and fluid layers, the general current speed V, the over-all gravitational stability ($g\partial ln\theta/\partial z$, where θ is potential temperature[1]), and in detail on the vertical profile of the current and the vertical variation of the stability. The latter two factors are discussed especially by Scorer (1953, 1954, 1955a,b). To a great extent the phenomena are not strongly influenced by friction and the theoretical investigations are all for frictionless fluids.

Experiments on internal wave motions have a long history; two important ones in geophysics being by Ekman (1904)[2] and (Schmidt 1908, 1910) together with a few others mentioned below. The important recent ones were initiated by

[1]This quantity for the atmosphere is essentially comparable to gE_z^*/δ of the earlier discussion.

[2]Ekman's (1904) work was stimulated by the problem of accounting for an instance of "dead water" observed during the voyage of the Fram in 1893–1896 off Taimur. As against a normal speed of five knots, the Fram was able to make only about one knot at full power for many hours. At the suggestion of Nansen and V. Bjerknes, Ekman proved conclusively by model experiments that the excessive resistance was produced by strong internal waves associated with an upper fresh-water layer several meters deep. The experiments were very carefully organized and quantitative even to giving reasonable resistance values. Typical F_i^* values were 0.2 to 0.3 and the phenomena, while three-dimensional, had many features resembling, for example, those in Fig. 31.

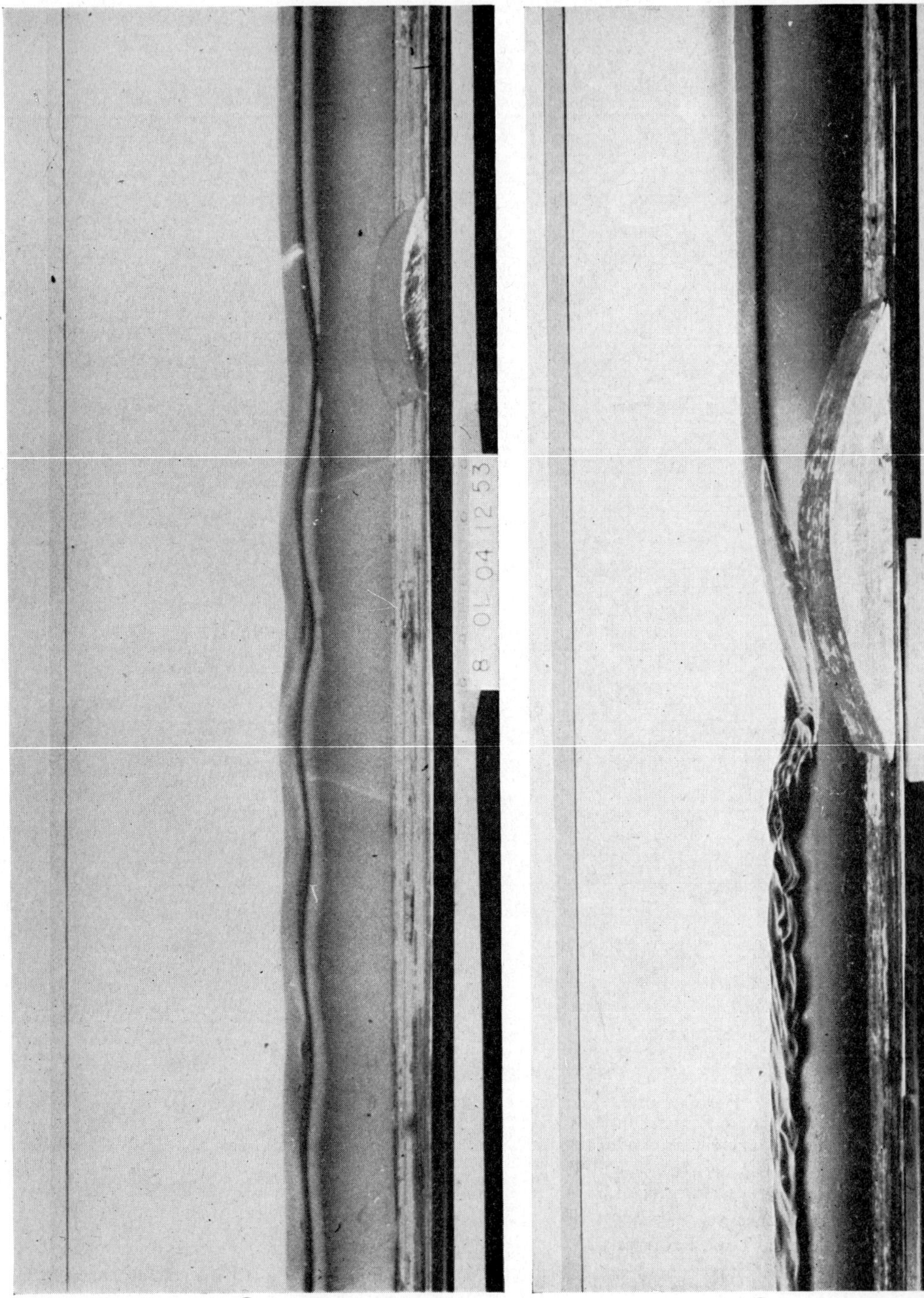

(a) (b)

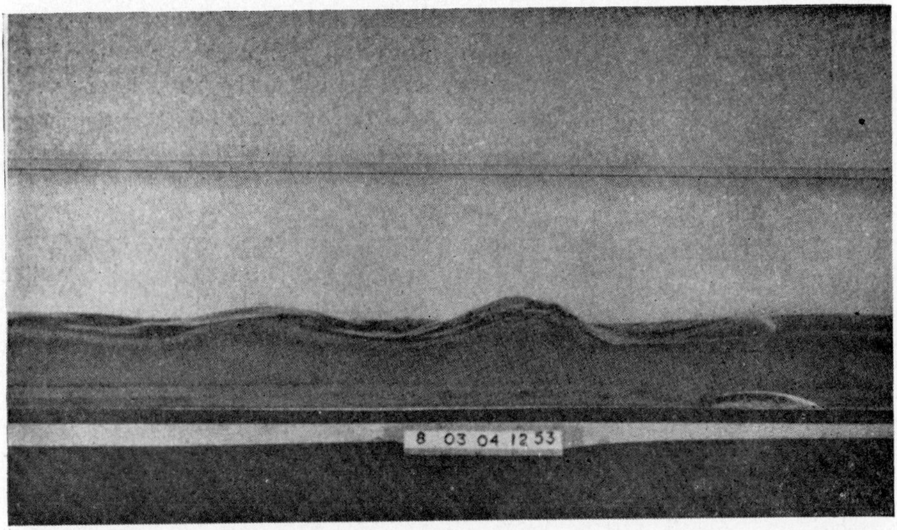

(c)

FIG. 31. Photographs from Long (1954b) of two-layer motions over a low rounded ridge. The lower layer (carbon tetrachloride and cleaning fluid) is dark and the clear (salt-water) upper layer ends at the straight dark line. At the speeds of motion involved the free top surface is practically undisturbed; that is, the external Froude number for the total layer is so small that the flow over the ridge would be like a potential flow in the absence of the lower layer. (Here, $\delta_1'' \equiv \delta_1/\delta$ below is the upstream depth of the lower layer in units of total depth and $h'' \equiv h/\delta$ the height of the obstacle in the same units.) a. Small regular internal lee waves at an internal Froude number $F_i^* = 0.220$. Motion is from right to left relative to the ridge. b. Moderate internal hydraulic jump in the lee of a considerably larger ridge. $F_i^* = 0.155$, $\delta_1'' = 0.33$, $h'' = 0.205$, $E_z^* \sim 0.02_3$, $F^* = 0.02_3$ (Fig. 4, Long, 1954b). c. Strong internal undular jump in the lee of the small ridge. Note the high amplitudes of crests compared to the size of the ridge. $F_i^* = 0.284$, $\delta_1'' = 0.33$, $h'' = 0.067$, $E_z^* \sim 0.02_3$, $F^* = 0.04_3$ (Fig. 5, Long, 1954b).

Long (1953b,c, 1954b, 1955, 1956a,b, 1957, 1958), partly with mountain-wave questions in mind. He has carried them out along with accompanying very important theoretical investigations. Long's initial work was with two- and three-layer systems of homogeneous liquids having slightly differing densities. The undisturbed velocity profiles are uniform with height and are obtained by moving a suitable obstacle relative to an initially resting set of layers. Aside from geometrical ratios such as ratios of layer depths and height of the obstacle to depths, the important nondimensional parameter of these systems is the internal Froude number $F_i^* \equiv V/(gE_z^*\delta)^{1/2}$. Here, F_i^* plays a role similar to the role of the ordinary Froude number $F^* \equiv V/(g\delta)^{1/2}$ [where $(g\delta)^{1/2}$ is the speed of a long gravitational wave in a layer of depth δ] in ordinary hydraulic channel flow. For a single-ridge object disturbing a two-dimensional two-layer motion in vertical planes perpendicular to the ridge, one may have a variety of distinct kinds of disturbance depending on the values of F_i^* and F^*. There may be motions similar to irrotational potential flows, small or large internal lee waves, internal hydraulic jumps or drops (shock waves), or, at very high speeds and high F^*, the same types with disturbances of the top surface as in a single-layer flow. Figure 31 gives photographs from Long (1954b) of two-layer motions over a ridge of both simple lee-wave and hydraulic-jump type The velocities are so low (small F^*)

FIG. 32. Theoretical streamlines and streak photograph from Long (1955) of the flow of stratified salt solution over a rounded ridge. The stable density distribution is approximately linear with height and the relative velocity approximately constant. The theoretical case is strictly for a flow with ρv^2 constant with height. Here, F_i^* lies between $(\pi)^{-1}$ and $(2\pi)^{-1}$. Experimental conditions: $F_i^* = 0.204$, $h'' = 0.200$, $b'' \equiv b/\delta = 0.86$ the breadth of the ridge. Theoretical conditions: $F_i^* = 0.200$, $h'' = 0.200$, $b'' = 1.00$.

that no appreciable top-surface effects are produced but only strong variations at the interface. In meteorological contexts, the interface corresponds to a temperature inversion. There is much observational evidence for such temperature inversions in association with many lee-wave occurrences. For any given ridge geometry and layer depth, beyond a critical F_i^*, hydraulic jumps and shocks are the prevalent response rather than simple lee waves and Long points out a number of resemblances of the internal-jump experiments to a number of observed features of mountain effects in strong winds, such as "rotor" clouds and others (Long, 1953b, 1954b, 1955; Corby, 1954) Typical F_i^* values for the troposphere are near those of Fig 31 as seen in Table I.

In later papers, Long (1955, 1956a, b, 1958) has considered continuous, stable, vertical density distributions that are approximately linear with height and are asscciated with approximately uniform velocity profiles. Here, a number of new phenomena appear that he has been able to account for by solutions of the governing equations that allow finite amplitudes when special restrictions

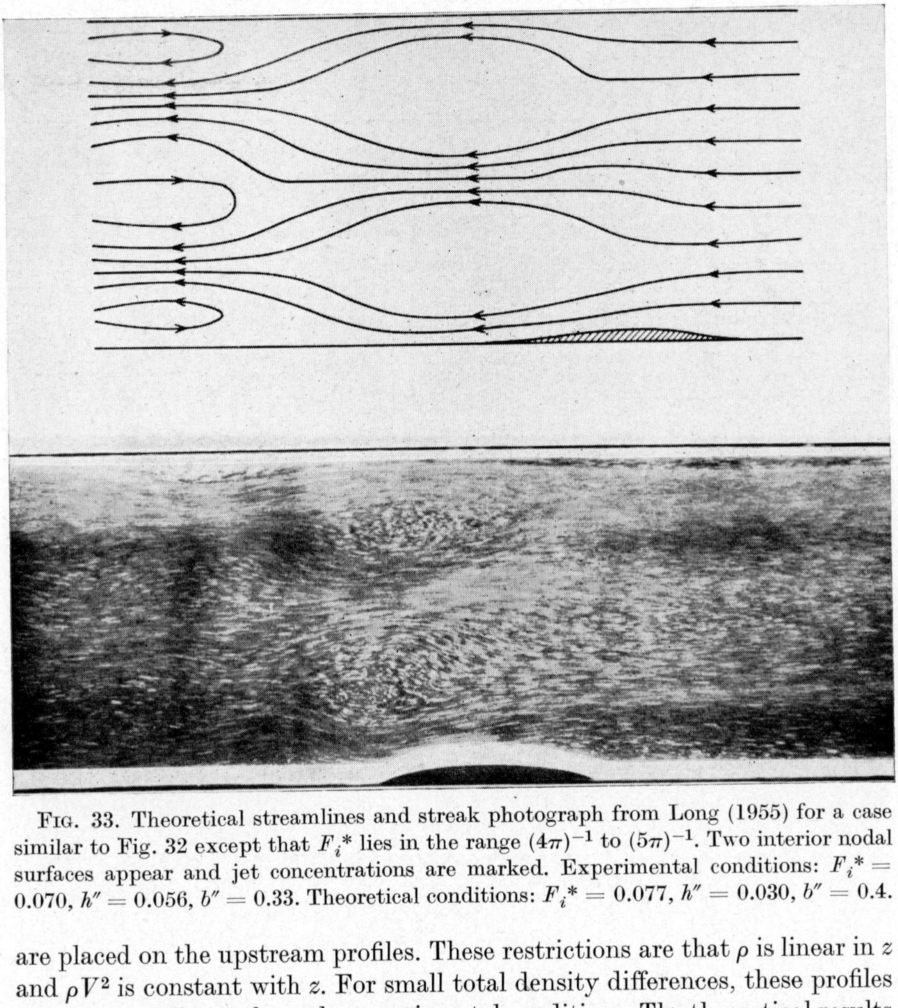

FIG. 33. Theoretical streamlines and streak photograph from Long (1955) for a case similar to Fig. 32 except that $F_i{}^*$ lies in the range $(4\pi)^{-1}$ to $(5\pi)^{-1}$. Two interior nodal surfaces appear and jet concentrations are marked. Experimental conditions: $F_i{}^* = 0.070$, $h'' = 0.056$, $b'' = 0.33$. Theoretical conditions: $F_i{}^* = 0.077$, $h'' = 0.030$, $b'' = 0.4$.

are placed on the upstream profiles. These restrictions are that ρ is linear in z and ρV^2 is constant with z. For small total density differences, these profiles are not too different from the experimental conditions. The theoretical results show that the solutions are singular at $F_i{}^* = (n\pi)^{-1}$, where n is an integer $\geqslant 1$, and differ strongly in character in the intervals between singular values. For $\pi^{-1} > F_i{}^* > (2\pi)^{-1}$, simple sinusoidal lee waves occur as in Fig. 32. The solutions indicated closed circulations (which imply static instability and turbulence because density is conserved on the steady streamlines) beyond certain maximum heights of the obstacle for given F_i^*. The changes in character of the flow as F_i^* just becomes less than $(n\pi)^{-1}$ are to introduce quasi-nodal streamlines in the vertical ($n - 1$ in number) and change phases of the vertical motion so that maximum currents (jets) alternate in the vertical (with n or $n - 1$ at various downstream stations). These general features are verified most remarkably as shown by the example in Fig. 33 for $(4\pi)^{-1} \gg F_i^* > (5\pi)^{-1}$.

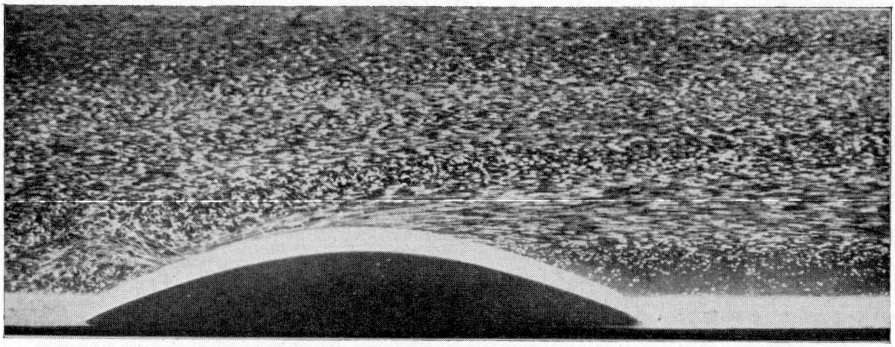

FIG. 34. Streak photograph from Long (1955) for a case similar to Figs. 32 and 33 except that F_i^* is considerably lower. (Full depth of the fluid is not shown.) Multiple jets are present in the vertical, four being visible in a depth of about 0.4δ at the ridge. The tendency for blocking the upstream flow up to the height of the obstacle increases through the three figures (see text). Experimental conditions: $F_i^* = 0.017$, $h'' = 0.150$, $b'' = 0.56$.

In Fig. 34, at a very low F_i^*, wavy perturbations are weak but a series of multiple jets appear in the vertical. Another feature, which increases in importance from Figs. 32 to 34 as F_i^* decreases, is blocking of the upstream layer ahead of the obstacle. This has the effect of actually changing the upstream approach profiles. It is very closely analogous to an effect discovered by Taylor (1921; Long, 1953a) in rotating fluids at very low Rossby numbers that is related to those discussed earlier. A small object translated very slowly parallel to the rotation axis ultimately pushes a lengthening column of the same cross section ahead of it. In both cases, at low R_0^* or low F_i^*, the strong transverse stability (rotational and gravitational, respectively) suppresses transverse motion and a column of fluid ultimately moves with the object. At still lower F_i^*, the same effect can occur on both leeward and windward sides of the object though Fig. 34 shows a lee wave.

The very deep analogies between the rotational and gravitational stability effects have been further clarified by a recent important paper of Ball (1959).

Ball shows that for steady inviscid flow with conservative density of the type considered by Long, the fundamental (vorticity) differential equations for two-dimensional stratified flow have exactly the same form as the equations of two-dimensional horizontal flow of the Rossby-wave type discussed above but in Cartesian coordinates with the β-parameter included (the so-called "β-plane"). The nondimensional equations are identical except that a parameter equivalent to $(F_i^*)^2$ occurs in one and, at the same position in the other, a "β-Rossby number" $R_{0\beta}^* \equiv V/L^2\beta$ that has been used by Long (1952). Thus, with similar boundary conditions in dimensionless variables, if $(R_{0\beta}^*)^{1/2} = F_i^*$ we will have identical solutions. Ball, as an illustrative example, shows a 500-mb map situation qualitatively like Long's Fig. 9 (1955) in which an $R_{0\beta}^*$ of about 0.02 compares with the experimental F_i^* of 0.13. The problems of jet formation and maintenance in the atmosphere and in Figs. 33 and 34 should be considerably clarified in the near future when the implications of these correspondences between the two types of systems are worked out in detail.

Returning, however, to the mountain flow problems themselves, Long (1957) has later tested the flow of stratified salt solutions over a generalized object with a vertically exaggerated profile like that of the Sierra Nevada range in California. Two cases of lee waves near Merced, California drawn from observations of the Sierra Wave Project (Holmboe and Klieforth, 1954) compare very well with experiments at comparable F_i^*. Some of the earlier work is summarized by Rouse (1951), and only one other example will be noted. Scorer (1956) and Wurtele (1957) have obtained solutions of the perturbation equations for the three-dimensional motion excited by an isolated mountain. These exhibited crescent-shaped crests and troughs in the internal lee waves very much as in the ship-wave problem. Wurtele draws attention to the fact that such crescent-shaped lee clouds had been photographed by Abe (1941) at Mount Fuji and produced in a stratified-current wind tunnel experiment (Fig. 35). Abe picked the experimental conditions on the basis of Reynolds number (using eddy viscosity in the manner mentioned in an earlier section), and does not give enough quantitative data to evaluate similarity other than qualitatively but clearly must have been reasonably correct in his choice of experimental parameters.

4.1.2. Free Cold Front Surges. The remaining two areas of work to be mentioned in connection with stable stratifications both involve situations where two fluid layers (sometimes more) are separated sufficiently sharply to approach a discontinuity of density at the internal boundary surface (as in Long's early work above). In large-scale problems, the polar front of the Norwegians is the most important example and was mentioned earlier but, whereas the large-scale properties are essentially influenced by rotation, at medium and small scales a number of very similar problems can be investigated

(a)

(b)

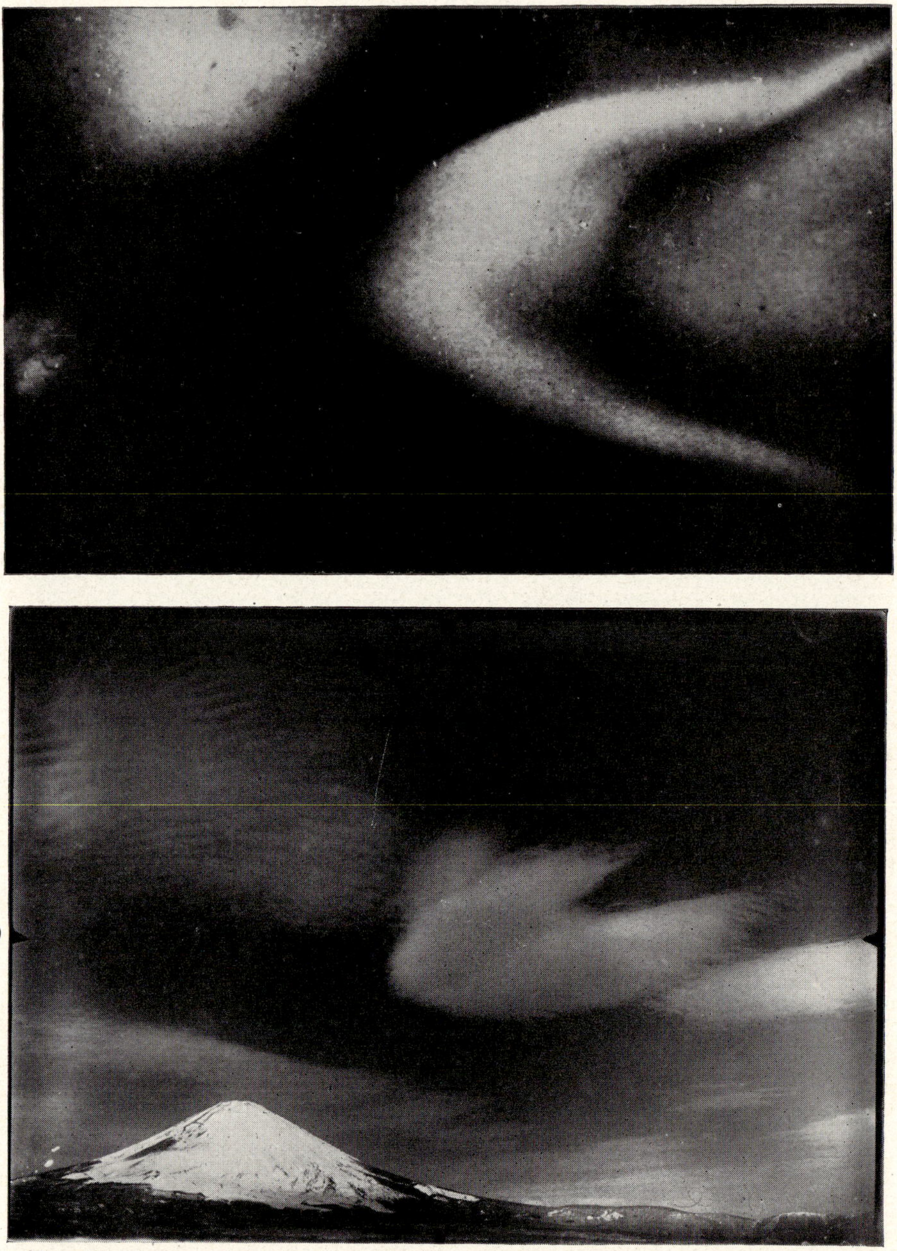

FIG. 35. Photographs from Abe (1941) of a crescent-shaped lee-wave cloud of Mt. Fuji. a. Crescent smoke pattern in lee of a model of Mt. Fuji produced in a wind-tunnel experiment with vertical stratification produced by a cold base plate. The lighting is by a horizontal slit at the level of the summit. Conditions: tunnel cross section 80 cm by 35 cm high, mountain height 7.6 cm, velocity probably 1 meter/sec or less, no temperature data. b. Photograph (February 16, 1938) of such a crescent cloud at Mt. Fuji.

without taking account of rotation (local Rossby numbers are high). Examples are local cold front motions (e.g., from thunderstorms or sea breezes), density or turbidity currents in lakes, rivers, and the oceans, and others. A situation that has given rise to a number of experimental investigations for geophysical, technical, or theoretical reasons is that where a denser layer is freely flowing under and displacing a lighter layer. For example, as in some experimental arrangements, when a gate is opened and a dense liquid allowed to flow freely through it into the bottom of a tank of light liquid. The leading edge region of dense liquid is a miniature cold front (with a "nose" or Böenkopf) and is essentially an internal counterpart of a surge, bore, or moving hydraulic jump in single-layer, channel flow or as in the problem of bursting of a dam. The analogy to squalls and cold fronts stimulated a number of early experiments (Schmidt, 1910, 1911, 1913; Ghatage, 1936; Prandtl, 1932, 1937) and a number of later quantitative ones have been done by people interested in small-scale density current problems in rivers or in general theory (Yih, 1947; Yih and Guha, 1955; Ippen and Harleman, 1952; von Kármán, 1940). In the above-described situation, the leading edge attains a characteristic steady velocity (Yih, 1947; Schmidt, 1910) as a function of the height of the nose and a characteristic profile (Ippen and Harleman, 1952). The most appropriate local nondimensional parameter is, in fact, an internal Froude number which can be constructed with the surge or cold front velocity V_c and the nose height h. The experimental results show that $F_i^* \equiv V_c/(gE_z^*h)^{1/2}$ is about constant at 0.9 (Schmidt, 1910; Ippen and Harleman, 1952) for reasonably high Reynolds numbers. A related result of Prandtl (1937) which states that the surge velocity should be one-half the fluid speed just to the rear in the denser layer and Ippen and Harleman's profile have recently been shown to agree roughly with fairly detailed atmospheric measurements in the regions up to about 1500 to 3000 ft during passage of squall lines in Australia (Berson, 1958). Recent interest in turbidity currents in the oceans has been stimulated by some cases of such strong currents that underwater cables were broken and by their apparent geological importance in redistributing ocean sediments. They are cold front or surge flows (beginning on an initial sloping bottom) in which the density excess is provided by suspended sediment and in which the velocities apparently can reach values as high as 25 meters/sec or more (Heezen and Ewing, 1955).

A very similar (though small-scale) phenomenon is the radially symmetric collapse of a heavy column of liquid that was studied by Penney and Thornhill (1952) and Martin and Moyce (1952a,b) in connection with the Bikini "base surge" at the underwater atomic bomb test of July, 1946 (Holzman, 1951). The base surge was essentially a circular cold front, similar to those associated with thunderstorms, which was produced by a density excess associated with the water and foreign material concentration in the explosion column.

4.1.3. Internal Seiches in Lakes. In contrast either to the free cold-front surge or the quasi-stationary waves induced by mountains, both freely progressing internal wave trains and standing oscillations should be possible in natural stratified media. Not too many such cases have been established observationally, mostly because of the lack of suitable and suitably frequent observations and because of the difficulty of separating effects of the waves from those due to other time-varying causes. In a very important case however, that of motions in lakes and other small bodies of water, oscillations of the standing type have been identified and shown to be highly important in the over-all behavior of the systems. This has led to extensive theoretical and experimental work. These motions are called seiches and were first identified quantitatively (though observed much earlier in many lakes) in Lake Geneva by Forel (1876), for a standing oscillation of the external surface-wave type. Internal seiches were first clearly identified in Loch Ness by Watson (1904). They, in common with ordinary seiches, are generally due to atmospheric wind and pressure variations. For small lakes, a strong wind changing fairly suddenly to weak values is one frequent exciting mechanism. The internal seiches involve considerably magnified vertical water displacements primarily because normal wind stresses can produce only small slopes of the free water surface (order of millimeters per kilometer) but for hydrostatic reasons can produce slopes of the thermocline exaggerated by about the reciprocal of $E_z^* \equiv -\Delta_z\rho/\rho$. Typically, $(E_z^*)^{-1}$ is of order 10^3 or greater, so that even in quite small lakes appreciable vertical displacements of the order of meters or tens of meters can occur at the thermocline and nearby levels.

Much of the early work on internal (and external) seiches has been concentrated on accounting for the observed oscillation periods. If T_n is the period of the nth mode of standing oscillation, h is the mean depth of the lake, and L a horizontal dimension, gravity wave theory for frictionless motion shows that

$$T_n(gh)^{1/2}/L = f \text{ (particular mode, shape, stratification characteristics)} \tag{4.1}$$

For the lowest simple (uninodal) external mode in a rectangular lake of uniform depth small compared to L, the oscillation is a long wave oscillation and the function f is a constant $= 2$. (Merian's formula; the nodal line must be parallel to a side and L the dimension normal to it.) A number of experimental studies have been made of external seiches in connection with theories that take account of depth and shape variations (e.g., White and Watson, 1905–1906; Kirchhoff and Hansemann, 1880; and recently Bottomley, 1956). Generally the experimental periods check the calculated values within a few per cent.

Following Watson's (1904) work on the internal seiche of Loch Ness, similar excellent experimental results on the periods of internal modes were obtained by Schmidt (1908) in a rectangular tank and by Wedderburn (1907) and Wedderburn and Williams (1911, for the Madüsee) in containers approximating the actual shape of the lakes involved. [Somewhat more complicated experiments on internal waves near boundaries with currents and wind stresses were carried out by Zeilon (1912, 1934).] A convenient form of the left-hand side of equation (4.1) then becomes of the nature of an internal Froude number (reciprocal). In recent years this line of investigation has been taken up again by Mortimer in a series of papers (1951a,b, 1952, 1953, 1954, 1955). In the experimental phases of his work, Mortimer used two or three layers of water, phenol and glycerine in a rectangular tank and generated motion by actual wind stresses on the free surface. Figure 36 (Mortimer, 1954) gives photographs of four stages in a two-layer experiment. The last two (*c*, *d*) show successive phases of oscillation after the wind is cut off. There are significant higher modes in *c* but the main motion is the fundamental internal mode shown in the opposite phase in *d*. Stages *a* and *b* in Fig. 36, while the wind is present, show the thickening of the upper layer downstream and the reverse effect in the lower layer that are responsible for exciting the internal seiche. The corresponding qualitative circulations are indicated and short shearing-instability waves on the interface are evident. Similar interesting displacement and waves occur in the three-layer experiments. Mortimer concludes from much observational evidence that very similar phenomena occur quite generally in the natural state and the resulting transfers of properties around and through the thermocline have profound effects on the biological and physical economy of the lake layers.

We cannot give a detailed discussion of his and other recent work but will comment only on some possible extensions of experimental work along the lines of the meteorological and oceanographic work surveyed earlier that appear perfectly feasible. The questions that might be raised concerning full similarity of Mortimer's wind generating mechanism to the natural occurrences have not yet been carefully checked and will depend on several fundamental questions. These involve proper similarity of the surface stresses and of the shearing-instability waves. A good deal of recent experimental work has been done on these essentially small-scale questions (e.g., Francis, 1954a,b,c; Keulegan, 1949; Van Dorn, 1953) and on the wind generation of waves. Much more can be expected in this area. The question of rotation influences on the motions is one that has often recurred and the answer has usually been equivalent to saying that lakes are usually small enough that local Rossby numbers are too large to allow significant rotational effects. However, for lakes of the size of the Great Lakes of North America the answer is probably quite the other way. Various agencies through the Great Lakes Research

Institute have recently begun obtaining essentially synoptic surveys for short periods of the motions and other properties in the Great Lakes (Ayers *et al.*, 1956, 1958). Lake Huron is one of those on which measurements have been made. A rough r_0 is about 150 km so that $r_0\Omega$ is about $7\frac{1}{2}$ meters/sec

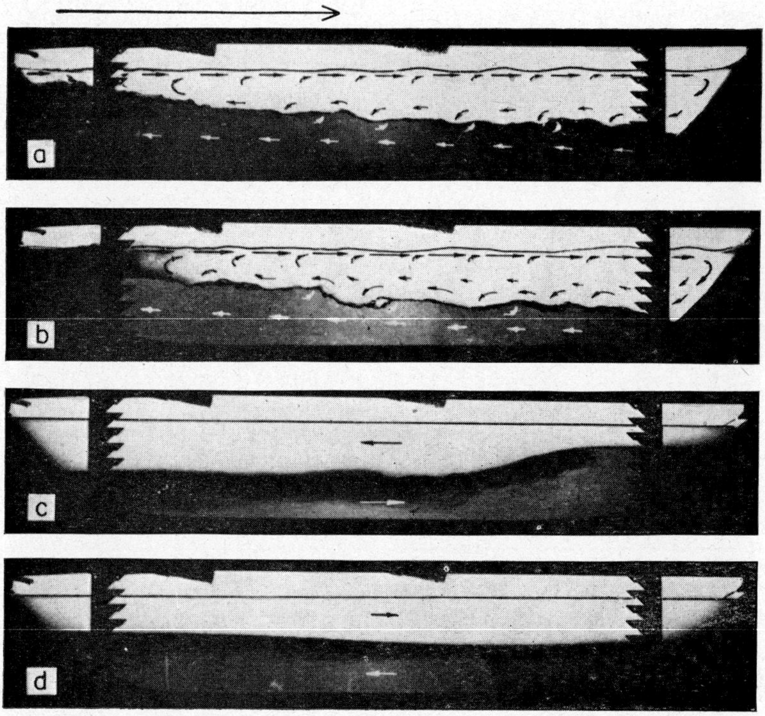

FIG. 36. Photographs from Mortimer (1954) showing wind-induced motions in a two-layer system of water (top) and a phenol-cresol mixture. While the wind is blowing from left to right (*a* and *b*) the thermocline tilts downward toward the right and small waves, some breaking, are induced on it by the circulation. The wind is cut off between *b* and *c* and a large-scale internal seiche begins which by *d* consists mainly of the fundamental mode. Conditions: average length 140 cm, width 7 cm, total depth 20 cm, height of interface 10 cm, upper density 1.00 gm/cm³, lower density 1.03–1.04 gm/cm³, $E_z^* \sim 0.03_4$, calculated and observed period of the fundamental internal seiche $\sim$ 22 sec.

using a local earth rotation value. The results of current, temperature, and other measurements made by Ayers *et al.* (1956) imply that kinematic Rossby numbers are of order 0.01 in the above characteristic unit, R_{0T}^*'s are of order 10^{-3}, S_z^*'s of order 10^{-3} to 10^{-2}, and F_i^*'s of order 10^{-1}. These values are all very similar to quite high rotation convection examples given in Fultz *et al.* (1959, e.g., Fig. 59) so that one is forced to the conclusion that these particular motions in Lake Huron are more strongly influenced by rotational effects than large-scale meteorological flow fields and only slightly less so than in the general oceanic flows. There thus appears to be a very strong ground for experiments of the sorts surveyed earlier on limnological problems, at least in the Great Lakes. Such experiments would also be of general interest in combining features of the meteorological experiments and of von Arx's wind-stress oceanographic experiments with effects of the closed boundaries and of imposed discharges through the inlets and outlets of such a lake as Huron.

4.2. Unstable Density Stratification

In contrast to phenomena like internal seiches, where some external agency must do mechanical work initially, a variety of important and familiar geophysical phenomena is produced by density arrangements which are vertically unstable so that buoyancy forces are more or less spontaneously capable of producing motion. An example is the small-scale vertical convection near the ground on a hot summer day. When the instability is at all vigorous, the results generally tend to involve systems having length scales of the order of the depth of convection. In consequence, the phenomena tend to be rather on the small-scale side from the point of view of this paper.

4.2.1. Bénard-Rayleigh Cellular Convection. At least passing mention must be made, however, of one type of vertically unstable convection even though most of the clear geophysical examples are small scale. This is the Bénard–Rayleigh cellular convection in horizontally uniform layers that occupies a crucial position because of the extensive and successful theoretical work that has been done since Rayleigh's pioneering paper of 1916. The important fact is that theory is able to predict the onset and to some extent the cellular form of the ensuing motion though this requires taking account both of viscous resistance and heat conduction and consequently of a complicated set of differential equations. A wide variety of cloud forms exhibit cellular structures. Detailed measurements by Mal (1930) have established the general reasonableness of the cellular interpretation for altocumulus in particular. A general survey of the cloud form aspects is given by Brunt (1951), and further experiments with this problem in mind have recently been begun by Tippelskirch (1957). Recent aircraft photographs of hurricane cloud systems and photographs from rockets even more strikingly exhibit the types of cloud element arrangements that are to be expected from the experimental work, especially when combined with a vertical shear of the velocity. Much has also been done on more complicated problems such as when the layer is rotating, where a vigorous development of theory has occurred. One group has

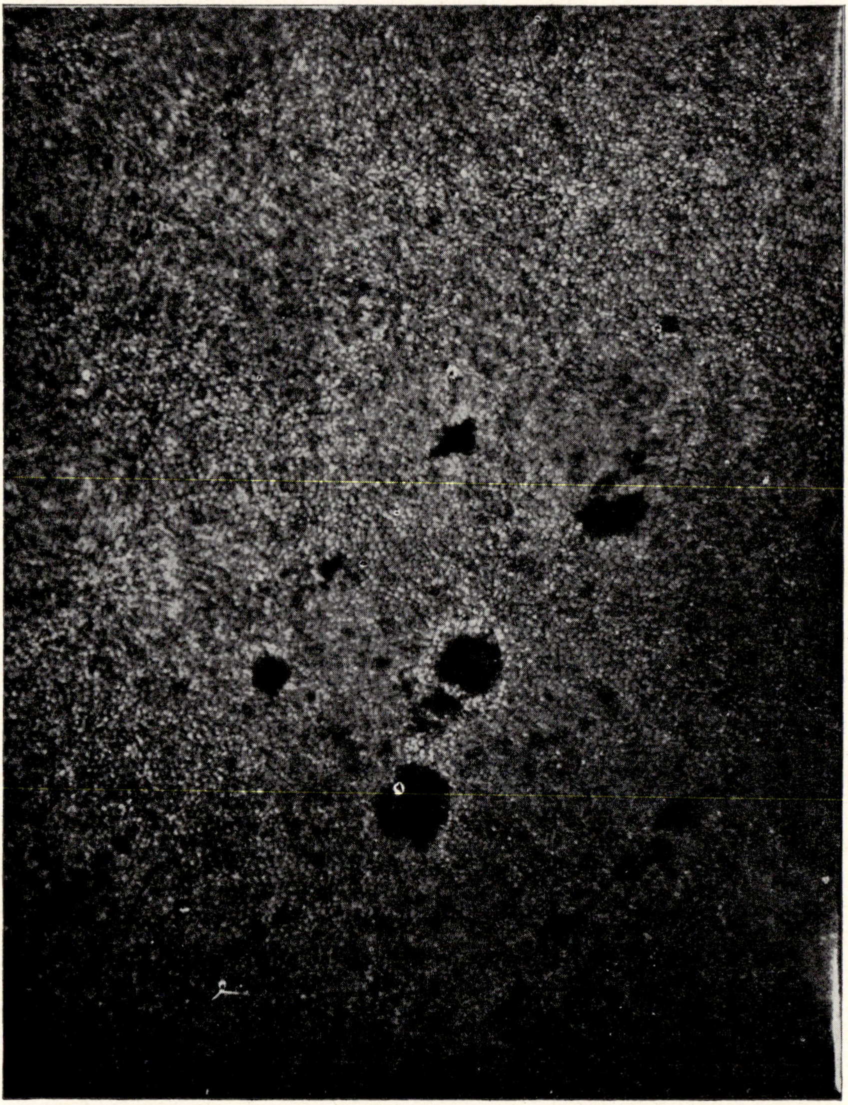

FIG. 37. Photograph of the solar photosphere in violet light (near Fraunhofer-G., taken by Janssen (1896) at the Meudon Observatory). In addition to the group of small sunspots, a network pattern is present that is highly reminiscent of experimental photographs of laboratory cellular convection taken later by Bénard and others. (Note especially the upper right-hand corner.) Diameters of elements of the réseau are of the order of a couple of seconds of arc or less; i.e., around 1 or 2000 km. Photograph: July 5, 1885, $8^h\ 23^m\ 12^s$ in central region about 11°S solar latitude, long side of plate about 500″ arc.

been mentioned earlier that forms a part of a systematic study of such hydrodynamic instabilities by Chandrasekhar (1957a).

A final, extremely interesting case where a cellular-convection interpretation has been given is that of the granulation in the solar photosphere (Wasiutynski, 1946). Extremely fine photographs (Fig. 37) of this granulation have been taken by Janssen (1896). The interpretation of his réseau as a net of Bénard-type convection cells has received recent support from photographs taken from high-altitude balloons (Loughhead and Bray, 1959; Schwarzschild and Schwarzschild, 1959) where the image quality is much better due to the reduction of atmospheric refraction effects.

4.2.2. Bubble Convection and Thermals. A considerably less regular phenomenon than cellular convection is the cumulus convection in the atmosphere which occurs on length scales at the lower end or below the distances mentioned at the beginning as medium scale. The classical meteorological view of this convection as a parcel process regards it as due essentially to the rise of lighter, limited portions of the fluid under positive buoyancy forces with passively induced motions in the surroundings. Considerable clarification of the detailed implications of this picture has resulted from a set of combined experimental and theoretical investigations of the behavior of bubbles and buoyant masses rising in liquids. A survey of the physical investigations has been given by Lane and Green (1956).

The work of geophysical interest has been done mainly in England and appears to have been stimulated, as in so many other fundamental questions of fluid mechanics, by some investigations of Sir G. Taylor. In two papers (Davies and Taylor, 1950; Taylor, 1950b) he established some very interesting properties of large bubbles (sufficiently large to make direct effects of surface tension small) rising in liquids. Davies and Taylor showed that the leading surface of such bubbles is very accurately a spherical cap with a turbulent wake and irregular, flat, lee surface. The bubbles very soon attain a constant vertical velocity W. Figure 38 gives two photographs by Davies and Taylor showing the spherical leading surface of an air bubble rising in nitrobenzene. The stagnation flow in the liquid relative to the bubble is very nearly an inviscid one and they show that the pressure condition on the interface implies that the terminal vertical velocity is related to the radius of curvature R of the spherical cap by

$$W/\sqrt{gR} = 2/3 \tag{4.2}$$

Their experimental data on rise velocities agree very closely with this relation which, it will be noted, is a special form of Froude criterion. Taylor (1945, 1950b) applied it also to give a very good estimate of the vertical velocity of the bubble of hot gas from the Trinity atomic explosion of 1945 in New Mexico.

Reynolds numbers of these experimental bubble motions are of the order of several thousand and inertia effects predominate in determining the rise

FIG. 38. Photographs from Davies and Taylor (1950) of an air bubble rising steadily in nitrobenzene. The upper surface is very accurately spherical. Conditions: interval between photographs 0.0103 sec, velocity of rise 36.7 cm/sec, radius of spherical cap 3.01 cm, bubble volume 8 + cm^3, $2/3\sqrt{gR} = 36.2$ cm/sec.

velocities. The applications to limited-volume cumulus convection growing out of or parallel with Taylor's work have proceeded along several lines. They have in common the feature of accepting inertial predominance, especially in the natural turbulent cases, and in applying dimensional arguments and governing equations restricted by various similarity hypotheses which lead to power-law dependences especially on the vertical coordinate. We cannot review details of the various developments but will confine ourselves to an example and descriptive comments on the principal types of investigations.

In the applications of bubble ideas to the interpretation of cumulus dynamics, modifications of Taylor's work above must be introduced to allow especially for the miscibility and mixing of the bubble with its environment and for the effects of stratification outside. Considering an individual buoyant bubble, Scorer (1957) starts from a hypothesis like equation (4.2) for the vertical velocity:

$$W/\sqrt{gE^*R} \equiv F \text{ (a number)} \tag{4.3}$$

where $E^* \equiv (\alpha_i - \alpha_0)/\alpha_0$ is a fractional expansion with the subscripts referring to specific volumes inside and outside the bubble (F is comparable to an internal Froude number). Here, E^* is a function of time due to mixing. With similarity hypotheses implying a conical expansion so that height $Z \propto R$, equation (4.3) can be related to the initial buoyancy or fractional expansion $E_0{}^*$ and leads to a prediction that the height Z, say of the apex of the bubble, is proportional to $t^{1/2}$ from a suitable time origin. Experiments in a neutrally stratified liquid mass have been carried out by releasing negatively buoyant masses (Fig. 39, Scorer and Ronne, 1956; Scorer, 1957). They lead to quite good verification of the $t^{1/2}$ relation and to a roughly constant value of $F \sim 1.2$. Similar results with $F \sim 1$, in spite of the radical differences of Reynolds numbers, have been obtained from measurements of atmospheric cumulus elements on time-lapse motion pictures by Malkus and Scorer (1955).

In the bubble problem, with either miscible or imiscible fluids, the local motions bear family resemblances to the motions around vortex rings. Some detailed experiments on buoyant rings were carried out by Turner (1957) in connection with questions raised by Dr. E. G. Bowen about the penetration upward in stable atmospheres of rings produced by demolition explosions. Some interesting and unexpected results were obtained by Turner which seem to scale roughly correctly to atmospheric conditions. Scorer's bubble experiments exhibited similar types of flow in a neutral environment (Scorer and Ronne, 1956; Scorer, 1957) which suggested vortex ring interpretations for thermals that agree with a considerable amount of gliding experience in both clear and cloudy air (Woodward, 1958, 1959).

The ideas in these bubble investigations have been much influenced and interconnected with earlier and concurrent studies of similarity solutions for continuous sources; thus, for example, a steady point heat source as contrasted with the impulsive release of buoyancy associated with bubble ideas. Analyses and experiments on the steady point source of buoyancy were apparently first carried out by Schmidt (1941) and then in a systematic series of investigations at the Iowa Institute of Hydraulic Research on both point and line sources by Rouse, Yih, and collaborators (Yih, 1951, 1952, 1956; Rouse *et al.*, 1952). A survey of the principal types of solutions was given

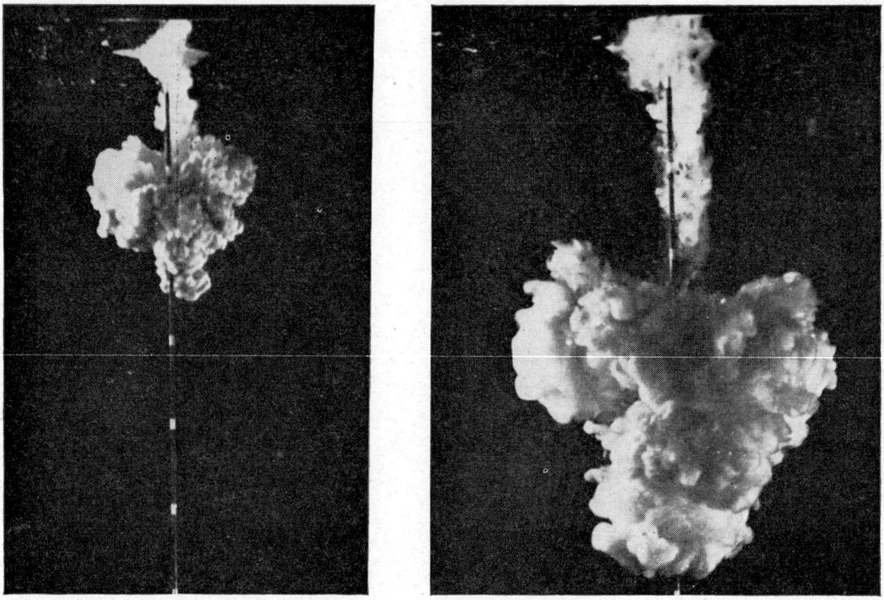

(a) (b)

FIG. 39. Two successive photographs from Scorer (1957) after the release of a volume of denser fluid from a pivoted hemispherical cup at the top surface of a tank of neutrally stratified water $2 \times 4 \times 3\frac{1}{2}$ ft high. The density excess is produced mainly by a suspension of white precipitate. As the motion proceeds, the apparent volume expands, the density excess is diluted by entrainment of surrounding liquid, and the vertical velocity decreased in magnitude. The stem of liquid left behind is more or less quiescent. Note the persistence of initial peculiarities despite the expansion and the strong resemblance to swelling cumulus forms (turn upside down). Conditions: initial volume $V_0 =$ 400 cm^3, initial volume deficiency ratio $E_0{}^* = 0.05$, time between photos 10.2 sec, distance between white markers 10 cm.

by Batchelor (1954) and later extensions, partly experimental and partly theoretical, by Morton *et al.* (1955), Priestley and Ball (1955), Morton (1959), and others. The especially interesting experimental parts of these studies are the quantitative results on the penetration of continuous plumes in stably stratified environments by Morton *et al.* (1955) and of vortex rings by Turner (1957). These in part represent the renewed taking up and quantifying of earlier work of Oberbeck (1877) and Czermak (1893) on mostly laminar source flows.

Finally, while they represent work somewhat outside the limits of this paper, some mention should be made of representative developments in the study of free-convection thermal turbulence on a small scale which, in part, stem from the previously mentioned work. These concern the flow above a uniform heated surface as in the meteorological surface layer. Many observational and theoretical studies are summarized by Priestley (1955, 1959), and recently important experimental studies on statistical properties of this type of free convection have been initiated by Townsend (Thomas and Townsend, 1957; Townsend, 1959). These are likely to have great consequences in deepening our understanding of atmospheric convection on many scales particularly when, as is extremely probable, detailed correspondences can be established with suitable natural observations.

4.2.3. Salt Domes. Returning now to some geological problems, a group of the most nearly quantitative and convincing experiments to date has revolved around the problem of the origin of the salt domes that are characteristic of many of the great petroleum fields such as those of the Texas and Gulf region (where more than two hundred are known) and Persia. The problem according to the currently accepted interpretation, as we will see, is in fact the creeping motion case of an instability problem lying somewhere between Bénard convection and the bubble or column convection of the last section.

The salt domes are columnar intrusions into upper sedimentary strata from deeply buried (as much as 10 to 20,000 ft) layers of salt. They have been extremely important because of the high frequency of associated trapped pools of oil and their consequent diagnostic use in prospecting. The salt columns take various rounded or flat-topped shapes extending from the source layer to near or above the present ground level and with average diameters of the order of two miles. The plar e sections and horizontal arrangements take many forms, often characteristic of the regional geology, but we will consider only the simple roughly circular type.

A variety of causes has been suggested for the origin of these structures in discussions since the mid-1800's (Nettleton, 1955). Since 1900, two general types of ideas have dominated and experiments have been actively pursued in both directions. The one trend ascribes the intrusion to various tectonic combinations of stresses on a weak layer, the salt, sandwiched among layers of different strengths and plasticities. Torrey and Fralich (1926), Escher and Kuenen (1929), and Link (1930) have published examples of experiments in which lateral squeezing, etc., was applied to various layer materials in rectangular boxes or a circular press, or a more plastic material was simply injected from below under high pressure.

The other group of investigations proceeds from the lower salt density and field observations of negative gravity anomalies near the domes to suggest that the rise of the domes is isostatic and is directly driven by the positive buoyancy of an initial disturbance of the salt in the surrounding heavier sediments. This was first definitely suggested by Arrhenius (1912) and has been pursued in a number of experimental investigations by Nettleton (1934), Dobrin (1941), and Parker and McDowell (1951, 1955). Nettleton and Dobrin both used pairs of viscous liquids such as heavy syrup over oil and with suitable initial disturbances obtained a number of different characteristic shapes much like those in the formation of drops from a dripping faucet. Dobrin made a variety of careful measurements of rise velocities of the column for different viscosities (10^3 to 10^6 poises) and found that the fully formed buoyant column reached a stage giving constant rise velocities just as with Taylor's bubbles (Davies and Taylor, 1950). The drag in this equilibrium is not that of a turbulent wake but of purely laminar viscous friction. For the container size (Fig. 40) and the very viscous liquids he used, a Froude number comparable to equation (4.3) was only 10^{-2} to 10^{-6}, the rise velocities lying between 0.3 and 10^{-5} cm/sec. A direct indicator of the viscous dominance is the range of Grashof numbers G_r^* from 10^{-1} to 10^{-7}.

Now the particular interest of these measurements is that the similarity arguments can be put in a form which, on quite plausible assumptions, gives a reasonable order-of-magnitude agreement with estimated times of formation of the salt domes. This is quite impressive evidence for the general physical picture and must involve one of the smallest ($\sim 10^{-12}$) time ratios at which a serious attempt has ever been made to obtain a physically similar model in addition to being almost unique in fitting time scaling with a geological experiment. If the radius $\bar{r}$ of a sphere equivalent to the salt volume in the dome is taken as length scale and the rise velocity W as a characteristic velocity, than a dimensionless ratio of buoyancy to viscous drag (the dominant effects) is

$$V^* \equiv g\Delta\rho\bar{r}^2/W\mu_t \tag{4.4}$$

Here μ_t is the viscosity coefficient of the upper layer (which Dobrin found, over a certain range, to dominate over effects of the lesser viscosity of the fluid layer playing the role of the salt) and $\Delta\rho$ is the density difference between the layers. Values of V^*, using the final steady W, clustered around 30 in the experiments or, using an average W up to the point where the liquid column had the proportions of a typical salt dome, scattered considerably more around 40–45 (some of the scatter is undoubtedly due to effects of the differing viscosities in the lower layer). Dobrin uses, as a typical salt dome configuration, a cylinder 20,000 ft deep with radius of one mile and densities of 2.2 and 2.4 gm/cm³ giving $\Delta\rho \sim 0.2$ gm/cm³. With a viscosity $\mu_t = 10^{21}$ poises, which is of the same order as those estimated from the glacial uplift measurements mentioned earlier, and using average $W \equiv \Delta z/t$ where t is the total time of rise of the salt dome, V^* as a similarity parameter implies:

$$t \sim \mu_t \Delta z(45)/g\Delta\rho r^2 \sim 6 \times 10^7 \text{ y}_r \tag{4.5}$$

This compares favorably with the estimates of formation time of about 10^8 years since the Cretaceous for the domes. Of course, the evidence for the

[*Gulf Research and Development Company photograph*

FIG. 40. Photograph from Dobrin (1941) of a salt-dome model using two viscous fluids. The upper denser fluid is boiled-down corn syrup and the lower is a heavy asphaltic oil. The container is a battery jar 7 in. in diameter and 9 in. high. The fluid layers were first thermostated for a day or so with the lighter layer on top. The jar was then inverted and the end (closed by a flexible diaphragm) placed on a flat dome to initiate a symmetrical mound on the bottom layer which then rises in a column in consequence of the buoyancy forces on the lighter layer. In this particular instance, the expanded head on the buoyant column is like one stage in the ordinary formation of bubbles. The velocity field must have some elements of the vortex-ring type of structure. Conditions (for a typical experiment, possible not precisely for the photograph): initial depth of light layer 1.9 cm, density of lower layer 0.970 gm/cm^3, density of upper layer 1.455 gm/cm^3 at room temperature (temperature 22.5°C), viscosity of lower layer $\mu_1 \sim 25{,}000$ poise, viscosity of upper layer $\mu_2 \sim 14{,}000$ poise, final steady rise velocity of dome peak $w \sim 0.021$ cm/sec, estimated $V^* \sim 38$, $R_e^* \sim 10^{-4}$.

choice of μ_t is very scanty, but the fact that the picture holds together even within several powers of 10 is encouraging.

A number of other detailed questions are discussed by Nettleton (1934) and Dobrin (1941) and in the very elaborate experiments conducted by Parker and McDowell (1951, 1955). The latter authors worked mainly with barite mud layers over asphalts. They devoted a great deal of attention to the horizontal patterns where the asphalt was not deliberately incited to rise in a single dome and to the complicated faulting patterns in the weak mud layers above the domes. Under reasonable similarity conditions for the viscosities and shear strengths of the muds, their results appear very satisfactorily realistic. As in the folding experiments discussed earlier, quantitative discussions do not appear to have proceeded very much farther but in view of the obvious gravitational instability nature of the problem, comparisons could be made in a number of respects with practicable theories. For example, in the two-viscous liquid approximation, horizontal scales of the dome and ridge experimental patterns could be checked against such theories as Hide's (1955) for this instability problem.

4.3. Seismic Waves

The last medium- or small-scale topic which will be briefly referred to is the geophysical one of seismic (elastic) waves in the earth's crust and interior. Only a few aspects of recent developments amidst a vast literature will be described because of the particularly interesting interplay with technological advances in this particular case.

Insofar as seismic propagation phenomena are governed by a simple wave equation, the question of modeling is extremely simple. It has been used in other contexts for over a century in physics teaching in the form of analogies between optical, acoustic, water waves, capillary waves, etc., for studying reflection, diffraction, and other effects. Essentially only length and velocity scales need be considered and these are set by the geometry of the region, wavelengths, and the wave phase velocity appearing in the wave equation. The complications in modeling for elastic waves arise from the possibility of both longitudinal (compressional) and transverse (shear) simple waves and of composite types, the limited range of elastic properties in modeling materials, and the complicated variation of wave propagation properties in the earth even when it is approximated by a series of homogeneous, isotropic layers.

A seismic wave model, because of the necessity for simultaneous compressional and shear-type motions, is pretty much forced to utilize elastic waves rather than any others. The fundamental compressional (P) and distortional (S) body wave velocities do not have great variations among the feasible metal, plastic, or stony solids: seismic P velocities range from 1.8 to 14 km/sec (at depth), S velocities from 2 to 7.3 km/sec while typical laboratory materials from aluminum alloy to wax have P velocities between 6.2 and 2 km/sec and S velocities between 3.2 and 1 km/sec. In most cases, consequently, velocity ratios are $O(1)$ although significant seismic experiments have been carried out with gelatines where wave velocities are a few meters per second (Terada and Tsuboi, 1927). The time-scale ratios must therefore be about the same as the model length ratios and frequencies are multiplied by the inverse. Put otherwise, since ratios of wavelengths to layer extent must be maintained, with models of a few tens of centimeters in size, wavelengths

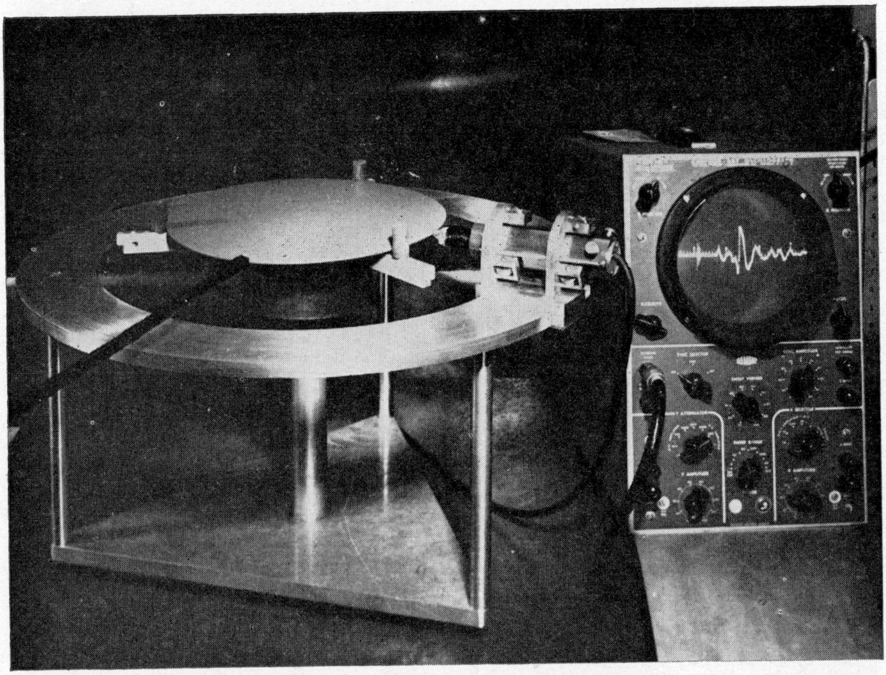

FIG. 41. Setup for two-dimensional (thin plate) model seismology (Oliver *et al.*, 1954) with oscilloscope display of a seismogram received at the edge of the disk. The source and detector are barium titanate crystals, the source operating at 2 to 100 pulses per second, each 15 μsec in length. The disk is of aluminum alloy 1/16 in. thick and 20 in. in diameter.

should be in the millimeter to centimeter range in the models and this implies frequencies in hundreds of kilocycles and in megacycles. This is in the ultrasonic range and much of the earlier work stems from purely physical investigations of ultrasonic vibrations. In fact, a good deal of the impetus from the experimental side in this work appears to have been the availability of high-quality piezoelectric transducers and the many high-frequency pulse and timing circuits developed in radar work during World War II.

A typical setup for the experiments uses a block or set of layers of elastic materials or a flat disk or sheet as in Fig. 41. The vibrations are excited and picked up by piezoelectric crystal transducers of barium titanate or other suitable salt. The source is excited by carefully shaped electric pulses of some thousand volts at repetition rates in the hundreds per second. This, if the cycling rate is chosen so that detected signals have decayed before the next pulse, allows the signal to be amplified and displayed with proper triggering and delays as a steady pattern on an oscilloscope.

A variety of types of problems has been successfully attacked experimentally. Some of the most significant have provided clear and quite detailed verification of theoretical calculations made long ago that had never received unequivocal confirmation because of the complexity of natural seismograms. The most important single instance is Lamb's (1904) calculation of the waves from a point or line source at the surface of a semi-infinite elastic solid. This and similar problems have been investigated by Kaufman and Roever (1951), Northwood and Anderson (1953), Tatel (1954), Knopoff (1955), and Shamina and Silayeva (1958). Tatel (1954), for example, who used a large steel block, was able to show that the seismograms at 5 to 10 cm from the source show just the simple structure calculated by Lamb of single pulse P and S arrivals (S rather obscure) followed by a larger Rayleigh surface wave oscillation. Tests with small drilled holes or a block contacting the surface as scattering centers immediately produce much more complicated seismograms partly as a result of mode conversions.

The complicated problems arising from velocity variations, reflections, refractions, etc., due to the horizontal layering in the earth's crust, have been attacked in a number of experimental investigations. These are especially relevant to the many problems of small- and medium-scale seismogram interpretation arising in the important techniques of refraction shooting with small explosions for oil and other types of exploration. Many of these experiments have involved generating acoustic waves with a spark source in air or water and following them into a plate or slab with piezoelectric detectors (Rieber, 1936; Howes *et al.*, 1953; Evans *et al.*, 1954; O'Brien, 1955; Sarrafian, 1956), and have included studies of the air-coupled flexural waves observed naturally on ice sheets (Press and Oliver, 1955). Others have worked with layered solid slabs to study the various direct and refracted arrivals (Press *et al.*, 1954; Levin and Hibbard, 1955). This general area has been very actively pursued in Russia (e.g., Riznichenko *et al.*, 1951; Riznichenko and Shamina, 1957). As an example of the scattering of detailed quantitative checks against theory, Fig. 42 gives results obtained by Clay and McNeil (1955) on a double layer of cement and marble. The quantities measured are amplitudes of reflected arrivals *versus* distance from the source of an initial P wave reflected respectively in a P or S mode. The agreement is very satisfactory.

A very interesting version of these experiments is that using thin plates described by Oliver *et al.* (1954), Fig. 41. They show that there are a number of advantages in replacing a scaled-down seismic problem by an equivalent plate problem where the original problem has sufficient symmetry. For example, most convenient laboratory materials have Poisson's ratios which are substantially higher than the value 1/4 taken as a best estimate for crust and mantle material whereas the equivalent plate waves in these materials, if wavelengths are large compared to the plate thickness, have an equivalent Poisson ratio considerably closer to 1/4. The Poisson ratio being a dimensionless ratio of elastic constants is, of course, one of the similarity parameters.

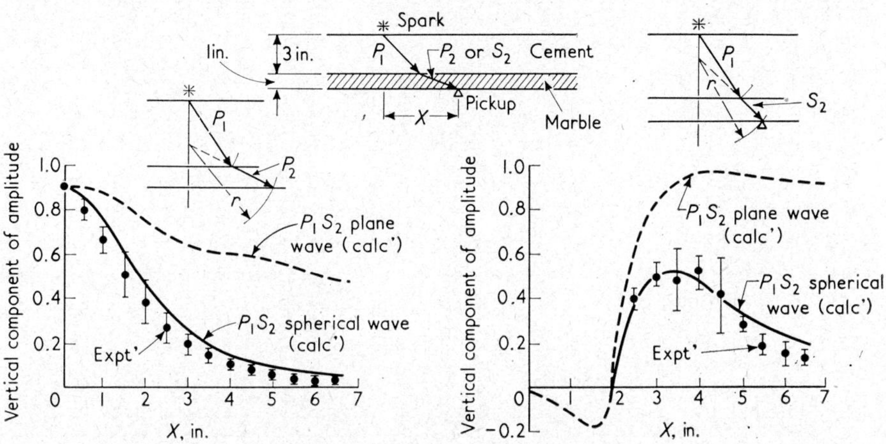

FIG. 42. Diagram of observed and theoretical amplitudes of P and S arrivals as functions of horizontal distance between source and detector at the bottom of a double layer of cement and marble (from Clay and McNeil, 1955).

The type of circular disk setup in Fig. 41 should in principle be applicable to a variety of global seismic problems if one represents the layering by cementing together a series of concentric rings of suitable sheet materials. Oliver *et al.* obtain, for example, good results for Rayleigh waves in a low-velocity layer over a high-velocity one and in Lamb's (1917) flexural wave problem which involved measuring group velocities.

At certain points, experiments of these types pushed in the direction of finite amplitude and inelastic effects will undoubtedly reach limits set by the practical problems of laboratory material properties, etc., but these limits have not yet been approached.

5. Conclusions

In concluding this brief and incomplete review of a rather wide variety of topics, it seems only necessary to restate a few of the already stated or implied lessons affecting future work. Particularly in connection with the large-scale problems, it is obvious that only the surface of the experimental possibilities has been scratched. It seems clear that the work carried out so far has already contributed to clarification of a number of geophysical and astrophysical problems in a degree that would have been generally dismissed as figments of a wild imagination had anyone predicted such developments, say, thirty or forty years ago. The consequence is certainly going to be a much healthier balance among theory, deliberate experiment, and observation in these fields and a much more effective and rapid development of the corresponding sciences.

The bad odor in which a number of the types of experiments we have reviewed has been held has resulted in great part from their earlier qualitative and therefore inconclusive character. Here the essentially new and promising feature, even though only the beginnings are at hand, is the systematic trend both in experiments and observational work toward intensive quantitative measurement and interpretation. The obverse attitudes from the theoretical side have been a combination of overvaluing the achievement of establishing general principles and selling short in despair the possibilities of obtaining reasonably conclusive theoretical conclusions from the principles with respect to some of the complicated physical problems that actually arise. Versions of these attitudes are, of course, common in all sciences and many more general developments than we have mentioned are contributing to their replacement by more philosophic, balanced, and optimistic points of view.

We may perhaps close with two relevant quotations. The first is from Daniel Bernoulli as quoted by Truesdell (1956): ". . . there is no philosophy which is not founded upon a knowledge of the phenomena, but to get any profit from this knowledge it is absolutely necessary to be a mathematician." The second is from George Boole (1872): "But while this consideration vindicates to (symbolical methods) a high position, it seems to me clearly to define that position. As discussions about words can never remove the difficulties that exist in things, so no skill in the use of those aids to thought which language furnishes can relieve us from a prior and more direct study of the things which are the subjects of our reasonings."

6. Acknowledgments

I am very much indebted to many colleagues and friends for extensive discussions, suggestions, and help in collecting material on the topics touched on. I hope they will excuse the lack of a full enumeration. That part of this survey which covers the experimental work at Chicago has essentially been made possible by the long-term financial support of that work received from the Geophysics Research Directorate of the Air Force Cambridge Research Center, U.S. Air Force.

List of Symbols

$R_0^* \equiv V/L\Omega$ —	Kinematic Rossby number
V	A representative velocity relative to a coordinate frame rotating with angular velocity Ω
L	A representative horizontal length scale
Ω	Angular velocity of rotation of a suitable coordinate frame
$R_{0T}^* \equiv gE_r^*\delta/fr_0\Omega(\Delta r)$	Thermal Rossby number
g	Magnitude of the gravity acceleration
E_r^*	Representative fractional expansion across horizontal strip of width Δr
δ	Fluid layer depth
f	Coriolis parameter (2Ω for a disk, $2\Omega \sin \phi$ for a sphere)
ϕ	Latitude on a sphere
r_0	Reference radius
Δr	Reference horizontal width
$S_z^* \equiv gE_z^*\delta/f^2(\Delta r)^2$	Vertical stability parameter
E_z^*	Representative fractional expansion through depth δ
$R_i^* \equiv S_z^*4(\Delta r)^2/(R_{0T}^*)^2 r_0^2$	Richardson number
$F_i^* \equiv V/(gE_z^*\delta)^{\frac{1}{2}}$	Internal Froude number
$R_e^* \equiv VL/\nu$	Reynolds number
ν	Kinematic viscosity
$P_e^* \equiv VL/\kappa$	Peclet number
κ	Thermometric conductivity
$\mathbf{V}$	Relative velocity vector
$(\mathbf{VV})$	Relative momentum transport tensor on unit mass basis
∇	Gradient differentiation operator
$\hat{\Omega}$	Unit vector in direction of vector angular velocity of the coordinate frame
$R^* \equiv \Omega r_0^2/\nu$	Rotation Reynolds number
$\delta_v \equiv (\nu/\Omega)^{\frac{1}{2}}$	Characteristic depth of layer of viscous influence
$R_\delta^* \equiv \Omega\delta^2/\nu$	Rotation Reynolds number on depth basis
$\overline{(\mathbf{VV})}$	Average momentum transport tensor by average velocity on unit mass basis
$\overline{(\mathbf{V'V'})}$	Eddy momentum transport tensor on unit mass basis (negative Reynolds' stress tensor)
$N_u^* \equiv H/kA\Delta_r T(\Delta r)^{-1}$	Nusselt number
H	Heat transfer per unit time across surface of area A
k	Thermal conduction coefficient
$\Delta_r T$	Representative horizontal temperature difference perpendicular to surface above across strip of width Δr
$G_r^* \equiv gE_r^*\delta^3/\nu^2$	Grashof number
$R_a^* \equiv -gE_z^*\delta^3/\nu\kappa$	Rayleigh number
$P^* \equiv \nu/\kappa$	Prandtl number
$\overline{(T\mathbf{V})}$	Vector proportional to mean convective heat transport
$\overline{(T'\mathbf{V}')}$	Vector proportional to eddy convective heat transport
d/dt	Material time differentiation operator

$\mathbf{V}_a$	Absolute velocity vector (with respect to inertial coordinate frame)
$\boldsymbol{\zeta}_a \equiv \nabla \times \mathbf{V}_a$	Absolute vorticity vector
$\mathbf{r}$	Position vector
$d\boldsymbol{\sigma}$	Vector element of area
$\overline{\zeta_a}$	Magnitude of absolute vorticity averaged with respect to area across cross section of a vortex tube
$\beta \equiv \frac{1}{r_0}\frac{\partial f}{\partial \phi}$	Rate of variation of f, the Coriolis parameter, in the meridional direction on a sphere
ζ	Vertical (radial) component of the vorticity of relative motion
ω_r	Relative angular velocity of a zonal current on a sphere
m	Longitudinal wave number on a sphere
$\overline{[u'v']}$	Eddy meridional transport of zonal momentum on unit mass basis averaged with respect to longitude and time
η	Ratio of inner to outer radius of an annulus between concentric circular cylinders
S	Yield strength of a solid
subscript M	Model values of a variable
subscript P	Prototype values of a variable
ρ	Density
μ	Magnetic permeability
$\mathbf{H}$	Magnetic field vector
$\mathbf{j}$	Current density vector
σ	Electrical conductivity
$R_m{}^* \equiv VL/(\mu\sigma)^{-1}$	Magnetic Reynolds number
$\mathbf{B} \equiv \mu\mathbf{H}$	Magnetic induction vector
$K_m{}^* \equiv V^2(\mu\rho)/B^2$	Characteristic ratio of inertia to magnetic force
B	Characteristic magnitude of magnetic induction vector
$V_A \equiv B/(\mu\rho)^{\frac{1}{2}}$	Alfvén velocity
$Q^* \equiv \sigma B^2 L/\rho\nu$	Characteristic ratio of magnetic to viscous force
$C_\sigma{}^* \equiv \sigma B^2/2\rho\Omega$	Characteristic ratio of magnetic to Coriolis force
$T^* \equiv (2R\delta^*)^2 = \frac{4\Omega^2\delta^4}{\nu^2}$	Taylor number
θ	Potential temperature
$F^* \equiv V/(g\delta)^{\frac{1}{2}}$	Froude number
$R_{0\beta}{}^* = V/\beta L^2$	β-Rossby number
h	Nose height of a cold-front type surge or mean depth of fluid layer
T_n	Period of the nth mode of standing gravity wave oscillation
W	Vertical velocity component of the apex of a bubble or fluid column
R	Radius of curvature at apex of a bubble or fluid column
$E^* \equiv (\alpha_i - \alpha_0)/\alpha_0$	Fractional expansion of a bubble relative to environment
α_i	Specific volume inside bubble
α_0	Specific volume in the environment

Z	Vertical coordinate
$V^* \equiv g\Delta\rho\bar{r}^2/W\mu_t$	Characteristic ratio of buoyancy to viscous force
$\Delta\rho$	Characteristic density difference between a fluid column and environment
$\bar{r}$	Radius of sphere equal in volume to a fluid column
μ_t	Dynamic viscosity of the top layer of two liquid layers

References

Abbé, C. (1907a). Comprehensive maps and models of the globe for special meteorological studies. *Monthly Weather Rev.* **35**, 559–564.

Abbé, C. (1907b). Projections of the globe appropriate for laboratory methods of studying the general circulation of the atmosphere. *Bull. Am. Math. Soc.* **13**, 502–506.

Abe, M. (1941). Mountain clouds and their connected air currents, Part II. *Bull. Cen. Meteorol. Obs. Tokyo* **7**, 93–145.

Adams, F. D. (1918). Experiment in geology. *Bull. Geol. Soc. Am.* **29**, 167–186.

Alfvén, H. (1942). On the existence of electromagnetic-hydrodynamic waves. *Arkiv. mat. ast. fys.* **29**B (2), 1–7.

Alfvén, H. (1950). "Cosmical Electrodynamics." Oxford Univ. Press, London and New York.

Arons, A. B., and Stommel, H. (1956). A β-plane analysis of free periods of the second-class in meridional and zonal oceans. *Deep-Sea Res.* **4**, 23–31.

Arrhenius, S. (1912). Zur Physik der Salzlagerstätten. *Meddel. K. Vetakad. Nobelinst.* **2** (20).

Avebury, Lord (1903). An experiment in mountain building. *Quart. J. Geol. Soc.* **59**, 348–355.

Ayers, J. C., Anderson, D. V., Chandler, D, C., and Lauff, G. H. (1956). Currents and water masses of Lake Huron (1954 Synoptic Surveys). *Ontario Dept. Lands and Forests Div. Res., Res. Rept. No.* **35**; *Univ. Michigan, Gt. Lakes Res. Inst., Tech. Pap. No.* **1**.

Ayers, J. C., Chandler, D. C., Lauff, G. H., Powers, C. F., and Henson, E. B. (1958). Currents and water masses of Lake Michigan. *Gt. Lakes Res. Inst., Publ. No.* **3**, 169 pp.

Bain, G. W., and Beebe, J. H. (1954). Scale model reproduction of tension faults. *Am. J. Sci.* **252**, 745–754.

Ball, F. K. (1959). Long waves, lee waves, and gravity waves. *Quart. J. Roy. Meteorol. Soc.* **85**, 24–30.

Batchelor, G. K. (1953). The conditions for dynamical similarity of motions of a frictionless perfect-gas atmosphere. *Quart. J. Roy. Meteorol. Soc.* **79**, 224–235.

Batchelor, G. K. (1954). Heat convection and buoyancy effects in fluids. *Quart. J. Roy. Meteorol. Soc.* **80**, 329–358.

Bennett, W. H. (1955). Self-focusing streams. *Phys. Rev.* **98**, 1584–1593.

Bennett, W. H. (1959). Solar proton stream forms with a laboratory model. *Rev. Sci. Instrum.* **30**, 63–69.

Bennett, W. H., and Hulburt, E. O. (1953). Magnetic self-focusing of auroral protons. *Phys. Rev.* **91**, 1562.

Berson, F. A. (1958). Some measurements on undercutting cold air. *Quart. J. Roy. Meteorol. Soc.* **84**, 1–16.

Bigelow, F. H. (1902a). IV Review of Ferrel's and Oberbeck's theories of the local and general circulations. *Monthly Weather Rev.* **30**, 163–171.

Bigelow, F. H. (1902b). Studies on the statics and kinematics of the atmosphere in the United States: V Relations between the general circulation and the cyclones and anticyclones. *Monthly Weather Rev.* **30**, 250–258.

Biot, M. A. (1957). Folding instability of a layered viscoelastic medium under compression. *Proc. Roy. Soc.* (*London*) **A242,** 444–454.

Birkeland, K. (1908, 1913). On the cause of magnetic storms and the origin of terrestrial magnetism. (Part II: Polar magnetic phenomena and terrella experiments). *The Norwegian Aurora Polaris Expedition* 1902–3 **1,** 801 pp.

Birkhoff, G. (1950). "Hydrodynamics: A Study in Logic, Fact, and Similitude." Princeton Univ. Press, Princeton, New Jersey.

Bjerknes, V., Bjerknes, J., Solberg, H., and Bergeron, T. (1933). "Physikalische Hydrodynamik." J. Springer, Berlin.

Block, L. (1955). Model experiments on aurorae and magnetic storms. *Tellus* **7,** 65–86.

Block, L. (1956). On the scale of auroral model experiments. *Tellus* **8,** 234–238.

Boole, G. (1872). "A Treatise on Differential Equations," 3rd ed., Macmillan, London.

Bostick, W. H. (1956). Experimental study of ionized matter projected across a magnetic field. *Phys. Rev.* **104,** 292–299.

Bostick, W. H., and Twite, O. A. (1957). Simulation of astrophysical processes in the laboratory. *Nature* **179,** 214–215.

Bottomley, G. A. (1956). The free oscillations of Lake Wakatipu, New Zealand. *Trans. Am. Geophys. Un.* **37,** 51–55.

Bridgman, P. W. (1931). "Dimensional Analysis." Yale Univ. Press, New Haven, Connecticut.

Brüche, E. (1930). Modellversuche mit sichtbaren Elektronenstrahlen zu Störmers Theorie des Polarlichtes und des 'Weltraumechos'. *Naturwissenschaften* **18,** 1085–1093.

Brunberg, E.-Å. (1953). Experimental determination of electron orbits in the field of a magnetic dipole. *Tellus* **5,** 135–156.

Brunberg, E.-Å. (1956). Cosmic rays in the terrestrial magnetic dipole field. *Tellus* **8,** 215–233.

Brunberg, E.-Å., and Dattner, A. (1953). Experimental determination of electron orbits in the field of a magnetic dipole II. *Tellus* **5,** 269–292.

Brunt, D. (1939). "Physical and Dynamical Meteorology," 2nd ed. Cambridge Univ. Press, London and New York.

Brunt, D. (1951). Experimental cloud formation. *Compendium of Meteorol.* **1951,** 1255–1262.

Bucher, W. H. (1951). Colloquium on plastic flow and deformation within the earth. *Trans. Am. Geophys. Un.* **32,** 516.

Bucher, W. H. (1956). Orogenic deformation as a symptom of subcrustal changes. (*Conf. Theor. Geophys.*), *J. Geophys. Res.* **61,** 374–377.

Bull, A. J. (1932). The pattern of a contracting earth. *Geol. Mag.* **69,** 73–75.

Bullard, E. C., and Gellman, H. (1954). Homogeneous dynamos and terrestrial magnetism. *Phil. Trans. Roy. Soc. London* **247,** 213–278.

Burhorn, F., Griem, H., and Lochte-Holtgreven, W. (1954). Auftreten von Magnetfeldern in zirkulierendem flüssigen Quecksilber. *Z. Phys.* **137,** 175–189.

Cadell, H. M. (1890). Experimental researches on mountain building; thrust faults without folding. *Phil. Trans. Roy. Soc. Edinburgh* **35,** 337–357.

Chamberlin, R. T., and Shepard, F. P. (1923). Some experiments on folding. *J. Geol.* **31,** 490–512.

Chandrasekhar, S. (1952a). The thermal instability of a fluid sphere heated within. *Phil. Mag.* (7) **43,** 1317–1329.

Chandrasekhar, S. (1952b). On the inhibition of convection by a magnetic field. *Phil. Mag.* (7) **43,** 501–532.

Chandrasekhar, S. (1954a). The stability of viscous flow between rotating cylinders in the presence of a radial temperature gradient. *J. Rat. Mech. Anal.* **3,** 181–207.

Chandrasekhar, S. (1954b). On the inhibition of convection by a magnetic field: II. *Phil. Mag.* (7) **45,** 1177–1191.

Chandrasekhar, S. (1954c). The instability of a layer of fluid heated below and subject to the simultaneous action of a magnetic field and rotation. *Proc. Roy. Soc.* (*London*) **A225,** 173–184.

Chandrasekhar, S. (1956a). The instability of a layer of fluid heated below and subject to the simultaneous action of a magnetic field and rotation II. *Proc. Roy. Soc.* (*London*) **A237,** 476–484.

Chandrasekhar, S. (1956b). Hydromagnetic oscillations of a fluid sphere with internal motions. *Astrophys. J.* **124,** 571–579.

Chandrasekhar, S. (1957a). Thermal convection. *Daedalus* **86**(4), 323–339.

Chandrasekhar, S. (1957b). The thermal instability of a rotating fluid sphere heated within. *Phil. Mag.* (8)**2,** 845–858.

Chandrasekhar, S., and Fermi, E. (1953). Magnetic fields in spiral arms. *Astrophys. J.* **118,** 113–115.

Clarke, B. L. (1937). Folding of the California coast range type illustrated by a series of experiments. *J. Geol.* **45,** 296–319.

Clay, C. S., McNeil, H. (1955). An amplitude study on a seismic model. *Geophysics* **20,** 766–773.

Cloos, H. (1930). Zur experimentellen Tektonik. *Naturwissenschaften* **18,** 741–747.

Cloos, H. (1931). Zur experimentellen Tektonik, Brüche, und Falten. *Naturwissenschaften* **19,** 242–247.

Cloos, E. (1955). Experimental analysis of fracture patterns. *Bull. Geol. Soc. Am.* **66,** 241–256.

Corby, G. A. (1954). The airflow over mountains. *Quart. J. Roy. Meteorol. Soc.* **80,** 491–521.

Corn, J., and Fultz, D. (1955). Synoptic analyses of convection in a rotating cylinder. *Geophys. Res. Pap. U.S.A.F. No.* **34,** 72 pp.

Cowling, T. G. (1933). The magnetic field of sunspots. *Monthly Not. Roy. Ast. Soc.* **94,** 39–48.

Czermak, P. (1893). Ueber warme Luft- und Flüssigkeitsströmungen. *Ann. Phys.* [3] **50,** 329–344.

Daubree, A. (1879a). "Études Synthétiques de Géologie Expérimentale." Dunod, Paris.

Daubree, A. (1879b). Expériences sur l'action et la réaction exercées sur un spheroide que se contracte, par une envellope adhérente et non contractile. *Bull. soc. géol. France* **7,** 152–157.

Davies, R. M., and Taylor, Sir G. (1950). The mechanics of large bubbles rising through extended liquids and through liquids in tubes. *Proc. Roy. Soc.* (*London*) **A200,** 375–390.

Davies, T. V. (1953). The forced flow of a rotating viscous liquid which is heated from below. *Phil. Trans. Roy. Soc. London* **A246,** 81–112.

Davies, T. V. (1956). The forced flow due to heating of a rotating liquid. *Phil. Trans. Roy. Soc. London* **A249,** 27–64.

De Chaucourtois, B. (1878). Moyens simples d'imiter la formation des chaînes de montagnes sur un globe et celle des cirques volcaniques sur un plan, conformément a la théorie des soulèvements. *Compt. rend. acad. sci.* **86,** 1091–1092.

DeLury, J. S. (1926). Experiments with model glaciers. *J. Geol.* **34,** 266–274.

Dive, P. (1932). Viscosité du fluide terrestre dans un modele reduit. *Compt. rend. soc. Phys. et hist. nat. Genève* **49,** 105.

Dobrin, M. B. (1939). Recreating geological history with models the experimental study of the earth's crustal deformation. *J. Appl. Phys.* **10,** 360–371.

Dobrin, M. B. (1941). Some quantitative experiments on a fluid salt-dome model and their geologic implications. *Trans. Am. Geophys. Un.* **22**, 528–542.

Dryden, H. L., Murnaghan, F. D., and Bateman, H. (1932). Report of the Committee on Hydrodynamics. *Bull. Natl. Res. Council (U.S.) No.* **84**, 634 pp.

Ekman, V. W. (1906). On dead-water: Being a description of the so-called phenomenon often hindering the headway and navigation of ships in Norwegian fjords and elsewhere, and an experimental investigation of its causes, etc. *The Norwegian North Polar Expedition 1893–1896* **5** (15), 1–152 (1906).

Ekman, V. W. (1905). On the influence of the earth's rotation on ocean-currents. *Arkiv. Mat. ast. fys.* **2**(11), 1–52.

Ekman, V. W. (1923). Über Horizontalzirkulation bei winderzeugten Meeresströmungen. *Arkiv. mat. ast. fys.* **17** (26), 1–74.

Elsasser, W. M. (1950). The earth's interior and geomagnetism. *Rev. Mod. Phys.* **22**, 1–35.

Elsasser, W. M. (1954). Dimensional relations in magneto-hydrodynamics. *Phys. Rev.* **95**, 1–5.

Elsasser, W. M. (1955). Hydromagnetism. I. A review. *Am. J. Phys.* **23** (9), 590–609.

Elsasser, W. M. (1956a). Hydromagnetism. II. A review. *Am. J. Phys.* **24**(2), 85–110.

Elsasser, W. M. (1956b). Hydromagnetic dynamo theory. *Rev. Mod. Phys.* **28**(2), 135–163.

Escher, B. G., and Kuenen, P. H. (1929). Experiments in connection with salt domes. *Leidse. Geol. Mededeel.* **3**, 151–182.

Evans, J. F., Hadley, C. F., Eisler, J. D., and Silverman, D. (1954). A three-dimensional seismic wave model with both electrical and visual observation of waves. *Geophysics* **19**, 220–236.

Ewing, W. M., Jardetzky, W. S., and Press, F. (1957). "Elastic Waves in Layered Media." McGraw-Hill, New York.

Exner, F. M. (1923). Über die Bildung von Windhosen und Zyklonen. *Sitzber. Akad. Wiss. Wien Abt. IIa* **132**, 1–16.

Faller, A. J. (1956). A demonstration of fronts and frontal waves in atmospheric models *J. Meteorol.* **13**, 1–4.

Faller, A. J. (1958). An optical method for the measurement of surface waves in the laboratory. *Trans. Am. Geophys. Un.* **39**, 716–720.

Faller, A. J., and von Arx, W. S. (1959). The modeling of fluid flow on a planetary scale. *Proc. 7th Hydraul. Conf., State Univ. Iowa, Iowa City 1958*, pp. 53–70.

Focken, C. M. (1953). "Dimensional Methods and Their Application." Arnold, London.

Forel, F. A. (1876). La formule des séiches. *Arch. sci. phys. et nat.* **57**, 278–292.

Francis, J. R. D. (1954a). Laboratory models of sea surface phenomena. *Weather* **9** (6), 163–168.

Francis, J. R. D. (1954b). Wave motions and the aerodynamic drag on a free oil surface. *Phil. Mag.* [7] **45**, 695–702.

Francis, J. R. D. (1954c). Wind stress on a water surface. *Quart. J. Roy. Meteorol. Soc.* **80**, 438–443.

Frenzen, P. (1955). Westerly flow past an obstacle in a rotating hemispherical shell. *Bull. Am. Meteorol. Soc.* **36**, 204–210.

Fuglister, F. C., and Worthington, L. V. (1951). Some results of a multiple ship survey of the Gulf Stream. *Tellus* **3**, 1–14.

Fultz, D. (1949). A preliminary report on experiments with thermally produced lateral mixing in a rotating hemispherical shell of liquid. *J. Meteorol.* **6**, 17–33.

Fultz, D. (1950a). Experiments combining convection and rotation and some of their possible implications. *Proc. Midwestern Conf. Fluid Dynamics 1st Conf.* pp. 297–304.

Fultz, D. (1950b). Experimental studies related to atmospheric flow around obstacles. *Geofyis. Pural Appl.* **17**, 89–93.

Fultz, D. (1951a). Experimental analogies to atmospheric motions. *Compendium Meteorol. American Meteorological Society, Boston*, 1235–1248.

Fultz, D. (1951b). Non-dimensional equations and modelling criteria for the atmosphere. *J. Meteorol.* **8**, 262–267.

Fultz, D. (1952). On the possibility of experimental models of the polar-front wave. *J. Meteorol.* **9**, 379–384.

Fultz, D. (1956a). A survey of certain thermally and mechanically driven fluid systems of meteorological interest. *Fluid Models in Geophysics, Proc. 1st Symposium on the use of Models in Geophys. Fluid Dynamics Baltimore 1953*, pp. 27–63.

Fultz, D. (1956b). A fluid convection experiment of special theoretical interest. *J. Geophys. Res.* **61**, 328–334.

Fultz, D. (1959). A note on overstability and the elastoid-inertia oscillations of Kelvin, Solberg and Bjerknes. *J. Meteorol.* **16**, 199–208.

Fultz, D., and Frenzen, P. (1955). A note on certain interesting ageostrophic motions in a rotating hemispherical shell. *J. Meteorol.* **12**, 332–338.

Fultz, D., and Kaylor, R. (1959). The propagation of frequency in experimental baroclinic waves in a rotating annular ring. *The Atmosphere and the Sea in Motion, Rossby Memorial Vol.* Rockefeller Inst. Press and Oxford Univ. Press 359–371.

Fultz, D., and Long, R. R. (1951). Two-dimensional flow around a circular barrier in a rotating spherical shell. *Tellus* **3**, 61–68.

Fultz, D., Long, R. R., Owens, G. V., Bohan, W., Kaylor, R., and Weil, J. (1959). Studies of thermal convection in a rotating cylinder with some implications for large-scale atmospheric motions. *Meteorol. Monographs* Vol. **4**, No. 21, 104 pp.

Galilei, Galileo (1638). Dialogues concerning two new sciences; Translated by Crew, H., and de Salvio, A. (1946). *Northwestern Univ. Studies* pp. 1–282.

Geitel, H. (1900). Eine Vorrichtung zur Demonstration von Luftwogen. *Meteorol. Z.* **17**, 426–427.

Ghatage, V. M. (1936). Modellversuche über die gegenseitige Bewegung von Luftmassen verschiedenen Temperaturen. Inaugural Dissertation, Göttingen.

Goldstein, S., ed. (1938). "Modern Developments in Fluid Dynamics," Vols. I and II. Oxford Univ. Press, London and New York.

Görtler, H. (1941). Neuere Beiträge zur Dynamik atmosphärischer und ozeanischer Strömungen. *Naturwissenschaften* **29**, 473–479.

Görtler, H. (1943). Über eine Schwingungserscheinung in Flüssigkeiten mit stabiler Dichteschichtung. *Z. agnew. Math. Mech.* [2] **23**, 65–71.

Görtler H. (1944). Einige Bemerkungen über Strömungen in rotierenden Flüssigkeiten. *Z. angew. Math. Mech.* **24**, 210–214.

Görtler H. (1957). On forced oscillations in rotating fluids. *Proc. 5th Midwestern Conf. Fluid Mech. Ann Arbor* pp. 1–10.

Griggs D. (1939). A theory of mountain building. *Am. J. Sci.* **237**, 611–650.

Gutenberg, B., ed. (1951). Colloquium on plastic flow and deformation within the earth. *Trans. Am. Geophys. Un.* **32**, 497–543.

Hadley, G. (1735). Concerning the cause of the general trade-winds. *Phil. Trans. Roy. Soc. London* **39**, 58–62.

Hall, Sir J. (1803). Experiments on whinstone and lava. *Trans. Roy. Soc. Edinburgh* **5**, 43–75.

Hall, Sir J. (1812). Account of a series of experiments, shewing the effects of compression in modifying the action of heat. *Phil. Trans. Roy. Soc. Edinburgh* **6,** 71–185.

Hall, Sir J. (1815). On the vertical position and convolutions of certain strata, and their relation with granite. *Phil. Trans. Roy. Soc. Edinburgh* **7,** 79–108.

Hartmann, J., and Lazaraus, F. (1937). Hg-dynamics II—Experimental investigations on the flow of mercury in a homogeneous magnetic field. *Kgl. Danske Vidensk. Selsk. Mat.-fys Meddel.* **15** (7), 1–45.

Haskell, N. A. (1935). The motion of a viscous fluid under a surface load. *Physics* **6,** 265–269.

Haurwitz, B. (1940). The motion of atmospheric disturbances. *J. Marine Res.* **3,** 35–50.

Heezen, B. C., and Ewing, M. (1955). Orleansville earthquake and turbidity currents. *Bull. Am. Assoc. Petrol. Geol.* **39,** 2505–2514.

Helmholtz, H. (1873). Über ein Theorem, geometrisch ähnliche Bewegungen flüssiger Körper betreffend, nebst Anwendung auf das Problem, Luftballons zu lenken. *Monatsber. K. Akad. Wiss. Berlin* pp. 501–504; Translated by Abbé, C. (1893). The mechanics of the earth's atmosphere. *Smithsonian Inst. Misc. Coll.* **34,** 67–77.

Helmholtz, H. (1888). Über atmosphärische Bewegungen I. *Preuss. Akad. Wiss. Berlin* **1,** 647–663; Translated by Abbé, C. (1891). The mechanics of the earth's atmosphere. *Smithsonian Inst. Misc. Coll.* **34,** 78–93.

Hide, R. (1953). Some experiments on thermal convection in a rotating liquid. *Quart. J. Roy. Meteorol. Soc.* **79,** 161.

Hide, R. (1955). The character of the equilibrium of an incompressible heavy viscous fluid of variable density: An approximate theory. *Proc. Cambridge Phil. Soc.* **51,** 179–201.

Hide, R. (1956a). Fluid motion in the earth's core and some experiments on thermal convection in a rotating liquid. *Fluid Models in Geophysics, Proc. 1st Symposium on the Use of Models in Geophys. Fluid Dynamics Baltimore 1953* pp. 101–116.

Hide, R. (1956b). The hydrodynamics of the earth's core. *Phys. and Chem. of the Earth* **1,** 94–137.

Hide, R. (1958). An experimental study of thermal convection in a rotating liquid. *Phil. Trans. Roy. Soc. London* **A250,** 441–478.

Høiland, E. (1950). On horizontal motion in a rotating fluid. *Geofys. Publik.* **17** (10), 1–26.

Holmboe, J., and Klieforth, H. (1954). Sierra Wave Project, Final Rept. Contract AF 19(122)-263. University of California, Los Angeles.

Holzman, B. G. (1951). Meteorological aspects of the atomic "Base Surge." *Bull. Am. Meteorol. Soc.* **32,** 251–258.

Hopkins, W. (1839). Researches in physical geology. *Phil. Trans. Roy. Soc. London* **A129,** 381–423.

Howes, E. T., Tejada-Flores, L. H., and Randolph, L. (1953). Seismic model study. *J. Acoust. Soc. Am.* **25,** 915–921.

Hubbert, M. K. (1937). Theory of scale models as applied to the study of geologic structures. *Bull. Geol. Soc. Am.* **48,** 1459–1520.

Hubbert, M. K. (1945). The strength of the earth. *Bull. Am. Assoc. Petrol. Geol.* **29,** 1630–1653.

Hubbert, M. K. (1951). Mechanical basis for certain familiar geologic structures. *Bull. Geol. Soc. Am.* **62,** 355–372.

Inglis, D. R. (1941). Motion of the earth's fluid core: A geophysical problem. *Phys. Rev.* **59,** 178–188.

Inglis, D. R. (1955). Theories of the earth's magnetism. *Rev. Mod. Phys.* **27,** 212–248.

Ippen, A. T., and Harleman, D. R. F. (1952). Steady-state characteristics of subsurface flow. *Gravity Waves, Nath. Bur. Stand. Circ.* **521,** 79–93.

Janssen, J. (1896). *Ann. Obs. ast. phys. Paris* **1,** 122 pp.

Jirlow, K. (1956). Experimental investigation of the inhibition of convection by a magnetic field. *Tellus* **8,** 252–253.

Kaufman, S., and Roever, W. L. (1951). Laboratory studies of transient elastic waves. *World Petrol. Congr. Proc. 3rd Congr. Hague 1951 Sec. I,* pp. 537–545.

Keulegan, G. H. (1949). Interfacial instability and mixing in stratified flows. *J. Res. Natl. Bur. Stand.* **43,** 487–500.

Kirchhoff, G., and Hansemann, G. (1880). Versuche über stehende Schwingungen des Wassers. *Ann. Phys. Chem.* **10,** 337-347; *Gesammelte Abhandlungen.* J. A. Barth Leipzig, 1882, pp. 442–454.

Knopoff, L. (1955). Small three-dimensional seismic models. *Trans. Am. Geophys. Un.* **36,** 1029–1034.

Königsberger, J., and Morath, O. (1913). Theoretische Grundlagen der experimentellen Tektonik. *Z. deut. geol. Ges.* **65,** 65–86.

Kuenen, P. H. (1936). The negative isostatic anomalies in the East Indies (with experiments). *Leidsl. Geol. Mededeel,* **8,** 169–214.

Kuo, H. L. (1953). On the production of mean zonal currents in the atmosphere by large disturbances. *Tellus* **5,** 475–493.

Kuo, H. L. (1954). Symmetrical disturbances in a thin layer of fluid subject to a horizontal temperature gradient and rotation. *J. Meteorol.* **11,** 399–411.

Kuo, H. L. (1955). On convective instability of a rotating fluid with a horizontal temperature contrast. *J. Mar. Res.* **14,** 14–32.

Kuo, H. L. (1956a). On convective instability of a rotating fluid. *Fluid Models in Geophysics, Proc. 1st Symposium on the Use of Models in Geophys. Fluid Dynamics Baltimore 1953,* pp. 65–72.

Kuo, H. L. (1956b). Forced and free axially-symmetric convection produced by differential heating in a rotating fluid. *J. Meteorol.* **13,** 521–527.

Kuo, H. L. (1956c). Energy-releasing processes and stability of thermally driven motions in a rotating fluid. *J. Meteorol.* **13,** 82–101.

Kuo, H. L. (1957). Further studies of thermally driven motions in a rotating fluid. *J. Meteorol.* **14,** 553–558.

Lamb, H. (1904). On the propagation of tremors over the surface of an elastic solid. *Phil. Trans. Roy. Soc. London* **A203,** 1–42.

Lamb, H. (1916). On waves due to a travelling disturbance, with an application to waves in superposed fluids. *Phil. Mag.* [6] **31,** 386–399.

Lamb, H. (1917). On waves in an elastic plate. *Proc. Roy. Soc.* (*London* **A93**) 114–128.

Lamb, H. (1932). "Hydrodynamics," 6th ed. Cambridge Univ. Press, London and New York.

Lance, G. N. (1958). The effect of a vertical lapse rate of temperature on the spiral flow of a fluid in a heated rotating cylinder. *J. Fluid Mech.* **3,** 523–530.

Lance, G. N., and DeLand, E. C. (1957). The steady, axially symmetric flow of a viscous fluid in a deep rotating cylinder heated from below. *Univ. Calif.* (*Berkeley*) *Publs. Eng.* **5**(6), 159–176.

Lane, W. R., and Green, H. L. (1956). The mechanics of drops and bubbles. *In* "Surveys in Mechanics; G. I. Taylor Anniversary Volume" (G. K. Batchelor and R. M. Davies eds.), pp. 162–215. Cambridge Univ. Press, London and New York.

Langhaar, H. L. (1951). "Dimensional Analysis and Theory of Models." Wiley, New York.

Lehnert, B. (1954). Magneto-hydrodynamic waves in liquid sodium. *Phys. Rev.* **94,** 815–824.

Lehnert, B. (1955). An instability of laminar flow of mercury caused by an external magnetic field. *Proc. Roy. Soc.* (*London*) **A233,** 299–302.

Lehnert, B. (1957). An experiment on axisymmetric flow of liquid sodium in a magnetic field. *Arkiv Fys.* **13,** 10.

Lehnert, B., and Little, N. C. (1957). Experiments on the effect of inhomogeneity and obliquity of a magnetic field in inhibiting convection. *Tellus* **9,** 97–103.

Levin, F. K., and Hibbard, H. C. (1955). Three-dimensional seismic model studies. *Geophysics* **20,** 19–32.

Link, T. A. (1930). Experiments relating to salt dome structures. *Bull. Am. Assoc. Petrol. Geol.* **14,** 483–508.

Lochte-Holtgreven, W., and Schilling P. (1953). Über das Auftreten von Magnetfeldern in rotierenden Plasmen. *Naturwissenschaften* **40,** 387.

Long, R. R. (1952). The flow of a liquid past a barrier in a rotating spherical shell. *J. Meteorol.* **9,** 187–199.

Long, R. R. (1953a). Steady motion around a symmetrical obstacle moving along the axis of a rotating fluid. *J. Meteorol.* **10,** 197–203.

Long, R. R. (1953b). A laboratory model resembling the "Bishop-wave" phenomenon. *Bull. Am. Meteorol. Soc.* **34,** 205–211.

Long, R. R. (1953c). Some aspects of the flow of stratified fluids. I. A theoretical investigation. *Tellus* **5,** 42–58.

Long, R. R. (1954a). Note on Taylor's "ink walls" in a rotating fluid. *J. Meteorol.* **11,** 247–249.

Long, R. R. (1954b). Some aspects of the flow of stratified fluids. II. Experiments with a two-fluid system. *Tellus* **6,** 97–115.

Long, R. R. (1955). Some aspects of the flow of stratified fluids. III. Continuous density gradients. *Tellus* **7,** 341–357.

Long, R. R. (1956a). Models of small-scale atmospheric phenomena involving density stratification. *Fluid Models in Geophysics, Proc. 1st Symposium on the Use of Models in Geophys. Fluid Dynamics Baltimore 1953,* pp. 135–147.

Long, R. R. (1956b). Solitary waves in the one- and two-fluid systems. *Tellus* **8,** 460–471.

Long, R. R. (1957). A laboratory model of air flow over the Sierra Nevada Mountains. U.S. Weather Bureau Contract CWB-8607, Tech. Rept. No. 8, 21 pp.

Long, R. R. (1958). Tractable models of steady state stratified flow with shear. *Quart. J. Roy. Meteorol. Soc.* **84,** 159–161.

Lorenz, E. N. (1955). Available potential energy and the maintenance of the general circulation. *Tellus* **7,** 157–167.

Lorenz, E. N. (1956). A proposed explanation for the existence of two regimes of flow in a rotating symmetrically-heated cylindrical vessel. *Fluid Models in Geophysics, Proc. 1st Symposium on the Use of Models in Geophys. Fluid Dynamics Baltimore 1953,* pp. 73–80.

Loughhead, R. E., and Bray, R. J. (1959). Turbulence and the photospheric granulation. *Nature* **183,** 240–241.

Lowell, S. C. (1958). Shear waves in stratified rotating liquids. Air Force Cambridge Research Center TN-58-484, Inst. Math. Sci., New York University.

Lundquist, S. (1949). Experimental investigations of magneto-hydrodynamic waves. *Phys. Rev.* **76,** 1805–1809.

Lundquist, S. (1952). Studies in magneto-hydrodynamics. *Arkiv Fys.* **5** (15), 297–347.

Lyra, G. (1940). Über den Einfluss von Bodenerhebungen auf die Strömung einer stabil geschichteten Atmosphäre. *Beitr. Phys. Atmos.* **26,** 197–206.

Lyra, G. (1943). Theorie der stationären Leewellenströmung in freier Atmosphäre. *Z. angew, Math. Mech.* **23,** 1–28.

Maillet, R., and Blondel, F. (1934). Sur la similitude en tectonique. *Bull. soc. géol. France* [5] **4,** 599–602.

Mal, S. (1931). Forms of stratified clouds. *Beitr. Phys. Atmos.* **17,** 40–68.

Malkus, J. S., and Scorer, R. S. (1955). The erosion of cumulus towers. *J. Meteorol.* **12,** 43–57.

Malmfors, K. G. (1945). Determination of orbits in the field of a magnetic dipole with application to the theory of the diurnal variation of cosmic rays. *Arkiv mat. ast. fys.* **32A**(8).

Malmfors, K. G. (1946). Experiments on the aurorae. *Arkiv mat. ast. fys.* **34**B(1) 8 pp.

Martin, J. C., and Moyce, W. J. (1952a). Pt. IV. An experimental study of the collapse of liquid columns on a rigid horizontal plane. *Phil. Trans. Roy. Soc. London* **244,** 312–324.

Martin, J. C., and Moyce, W. J. (1952b). Pt. V. An experimental study of the collapse of fluid columns on a rigid horizontal plane, in a medium of lower, but comparable, density. *Phil. Trans. Roy. Soc. London* **244,** 325–334.

Mead, W. J. (1920). Notes on the mechanics of geologic structures. *J. Geol.* **28,** 505–523.

Michelson, A. A., and Gale, H. G. (1919). The rigidity of the earth. *J. Geol.* **27,** 585–601.

Miles, J. W. (1959). Free surface oscillations in a rotating liquid. *Phys. Fluids* **2,** 297–305.

Mintz, Y. (1955). Final computation of the mean geostrophic poleward flux of angular momentum and of sensible heat in the winter and summer of 1949. Final Report, Contract AF 19(122)-48, UCLA.

Morgan, G. W. (1951). A study of motions in a rotating liquid. *Proc. Roy. Soc.* (*London*) **A206,** 108–130.

Morgan, G. W. (1956). Remarks on the problem of motions in a rotating fluid. *Fluid Models in Geophysics, Proc. 1st Symposium on the Use of Models in Geophys. Fluid Dynamics Baltimore 1953*, pp. 19–26.

Mortimer, C. H. (1951a). Water movements in stratified lakes, deduced from observations in Windermere and model experiments. *Un. geod. geophys. intern. Assemblée de Bruxelles* **111,** 335–349.

Mortimer, C. H. (1951b). The use of models in the study of water movement in stratified lakes. *Proc. Intern. Assoc. Theo. Appl. Limnol.* **11,** 254–260.

Mortimer, C. H. (1952). Water movements in lakes during summer stratification; evidence from the distribution of temperature in Windermere. *Phil. Trans. Roy. Soc. London* **236,** 355–404.

Mortimer, C. H. (1953). The resonant response of stratified lakes to wind. *Schweiz. Z. Hydrol.* **15** (1), 94–151.

Mortimer, C. H. (1954). Models of the flow-pattern in lakes. *Weather* **9,** 177–184.

Mortimer, C. H. (1955). Some effects of the earth's rotation on water movements in stratified lakes. *Proc. Intern. Assoc. Theo. Appl. Limnol.* **12,** 66–77.

Morton, B. R. (1959). Forced plumes. *J. Fluid Mech.* **5,** 151–163.

Morton, B. R., Taylor, Sir G., and Turner, J. S. (1955). Turbulent gravitational convection from maintained and instantaneous sources. *Proc. Roy. Soc.* (*London*) **A234,** 1–23.

Munk, W. H. (1950). On the wind-driven ocean circulation. *J. Meteorol.* **7,** 79–93.

Nakagawa, Y. (1955). An experiment in the inhibition of thermal convection by a magnetic field. *Nature* **175,** 417.

Nakagawa, Y. (1957a). Experiments on the inhibition of thermal convection by a magnetic field. *Proc. Roy. Soc.* (*London*) **A240,** 108.

Nakagawa, Y. (1957b). Experiments on the instability of a layer of mercury heated from below and subject to the simultaneous action of a magnetic field and rotation. *Proc. Roy. Soc.* (*London*) **A242**, 81–88.

Nakagawa, Y. (1959). Experiments on the instability of a layer of mercury heated from below and subject to the simultaneous action of a magnetic field and rotation II. *Proc. Roy. Soc.* (*London*) **A249**, 138–145.

Namias, J. (1954). Quasi-periodic cyclogenesis in relation to the general circulation. *Tellus* **6**, 8–22.

Nettleton, L. L. (1934). Fluid mechanics of salt domes. *Bull. Am. Assoc. Petrol. Geol.* **18**, 1175–1204.

Nettleton, L. L. (1955). History of concepts of Gulf Coast salt-dome formation. *Bull. Am. Assoc. Petrol. Geol.* **39**, 2373–2383.

Nettleton, L. L., and Elkins, T.A. (1947). Geologic models, made from granular materials. *Trans. Am. Geophys. Un.* **28**, 451–465.

Northwood, T. D., and Anderson, D. V. (1953). Model seismology. *Bull. Seismol. Soc. Am.* **43**, 239–245.

Oberbeck, A. (1877). Ueber discontinuierliche Flüssigkeitsbewegungen. *Ann. Phys.* [3] **2**, 1–16.

O'Brien, P. N. S. (1955). Model seismology—the critical refraction of elastic waves. *Geophysics* **20**, 227–242.

Oliver, J., Press, F., and Ewing, M. (1954). Two-dimensional model seismology. *Geophysics* **19**, 202–219.

Oser, H. (1957). Erzwungene Schwingungen in rotierenden Flüssigkeiten. *Arch. Rat. Mech. Anal.* **1**, 81–96.

Oser, H. (1958). Experimentelle Untersuchung über harmonische Schwingungen in rotierenden Flüssigkeiten. *Z. angew. Math. Mech.* **38**, 386–391.

Palm, E. (1953). On the formation of surface waves in a fluid flowing over a corrugated bed and on the development of mountain waves. *Astrophys. Norveg.* **5**(3), 129 pp.

Palmén, E., and Nagler, K. (1948). An analysis of the wind and temperature over North America in a case of approximately westerly flow. *J. Meteorol.* **5**, 58–64.

Parker, E. N. (1955). Hydromagnetic dynamo models. *Astrophys. J.* **122**, 293–314.

Parker, T. J., and McDowell, A. N. (1951). Scale models as guide to interpretation of salt dome faulting. *Bull. Am. Assoc. Petrol. Geol.* **35**, 2076–2086.

Parker, T. J., and McDowell, A. N. (1955). Model studies of salt-dome tectonics. *Bull. Am. Assoc. Petrol. Geol.* **39**, 2384–2470.

Pekeris, C. L. (1935). Thermal convection in the interior of the earth. *Monthly Not. Roy. Ast. Soc. Geophys. Suppl.* **3**, 343–367.

Penney, W. G., and Thornhill, C. K. (1952). Pt. III. The dispersion, under gravity, of a column of fluid supported on a rigid horizontal plane. *Phil. Trans. Roy. Soc. London* **A244**, 285–311.

Phillips, N. A. (1956). The general circulation of the atmosphere: A numerical experiment. *Quart. J. Roy. Meteorol. Soc.* **82**, 123–164.

Prandtl, L. (1904). Über Flüssigkeitsbewegung bei sehr kleiner Reibung. *Proc. 3rd Intern. Math. Cong. Heidelberg.*

Prandtl, L. (1932). Meteorologische Anwendungen der Strömungslehre. *Beitr. Phys. Atmos.* **19**, 188 pp.

Prandtl, L. (1937). Betrachtungen zur Mechanik der freien Atmosphäre. *Abh. Ges. Wiss, Göttingen Math. Phys. Kl.* **3** (18), 75–84.

Pratt, J. H. (1855). II. On the attraction of the Himalaya Mountains, and of the elevated regions beyond them, upon the plumb-line in India. *Phil. Trans. Roy. Soc. London* **A145**, 53–100.

Press, F., and Oliver, J. (1955). Model study of air-coupled surface waves. *J. Acoust. Soc. Am.* **27**, 43–46.

Press, F., Oliver, J., and Ewing, M. (1954). Seismic model study of refraction from a layer of finite thickness. *Geophysics* **19**, 388–401.

Priestley, C. H. B. (1955). Free and forced convection in the atmosphere near the ground. *Quart, J. Roy. Meteorol. Soc.* **81**, 139–143.

Priestley, C. H. B. (1959). "Turbulent transfer in the Lower Atmosphere." Univ. Chicago Press, Chicago, Illinois.

Priestley, C. H. B., and Ball, F. K. (1955). Continuous convection from an isolated source of heat. *Quart. J. Roy. Meteorol. Soc.* **81**, 144–157.

Proudman, J. (1916). On the motion of solids in a liquid possessing vorticity. *Proc. Roy. Soc.* (*London*) **A92**, 408–424.

Queney, P. (1947). Theory of perturbations in stratified currents with application to airflow over mountain barriers. Dept. Meteorol., Univ. of Chicago, Misc. Rept. No. 23.

Raethjen, P. (1958). Ähnlichkeitsbedingungen für geohydrodynamische Modellexperimente in rotierender Schale. *Arch. Meteorol. Geophys. Biokl.* **A10**, 178–193.

Ramberg, H. (1955). Natural and experimental boudinage and pinch-and-swell structures. *J. Geol.* **63**, 512–526.

Rayleigh, Lord (1916). On convection currents in a horizontal layer of fluid when the higher temperature is on the under side. *Phil. Mag.* [6] **32**, 529–546.

Reynolds, O. (1895). On the dynamical theory of incompressible viscous fluids and the determination of the criterion. *Phil. Trans. Roy. Soc. London* **A186**, 123–164; (1901). *Pap. Mech. Phy. Subjects* **2**, 535–577.

Rieber, F. (1936). Visual presentation of elastic wave patterns under various structural conditions. *Geophysics* **1**, 196–218.

Riehl, H., Badner, J., Hovde, S. E., and others (1952). Forecasting in middle latitudes. *Meteorol. Monographs* **1** (5).

Riehl, H., and Fultz, D. (1957). Jet stream and long waves in a steady rotating-dishpan experiment: Structure of the circulation. *Quart. J. Roy. Meteorol. Soc.* **83**, 215–231.

Riehl, H., and Fultz, D. (1958). The general circulation in a steady rotating-dishpan experiment. *Quart. J. Roy. Meteorol. Soc.* **84**, 388–417.

Rimbach, C. (1913). Versuche der Gebirgsbildung. *N. Jahr. Mineral. Geol. Paläontol. B.B.* **35**, 689–722.

Riznichenko, Y. V., and Shamina, O. G. (1957). Elastic waves in a laminated solid medium, as investigated on two-dimensional models. *Bull. Acad. Sci. U.S.S.R. Geophys. Ser.* **7**, 17–37 (translated by K. Syers from *Izvest. Akad. Nauk S.S.S.R. Ser. Geofiz. No.* **7**, 855–874).

Riznichenko, Y. V., Ivakin, B. M., and Bugroff, V. R. (1951). The modelling of seismic waves. *Izvest. Akad. Nauk S.S.S.R. Geophys. Ser.* **5**, 1–30.

Rogers, M. H. (1954). The forced flow of a thin layer of viscous fluid on a rotating sphere. *Proc. Roy. Soc.* (*London*) **A224**, 192–208.

Rogers, R. H. (1959). The structure of the jet-stream in a rotating fluid with a horizontal temperature gradient. *J. Fluid Mech.* **5**, 41–59.

Rossby, C.-G. (1926). On the solution of problems of atmospheric motion by means of model experiments. *Monthly Weather Rev.* **54**, 237–240.

Rossby, C.-G. (1928). Studies in the dynamics of the stratosphere. *Beitr. Phys. Atmos.* **14**, 240–265.

Rossby, C.-G. (1939). Relation between variations in the intensity of the zonal circulation of the atmosphere and the displacements of the semi-permanent centers of action. *J. Mar. Res.* **2**, 38–55.

Rossby, C.-G. (1947). On the distribution of angular velocity in gaseous envelopes under the influence of large scale horizontal mining processes. *Bull. Am. Meteorol. Soc.* **28,** 53–68.

Rossby, C.-G. (1949). On the nature of the general circulation in the lower atmosphere. *In* "The Atmospheres of the Earth and Planets" (G. P. Kuiper, ed.), Univ. Chicago Press, Chicago, Illinois. pp. 16–48.

Rouse, H. (1951). Model techniques in meteorological research. *Compendium Meteorol.* **1951,** 1249–1254.

Rouse, H., Yih, C. S., and Humphreys, H. W. (1952). Gravitational convection from a boundary source. *Tellus* **4,** 201–210.

Rumford, Sir B. T. (1800). Of the propagation of heat in fluids. "Essays, Political, Economical and Philosophical," Vol. II, pp. 199–465. *London*; (1870). "Complete Works," Vol. I, Am. Acad. Arts and Sci. Boston, Massachusetts, pp. 239–404.

Runcorn, S. K. (1954). The earth's core. *Trans. Am. Geophys. Un.* **35,** 49–63.

Sabin, B. A. (1954). A qualitative study of some of the critical parameters for the "dish-pan" model of the atmosphere. Mass. Inst. of Tech. (unpublished B.S. thesis).

Sandström, J. W. (1914). Über die Ablenkung des Windes infolge der Erddrehung und der Reibung. *Arkiv mat. ast. fys.* **9** (31), 8 pp.

Sarrafian, G. P. (1956). A marine seismic model. *Geophysics* **21,** 320–336.

Schilling, P., and Lochte-Holtgreven, W. (1953). Magnetic fields measured within flames brought mechanically into rotational motion. *Nature* **172,** 1053.

Schilling, P. O., and Lochte-Holtgreven, W. (1954). Magnetfelder in turbulent strömenden Plasmen. *Z. Naturforsch.* **9**a, 520–526.

Schlichting, H. (1955). "Boundary Layer Theory." McGraw-Hill, New York.

Schmidt, W. (1908). Stehende Schwingungen in der Grenzschicht zweier Flüssigkeiten. *Sitzber. K. Akad. Wiss. Wien (IIa)* **117,** 91–102.

Schmidt, W. (1910). Gewitter und Böen, rasche Druckanstiege. *Sitzber. K. Akad. Wiss. Wien (IIa)* **119,** 1101–1214.

Schmidt, W. (1911). Zur Mechanik der Böen. *Meteorol. Z.* **28,** 355–362.

Schmidt, W. (1913). Weitere Versuche über den Böenvorgang und das Wegschaffen der Bodeninversion. *Meteorol. Z.* **30,** 441–447.

Schmidt, W. (1941). Turbulente Ausbreitung eines Stromes erhitzter Luft. *Z. angew. Math. Mech.* **21,** 265–278, 351–363.

Schwarzschild, M., and Schwarzschild, B. (1959). Balloon astronomy. *Sci. American* **200** (5), 52–59.

Scorer, R. S. (1949). Theory of waves in the lee of mountains. *Quart. J. Roy. Meteorol. Soc.* **76,** 41–56.

Scorer, R. S. (1951). Gravity waves in the atmosphere. *Arch. Meteorol. Geophys. Biokl.* **A4,** 176–193.

Scorer, R. S. (1953). Theory of airflow over mountains: II—The flow over a ridge. *Quart. J. Meteorol. Soc.* **79,** 70–83.

Scorer, R. S. (1954). Theory of airflow over mountains: III—Airstream characteristics. *Quart. J. Roy. Meteorol. Soc.* **80,** 417–428.

Scorer, R. S. (1955a). Theory of airflow over mountains: IV—Separation of flow from the surface. *Quart. J. Meteorol Roy. Soc.* **81,** 340–350.

Scorer, R. S. (1955b). Theory of non-horizontal adiabatic flow in the atmosphere. *Quart. J. Roy. Meteorol. Soc.* **81,** 551–561.

Scorer, R. S. (1956). Airflow over an isolated hill. *Quart. J. Roy. Meteorol. Soc.* **82,** 75–81.

Scorer, R. S. (1957). Experiments on convection of isolated masses of buoyant fluid. *J. Fluid Mech.* **2,** 583–594.

Scorer, R. S., and Ronne, C. (1956). Experiments with convection bubbles. *Weather* **11**, 151–153.

Shamina, O. G., and Silayeva, O. I. (1958). The propagation of elastic pulses in free boundary layers of finite thickness. *Bull. Acad. Sci. U.S.S.R. Geophys. Ser.* (*English Translation*) *No.* **3**, 168–174.

Skeib, G. (1953). Modellversuche zur thermischen Konvektion. *Abhandlungen des Meteorologischen und Hydrologischen Dienstes der Deutschen Demokratischen Republik*, **3** (20), 1–56.

Squire, H. B. (1956). Rotating fluids. *In* "Surveys in Mechanics: G. I. Taylor 70th Anniversary Volume" (G. K. Batchelor and R. M. Davies, eds.), Cambridge Univ. Press, London and New York. pp. 139–161.

Starr, V. P., and Long, R. R. (1953). The flux of angular momentum in rotating model experiments. *In* "Contributions to the Study of Planetary Atmospheric Circulations" (R. M. White, ed.) pp. 103–113, *Geophys. Res. Pap. U.S.A.F. No.* **24.**

Starr, V. P., and White, R. M. (1954). Balance requirements of the general circulation. *Geophys. Res. Pap. U.S.A.F. No.* 35, 57 pp.

Stommel, H. (1948). The westward intensification of wind-driven ocean currents. *Trans. Am. Geophys. Un.* **29**, 202–206.

Stommel, H. (1957). A survey of ocean current theory. *Deep-Sea Res.* **4**, 149–184.

Stommel, H., Arons, A. B., and Faller, A. J. (1958). Some examples of stationary planetary flow patterns in bounded basins. *Tellus* **10**, 179–187.

Störmer, C. (1955). "The Polar Aurorae." Oxford Univ. Press, London and New York.

Summers, H. S. (1933). Experimental tectonic geology. *Rept. Australian New Zealand Assoc. Advance. Sci.* **21**, 49–75.

Swallow, J. C., and Worthington, L. V. (1957). Measurements of deep currents in the western North Atlantic. *Nature* **179**, 1183–1184.

Tatel, H. E. (1954). Note on the nature of a seismogram II. *J. Geophys. Res.* **59**, 289–294.

Taylor, Sir G. I. (1917). Motion of solids in fluids when the flow is not irrotational. *Proc. Roy. Soc.* (*London*) **A93**, 99–113.

Taylor, Sir G. I. (1921). Experiments with rotating fluids. *Proc. Roy. Soc.* (*London*) **A100**, 114–121.

Taylor, Sir G. I. (1922). The motion of a sphere in a rotating liquid. *Proc. Roy. Soc.* (*London*) **A102**, 180–189.

Taylor, Sir G. I. (1923). Experiments on the motion of solid bodies in rotating fluids. *Proc. Roy. Soc.* (*London*) **A104**, 213–218.

Taylor, Sir G. I. (1945). Dynamics of a mass of hot gas rising in air. U.S. Atomic Energy Comm., MDDC-919, LADC 276.

Taylor, Sir G. I. (1950a). The formation of a blast wave by a very intense explosion I. Theoretical discussion. *Proc. Roy. Soc.* (*London*) **A201**, 159–174.

Taylor, Sir G. I. (1950b). The formation of a blast wave by a very intense explosion II. The atomic explosion of 1945. *Proc. Roy. Soc.* (*London*) **A201**, 175–186.

Terada, T., and Miyabe, N. (1928). Experiments on the modes of deformation of a layer of granular mass floating on a liquid and c. *Bull. Earthquake Res. Inst. Tokyo Univ.* **4**, 21–32.

Terada, T., and Tsuboi, C. (1927). Experimental studies on elastic waves Pts. I and II. *Bull. Earthquake Res. Inst. Tokyo Univ.* **3**, 55–64; **4**, 9–20 (1928).

Thiriot, K. H. (1940). Über die laminare Anlaufströmung einer Flüssigkeit über einem rotierenden Boden bei plötzlicher Änderung des Drehungszustandes. *Z. angew. Math. Mech.* **20**, 1–13.

Thomas, D. B., and Townsend, A. A. (1957). Turbulent convection over a heated horizontal surface. *J. Fluid Mech.* **2,** 473–492.

Thomson, J. (1892). On the grand currents of atmospheric circulation. *Phil. Trans. Roy. Soc. London* **A183,** 653–684.

Thomson, W. (Lord Kelvin) (1886). On stationary waves in flowing water. *Phil. Mag.* **22** (5), 353–357, 445–452, 517–570 (Papers of W. Thomson (Lord Kelvin) **4,** 270 ff.).

Tippelskirch, H. von (1957). Über die Bénard-Strömung in Aerosolen. *Beitr. Phys. Atmos.* **29,** 219–233.

Torrey, P. D., and Fralich, C. E. (1926). An experimental study of the origin of salt domes. *J. Geol.* **34,** 223–234.

Townsend, A. A. (1959). Temperature fluctuations over a heated horizontal surface. *J. Fluid Mech.* **5,** 209–241.

Truesdell, C. (1956). Experience, theory, and experiment. *Proc. 6th Hydraul. Conf. Iowa Inst. Hydraul. Res.* Iowa City, June 13–15, 1955 pp. 3–18.

Turner, J. S. (1957). Buoyant vortex rings. *Proc. Roy. Soc.* (*London*) **239,** 61–75.

University of Chicago, Department of Meteorology. (1947). On the general circulation of the atmosphere in middle latitudes. *Bull. Am. Meteorol. Soc.* **28,** 225–280.

Van Dorn, W. G. (1953). Wind stress on an artificial pond. *J. Mar. Res.* **12** (3), 249–276.

Vettin, F. (1857). Über den aufsteigenden Luftstrom, die Entstehung des Hagels und der Wirbel-Stürme. *Ann. Physik.* [2] **102,** 246–255.

Vettin, F. (1884). Experimentelle Darstellung von Luftbewegungen unter dem Einfluss von Temperatur-Unterschieden und Rotations-Impulsen. *Meteorol. Z.* **1,** 227–230, 271–276; (1885) **2,** 172–183.

Vettin, F. (1886). Experimentelle Darstellung der Schäfchenwolken (Cirro-Cumuli). *Meteorol. Z.* **3,** 38–39.

von Arx, W. S. (1952). A laboratory study of the wind driven ocean circulation. *Tellus* **4,** 311–318.

von Arx, W. S. (1955). An experimental study of the dependence of the primary ocean circulation on the mean zonal wind field. MIT Sc.D. thesis.

von Arx, W. S. (1956). Some techniques for laboratory study of the ocean circulation. *Fluid Models in Geophysics, Proc. 1st Symposium on the Use of Models in Geophys. Fluid Dynamics Baltimore 1953,* pp. 89–99.

von Arx, W. S. (1957). An experimental approach to problems in physical oceanography. *Progr. in Phys. and Chem. Earth.* **2,** 1–29.

von Kármán, T. (1940). The engineer grapples with nonlinear problems. *Bull. Am. Math. Soc.* **46,** 615–683.

Wasiutynski, J. (1946). Hydrodynamics and structure of stars and planets. *Astrophys. Norveg.* **4,** 1–497.

Watson, E. R. (1904). Movements of the waters of Loch Ness as indicated by temperature observations. *Geograph. J.* **24,** 430–437.

Webster, H. F. (1957). Structure in magnetically confined electron beams. *J. Appl. Phys.* **28,** 1388–1397.

Wedderburn, E. M. (1907). An experimental investigation of the temperature changes occurring in fresh water lochs. *Proc. Roy. Soc. Edinburgh* **28,** 2–20.

Wedderburn, E. M., and Williams, A. M. (1911). The temperature seiche. IV Experimental verification of the hydrodynamical theory of temperature seiches. *Phil. Trans. Roy. Soc. Edinburgh* **47,** 636–642.

Weickmann, L. (1929). Kapitel 80: Hydrodynamische Analogien. "Lehrbuch der Geophysik, Mechanik und Thermodynamik der Atmosphäre," pp. 917–937. Borntraeger, Berlin.

White, P., and Watson, W. (1905–1960). Some experimental results in connection with the hydrodynamical theory of seiches. *Proc. Roy. Soc. Edinburgh* **26,** 143–156.

Willis, B. (1891–1892). The mechanics of Appalachian structure. *U.S. Geol. Surv. Ann. Rept.* **13,** Pt. 2, 211–289.

Woodward, B. (1958). A theory of thermal soaring. *Schweizer Aero-rev.* **33,** *No.* **6,** 1–4.

Woodward, B. (1959). The motion in and around isolated thermals. *Quart. J. Roy. Meteorol. Soc.* **85** (364), 144–151.

Wurtele, M. G. (1957). The three-dimensional lee wave. *Beitr. Phys. Atmos.* **29,** 242–252.

Yih, C. S. (1947). A study of the characteristics of gravity waves at a liquid surface. S.M. thesis, Iowa Inst. Hydraul. Res. Iowa City.

Yih, C. S. (1951). Free convection due to a point source of heat. *Proc. 1st U.S. Natl. Congr. Appl. Mech.* pp. 941–947.

Yih, C. S. (1952). Laminar free convection due to a line source of heat. *Trans. Am. Geophys. Un.* **33,** 669.

Yih, C. S. (1956). Free convection due to boundary sources. *Fluid Models in Geophysics, Proc. 1st Symposium on the Use of Models in Geophys. Fluid Dynamics Baltimore 1953,* pp. 117–133.

Yih, C. S., and Guha, C. R. (1955). Hydraulic jump in a fluid system of two layers. *Tellus* **7,** 358–366.

Zeilon, N. (1912). On tidal boundary-waves and related hydrodynamical problems. *Kgl. Svensk. Vetakad. Handl.* **47,** pp. 1–46.

Zeilon, N. (1934). Experiments on boundary tides. *Göteborg Kgl. Vetakad. Vitterhets–Samhäll. Handl. Ser. B* **3**(10) 8 pp.

ATMOSPHERIC TIDES

Manfred Siebert

Geophysikalisches Institut der Universität Göttingen, Germany

1. Outline of History and Present State

Atmospheric tides are small, regular, world-wide air pressure oscillations which occur with periods of λ^{-1} solar or lunar day ($\lambda = 1, 2, 3, \ldots$). They can be generated by gravitational tidal forces of the moon and the sun and by the thermal forces caused by a periodic absorption and emission of heat connected in some way with the periodic incoming solar radiation.

These pressure oscillations are generally so small that they are totally masked by pressure variations associated with the so-called normal weather processes. A very regular pattern of the barogram, showing a semidiurnal variation, is immediately apparent only at equatorial stations. By means of harmonic analysis (see Section 2), however, the semidiurnal as well as other periodic pressure oscillations were detected all over the world. The largest of all is that oscillation which appears on equatorial barograms with maxima at about 10 A.M. and 10 P.M. local solar time. It is surprising to find that the major period of the atmospheric tidal oscillations is half a solar day, rather than half a lunar day, as would be expected from the application of Newton's theory of gravitation to the tidal phenomenon. According to Newton's theory the tidal force of the moon near the earth is 2.2 times as large as that of the sun; and, indeed, half a lunar day is observed as the predominant period of the oceanic tides. Although much work has been done to explain the unexpected behavior of the outstanding air tide, there is as yet no completely satisfactory answer.

In order to describe clearly the following historical development, it is necessary to introduce a few details from subsequent sections: Let S_λ represent the λ^{-1}-diurnal solar pressure oscillation. Its amplitude and phase derived from observational data of stations all over the world are arbitrary functions of geographical longitude and colatitude. It is possible to resolve S_λ into terms of S_λ^s ($s = 0, 1, 2, \ldots$), the amplitudes and phases of which depend on colatitude only. Here, S_λ^s is called a *wave family*. It can be expanded by means of spherical harmonics or *Hough's functions* (see Section 5) which are more appropriate to theoretical investigations. Each term of these expansions has constant amplitude and constant phase and is the simplest of the expressions of planetary character. It may be called *wave type* and denoted by $S_{\lambda,n}^s$ ($n = 1, 2, 3, \ldots$) if Hough's functions are used. The same resolution can also be carried out for the lunar pressure variation L_λ. Each term $S_{\lambda,n}^s$ or $L_{\lambda,n}^s$ has its own resonance magnification which depends on the three parameters λ, s, n. Contrary to other simpler vibration problems, the resonance of a tidal wave type cannot be determined by its period alone. The appropriate resonance-parameter is the so-called *equivalent depth* h. The name comes from the appearance of the same quantity in Laplace's tidal equation (5.5) as the depth of an ocean covering the earth and capable of free oscillations of a given period and geographical distribution. The quantity h depends on all three parameters λ, s, n, so that each wave type has its equivalent depth. It follows from the theoretical formulism that free oscillations of the earth's atmosphere are possible only for one or more fixed values of h, which will be denoted by $\hat{h}$ and will be called *eigenvalues* of the earth's atmosphere or of the model atmosphere used. The numerical values and the number of the $\hat{h}$ values depend on the structure of the atmosphere under consideration. The closer an h value is to an $\hat{h}$ value the higher is the resonance magnification of the corresponding wave type. Hence, when h and $\hat{h}$ are used, factors of resonance can be described and resonance curves can be given analogous to simpler vibration processes.

The quoted large semidiurnal solar pressure oscillation is obviously within the class S_2. Further analysis shows that the predominant wave family of S_2 is S_2^2, a westward migrating semidiurnal solar pressure wave. Its main term is the wave type $S_{2,2}^2$ having an equivalent depth of $h = 7.85$ km. Hence, the original question, leading to theoretical investigations of atmospheric tides, is in terms of the given classification: why do S_2^2 and especially $S_{2,2}^2$ show an outstanding behavior?

Laplace [1] developed the first dynamical theories of oceanic and atmospheric tides. At the same time he gave the first answer to our problem. From the observed period of half a solar day he concluded that this pressure oscillation is not due to tidal forces but is due rather to the thermal action of the sun.

The next step was taken by Kelvin [2] in 1882. He agreed with Laplace that the semidiurnal pressure variation must be due to a variation of atmospheric temperature. If this is true, however, the diurnal should be larger than the semidiurnal pressure variation because the diurnal term of the temperature variation is appreciably larger than the semidiurnal one as has been shown by harmonic analysis. A way out of this situation was Kelvin's idea that the semidiurnal pressure oscillation is selected by resonance if the atmosphere, as a whole, is regarded as an oscillating system. This interpretation of the observations was the beginning of the so-called *resonance theory* of atmospheric tides. Many authors were stimulated by this idea and attempted its quantitative proof.

Following Kelvin, Rayleigh [3] and Margules [4] (see also Trabert [5]) investigated the period of free oscillations of an atmosphere covering a plane or spherical earth. Rayleigh oversimplified the problem by some of his assumptions, among which was the important omission of the earth's rotation. This factor was taken into account by Margules, who computed free and also forced atmospheric oscillations on the basis of Laplace's theory. That is, he made the assumptions that the vertical acceleration is negligible, that the atmosphere is of uniform temperature and constant composition, and that the pressure changes of the atmospheric tides occur isothermally. Although Margules' results support Kelvin's suggestion of resonance, they do not prove the correctness of the resonance theory since the assumptions used are not sufficiently realistic. Margules also investigated the oscillations of a periodically heated atmosphere using various heating models. He considered the influence of friction, assuming that friction is proportional to the tidal wind velocity, and gave a general classification of these oscillations.

In 1910, Lamb [6] succeeded in extending Laplace's theory while he was studying the propagation of long horizontal waves in a plane atmosphere. He found that the velocity of such waves and, hence, the period of a free atmospheric oscillation, is the same in both an isothermal atmosphere with isothermal changes of state (Laplace) and an atmosphere in convective equilibrium with adiabatic changes of state or, more general, an autobarotropic atmosphere. Later Lamb [7] proved the validity of this result also considering the sphericity and rotation of the earth. The eigenvalue he obtained for autobarotropic atmospheres is $\hat{h} = H(0)$, where $H(0)$ is the scale height of the atmosphere at ground level. Its mean value on the earth's surface is 8.4 km, varying between about 7.3 km at the poles and 8.7 km at the equator. Hence, this eigenvalue is not far from the equivalent depth $h = 7.85$ km of the wave type $S_{2,2}^2$ which should be magnified by resonance. On the basis of his atmosphere in convective equilibrium, Lamb estimated how large the magnification must be and how near h and $\hat{h}$ or the periods of forced and free oscillations must be in order that the magnitude of the observed amplitude be understandable. Assuming gravitational excitation, he found that $S_{2,2}^2$ must be magnified by dynamical action some eightyfold or ninetyfold compared to the *equilibrium tide* (see Section 6), and that such a strong resonance requires a free period of the earth's atmosphere (in regard to the geographical distribution of $S_{2,2}^2$, determined by $n = 2$, $s = 2$) which differs from half a solar day, the period of the forced oscillation, by not more than two or three minutes. This implies a sharp resonance maximum and could qualitatively explain the striking difference of the amplitudes of the semidiurnal solar and lunar (only 25.2 min longer) oscillations even if gravitational generation is assumed. On the other hand, the observed time of maximum of the semidiurnal solar oscillation occurs before noon, whereas the theoretically expected time of maximum is at noon or even after noon if friction is effective. Hence, thermal action appears to be involved.

In his 1910 paper, Lamb showed, moreover, that an infinite number of different velocities of long waves appears if the plane atmosphere has a uniform (but not adiabatic) lapse rate and if the pressure changes occur adiabatically. More than twenty years later, Taylor [8] extended this result to an equivalent atmosphere over a spherical, rotating earth, thus demonstrating that atmospheres may have more than one eigenvalue. This was an important step toward the modern version of the resonance theory. In the meantime, however, all conclusions were drawn from simple model atmospheres involving only one resonance maximum.

Strong support was given to the resonance theory by Chapman [9] in 1924 when he succeeded in explaining the phase of the maximum (about 155°) of S_2^2. Taking again thermal tidal forces into consideration, Chapman assumed that these forces are due to that temperature wave which spreads out from the earth's surface into the atmosphere by turbulent mass exchange (*eddy conductivity*). He used a constant *austausch* coefficient to derive a semidiurnal temperature oscillation from the data of some equatorial stations, and from these he computed the corresponding pressure wave generated in a plane, nonrotating model atmosphere. The phase angle found in this way for the pressure wave was 195°. Because a gravitationally caused semidiurnal solar pressure wave should have the phase of 90°, Chapman inferred from both results that the observed pressure oscillation is composed of two parts, one thermally and the other gravitationally generated, both having nearly equal amplitudes. Moreover, his conclusions were confirmed when he calculated the amplitudes of both parts from his formulas and data. The resonance magnification necessary for the understanding of the magnitude of the observed amplitude was estimated by Chapman to be about one hundredfold when again the equilibrium tide is taken as unity. Hence, Chapman's results are in agreement with Lamb's resonance considerations. The restriction of Chapman's calculations to a plane, nonrotating atmosphere was removed by Wilkes [10] nearly thirty years later without any change in the results.

After the appearance of Chapman's paper the resonance theory seemed quite well established, although it is difficult to understand "the *a priori* improbability of so very close an agreement between the two periods" (Lamb) of forced and free oscillations. Meanwhile knowledge of the earth's atmosphere increased, and the structure of the atmosphere turned out to be more complicated than the simple models used in theory. Again, therefore, the compatibility of the resonance hypothesis with the actual earth's atmosphere was discussed. In 1927, Bartels [11] introduced a two-layer model with constant temperature gradient in the lower layer, the troposphere, and uniform temperature in the upper layer, the stratosphere. The eigenvalue derived from this model by Bartels is about $\hat{h} = 10$ km. Two years later this result was confirmed by Taylor [12], who moreover showed that this value agrees with the propagation of the famous Krakatoa pressure wave of 1883, to which he applied the formula $V^2 = g\hat{h}$ (where V is the velocity of the disturbance and g the acceleration of gravity). It should be noticed that the pressure wave of the great Siberian meteor, often quoted in this connection, cannot be regarded as a free oscillation of the earth's atmosphere because its wavelength is not large enough (see Section 4.2).

The papers of Bartels and Taylor questioned the resonance theory seriously. In the case of a geographical distribution like that of $S_{2,2}^2$ the eigenvalue $\hat{h} = 10$ km implies a free period of about 10.5 hr instead of about 12 hr. In discussing the question of whether or not this result is compatible with the resonance theory, Bartels pointed to the influence of the Cordilleras in Western America. This north-south chain of mountains must have an effect, especially on eastward or westward migrating waves. Because the equivalent depths are usually computed under the assumption of a smooth earth's surface it might be possible that the h values of the predominant wave types change into values near $\hat{h}$ when the influence of the Cordilleras barrier is considered. Kertz [13] (see also Kertz [14] Section 24) investigated this effect quantitatively by means of the perturbation theory and obtained the result that the modification of the equivalent depths is not large enough to get a close agreement to the eigenvalue $\hat{h} = 10$ km and to explain these oscillations by resonance.

The possibility of another explanation appeared in the 1936 paper by Taylor [8] where it was shown that model atmospheres exist with more than one eigenvalue. Hence, the same property might be true as well for the actual earth's atmosphere. Taylor's investigation was continued by Pekeris [15], who succeeded in determining an additional eigenvalue of about 8 km. For his calculations Pekeris used five-layer models in order to take into account the temperature maximum near 50 km height which was deduced from studies of the anomalous propagation of sound and from the existence of noctilucent clouds. In this outstanding paper of 1937, Pekeris also brought the general mathematical treatment of the problem (restricted to gravitational excitation) into a final form.

Thus a quantitative explanation of the phase and the large amplitude of $S_{2,2}^2$ was obtained by Chapman and Pekeris. The resonance theory seemed proved and the problem finally solved.

The growing knowledge of the structure of the upper atmosphere gave subsequently rise to tests of more and more complicated model atmospheres with respect to their resonance properties. With the aid of electronic differential analyzers, Weekes and Wilkes [16] (see also [17]) as well as Jacchia and Kopal [18] computed complete resonance curves for different models. Thus the conditions became well known under which an atmosphere such as that of the earth can have the two eigenvalues of about $\hat{h}_1 = 10$ km and $\hat{h}_2 = 8$ km.

Considering critically this most recent state of the resonance theory, one has to observe that the theory has been developed with hardly another aim in mind than to explain the large semidiurnal solar oscillation. Other planetary pressure waves, however, are observed. As early as 1918 Hann [19] gave details of a terdiurnal solar oscillation, and in 1926 Pramanik [20] investigated the six-hourly solar variations of atmospheric pressure and temperature. There is also a semidiurnal lunar pressure oscillation, whose existence in tropical latitudes has been known since 1847. Its world-wide character could be proved after Chapman [21] showed in 1918 how to determine this oscillation from pressure readings in middle and high latitudes (see also [22]).

Although these other oscillations were described in reviews of atmospheric tides, e.g., by Bartels [23, 24] and Chapman *et al.* [25] in the 1920's and 1930's, only recently Kertz [26] and Siebert [27] used these additional observational data to study empirically the resonance properties of the earth's atmosphere. They compared amplitudes and phases of corresponding pressure and temperature wave types at the earth's surface, using as many as available 12-, 8-, and 6-hourly wave types. However, the results are not open to a simple interpretation. Restriction to the most important wave types which doubtless are planetary phenomena and caused in the same way (unless the improbable possibility of a coupling by nonlinear terms should prove correct) leads to the result that the resonance magnification of $S_{2,2}^2$ is of the same order of magnitude as that of the other oscillations. Therefore, $S_{2,2}^2$ cannot be favored by resonance. When such pressure wave types are considered, whose equivalent depths include an eigenvalue of the atmosphere ($h_1 < \hat{h} < h_2$), the phase differences between corresponding pressure and temperature oscillations should differ by 180°. But no simple feature of this kind could be read from the analysis.

Moreover, the high temperature maximum at about 50 km height required by the resonance theory for $\hat{h} = 8$ km was not confirmed by recent observations in the upper atmosphere obtained mainly by rockets. It is true that a temperature maximum exists in this height, but its magnitude of 300°K or less is at least 50°C lower than that required by the theory. It follows as a result of the computations of Jacchia and Kopal [18] that the second resonance maximum $\hat{h} = 8$ km disappears (see Fig. 7 in Section 6.2). The resonance curves based on temperature profiles which were derived from observations show only one maximum near $\hat{h} = 10$ km and differ only a little from the resonance curve of a two-layer model.

This sensitivity of the second resonance maximum to temperature changes in the middle mesosphere strengthens an old objection of Whipple [28] to the concept of the need of such a sharp tuning of a forced oscillation to the conditions of resonance. Because of the seasonal change of the mesospheric temperature a significant seasonal change should be expected for the amplification of $S_{2,2}^2$. It is not observed (see Chapman [29]).

These considerations and empirical facts suggest abandonment of the resonance theory. If this position is adopted, Chapman's explanation of the excitation of S_2^2 does no longer hold. Instead S_2^2 and also the other solar oscillations must be considered as caused only thermally, but the heating effect cannot be due to eddy conductivity. Hence, another and more effective kind of thermal excitation must exist, and so the problem arose to examine other periodical heating processes with regard to their effectiveness of generating air tides [30]. This has been done for the direct absorption of insolation by water vapor in the troposphere. The first quantitative results are given in Section 7 of this article. The amplitude of the temperature variation which is due to this periodical heating process, is, of course, small; but contrary to that due to eddy conductivity it decreases very slowly with height up to the tropopause. The phase is constant throughout the troposphere. It follows from the computations that, indeed, this new tide-generating force is stronger than that of the temperature variation produced by turbulent mass exchange. In the case of $S_{2,2}^2$, for example, the effectiveness increases by a factor of about 10 under conditions as they are found in the earth's atmosphere. However, even this thermal force is not quite sufficient since the resonance magnification of $S_{2,2}^2$ is not even fourfold after abandonment of the second resonance maximum. Because the magnification required by the resonance theory is about one hundredfold (when for the purpose of comparison the thermal action by eddy conductivity is assumed to be the only cause for $S_{2,2}^2$), a new thermal force is needed which must be at least thirty times as strong as the old one. Hence, there remains a factor of approximately 3 by which the thermal force due to absorption of insolation by water vapor is still too small to explain the magnitude of $S_{2,2}^2$ without pronounced resonance. Similar results are also obtained for other important wave types.

In search for a resolution of this remaining difference between observation and theoretical result, three possible explanations can be presented: First is the possibility that the model atmosphere employed for the numerical calculations is too simple and, therefore, the resonance magnifications derived from it are too small. Contrary to this assumption, the main lunar wave type $L_{2,2}^2$ does not require a larger magnification than that computed on the basis of the same model. Now the lunar gravitational tidal force is known exactly, but the same is not true for the solar thermal forces. Therefore, the second possibility is that the empirical data used for determination of the thermal action are not sufficiently reliable.

As a third possibility, a still more effective but unknown thermal tidal force might exist. In order to restrict this possibility, the tide-generating forces in the ozonosphere were estimated regarding $S_{2,2}^2$ only. The somewhat surprising result is that the periodic heating of this layer causes an $S_{2,2}^2$-pressure oscillation at the ground whose amplitude is a third to a fourth of that of the oscillation generated by the absorption of insolation in the troposphere. Hence, the effectiveness of the ozonosphere as a tide-generating source is even at ground level greater than that of the *austausch*-mechanism. Probably this effectiveness becomes more and more important with increasing height up to the ozonosphere. It does not make much sense, therefore, to calculate the dependence of the solar tidal oscillations on height without considering the thermal tidal forces of the ozonosphere. On the other hand, the empirical data necessary for these calculations are as yet so uncertain that exact results cannot be given.

Another thermal tidal force is the long-wave terrestrial radiation. Because it is absorbed nearly completely by only thin layers of water vapor—then emitted, reabsorbed, and so on—its spreading out is very similar to the heat transfer by eddy conductivity and its effectiveness should not exceed that of eddy conductivity. There remains only a window for wavelengths around $10\,\mu$ for which the troposphere is transparent. For this range absorption probably takes place in the lower stratosphere. The tidal oscillations due to this absorption have not yet been computed.

Hence, the most effective thermal tidal force which is known hitherto, is that due to absorption of incoming radiation by water vapor. Let us briefly discuss the theoretically obtained phases for this kind of excitation (see Table

IV). The phase angle of $S_{2,2}^2$ was found to be 180°; thus, the observed time of maximum occurs 48 min later than computed. An interesting point is that the observed time of maximum of the main semidiurnal lunar wave type $L_{2,2}^2$ occurs on the average 36 min later than computed, and in the case of $L_{2,2}^2$ the phase of the generating gravitational force is known exactly. Possibly these differences arise from surface friction which is not considered in the theory. This assertion is strengthened by the observation that the phase retardation of $L_{2,2}^2$ is larger in the northern than in the southern hemisphere (see [29], Fig. 9b) corresponding to the land-water distribution. The phase relations of other important wave types are quite satisfactory. The reliability of the observed phases, however, is not good enough to study the influence of possible surface friction.

Another point of interest is the influence of the land-water distribution on the earth's surface. It is almost certain that some observed solar wave families are influenced in some way by the land-water distribution which very probably affects the distribution of the thermal forces. The most important and best known of those wave families is S_2^0 with its main term $S_{2,2}^0$. According to thermal excitation by eddy conductivity the land areas have to be weighted much more than the ocean areas because of the much smaller amplitude of temperature variation above the oceans. However, it is not possible in this way to understand the 137° phase of $S_{2,2}^2$. When thermal excitation by absorption of insolation is assumed, the ocean areas have to be weighted more than the land areas because of the generally larger density of water vapor above the oceans compared to the land areas of the same latitude. The phase theoretically found in this way is 155°. That is, the observed time of maximum of $S_{2,2}^0$ occurs again 36 min later than computed. Also phases of other waves types can be explained in this way, but not by combining the land-water distribution with thermal excitation by eddy conductivity. A comparison of the theoretical amplitudes with the observational results is still impossible because of the nonavailability of very reliable data on the difference between the water vapor density above continents and oceans.

We can see from the last paragraphs that the development of a theory of atmospheric tides not based on the assumption of strongly resonant $S_{2,2}^2$ is possible. In spite of some clear shortcomings which still exist, the new theory seems to have a larger validity; that is, it seems to be able to explain more observational facts than the resonance theory. However, there is still one important objection. Kelvin's idea of resonance of the semidiurnal solar oscillation was essentially based on the smallness and irregularity of the diurnal solar oscillation. When now the assertion is made that $S_{2,2}^2$ and the other significant wave types are not greatly magnified by resonance, then a new explanation must be given for the absence of a large diurnal pressure oscillation which one would expect to be generated by the diurnal temperature variation (which is also the largest one among those variations caused by direct absorption of isolation). The only possible answer is that the diurnal oscillation is suppressed in the earth's atmosphere. There are, indeed, indications for such a behavior. If a model atmosphere with an isothermal top is used and thermal excitation is assumed, the resonance due to this model tends to zero as h^{-1} when $h \to \infty$ and as $h^{-1/2}$ when $h \to 0$. It would be expected from comparison with regular resonance curves that the amplitude of a disturbance should disappear when $h \to \infty$ and that it should tend to the unit of the resonance magnification when $h \to 0$. A resonance curve of this kind is also obtained for atmospheric tides if gravitational excitation is assumed. On the contrary, a pressure amplitude caused by thermal forces also vanishes when $h \to 0$, if the extrapolation to $h \to 0$ is allowed. This unusual result might explain the suppression of the migrating diurnal solar wave family S_1^1; for its wave types have only small equivalent depths, the largest of which is $h = 0.63$ km due to $S_{1,1}^1$ (see Fig. 10 in Section 7.2). The complete theoretical treatment of S_1^1 has not yet been given.

Summarizing the conclusions which follow from recent analyses of observational data and which are most compatible with the present, still incomplete state of the theory, we can characterize the situation by saying that, contrary to Kelvin's suggestion, not the semidiurnal but the diurnal solar pressure oscillation is the extraordinary phenomenon. The migrating semidiurnal solar air-tide S_2^2 is predominant because the diurnal air tide is suppressed in the earth's atmosphere in spite of its stronger excitation while the other 12-, 8-, and 6-hourly pressure waves which do not differ from S_2^2 in regard to the magnitude of resonance, are due to less strong thermal tidal forces.

In order to obtain a final solution of our problem, theoretical and observational work has to be done which goes beyond the present state in some directions. It does not seem necessary to search for further wave families because now many of them and probably the most significant ones are known. Their behavior, however, needs further investigation, especially their seasonal variation and their variation with height up to the ionosphere. First results are already available (see, for example, references [31–34] [103]), even some from the daily variation of the cosmic ray intensity [35]. The analysis of observational data of additional stations would bring out regional anomalies on the earth's surface much clearer than known at present. As such additional empirical material becomes available it should be possible to restrict greatly the ambiguity of the present theoretical concepts about the generation of atmospheric tides. The quantitative explanation of the details will require the development of a more and more complicated theoretical formulism.

2. Application and Results of Harmonic Analysis

2.1. Outline of Harmonic Analysis

Atmospheric pressure and temperature are given by recorders as functions of time. The shapes of the curves are determined mainly by the daily variation, the seasonal variation, and the irregular variation of the observed elements. The irregular variation is assumed to disappear when a large number of observations is averaged. Regarding this nonperiodic variation as if it consisted of random errors, the standard deviation vanishes as $1/\sqrt{N}$, where N is the number of data from which any mean is computed. A periodic variation, on the contrary, is preserved when the superposed-epoch method is used so that only values of the same phase are summed. The successful application of this method to pressure and temperature variations is based on the exact knowledge of the fundamental periods of the periodic variations. These are the length of the solar day and the length of the lunar day.

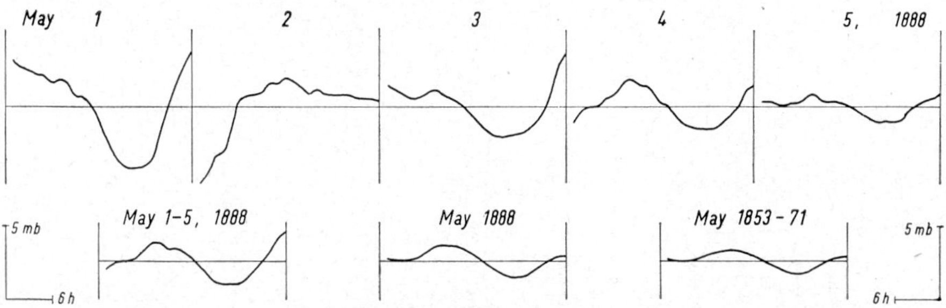

Fig. 1. Observed pressure deviations from the daily mean for May 1–5, 1888, at the observatory Hohe Warte, Vienna [36] (above). Mean daily pressure variation at the same observatory determined by using the superposed-epoch method for May 1–5, 1888 (lower left), May 1888 (lower middle), and 19 months of May after Hann [42] (lower right).

Instead of the absolute readings, the deviations from the daily mean are generally used. They are computed for the 24 hr of solar or lunar local mean time t^*. The next step, the elimination of the nonperiodic variation, is illustrated in Fig. 1. The observatory, Hohe Warte, Vienna, and the days, May 1–5, 1888 [36], were arbitrarily chosen. In this special case the observations of only one month were adequate to obtain a mean daily pressure variation nearly free of nonperiodic influences. Even this unanalyzed variation shows two maxima and two minima, so that the period of half a solar day turns out to be predominant also in middle latitudes (see also [106]). When the mean daily variation has been obtained, it generally shows a difference between 0^h and 24^h. This noncyclic variation is eliminated by representing it by a linear change during the whole day and subtracting it (Lamont's correction). Thus, a completely periodic variation remains. The errors which can arise by the linear elimination of the noncyclic variation were studied by Bartels [37] (effect of curvature).

To the corrected mean daily variation the Fourier analysis of equidistant ordinates is applied. In this manner, the variation is represented by the sum of a series of sine and cosine terms; for example, the pressure variation δp at any station is expressed by

$$\delta p = \sum_{\lambda=1}^{12} (a_\lambda \cos \lambda t^* + b_\lambda \sin \lambda t^*) \tag{2.1}$$

$$= \sum_{\lambda=1}^{12} c_\lambda \sin (\lambda t^* + \epsilon_\lambda) \tag{2.2}$$

with

$$c_\lambda^2 = a_\lambda^2 + b_\lambda^2 \text{ and } \tan \epsilon_\lambda = a_\lambda / b_\lambda \tag{2.3}$$

The coefficients in (2.1) are given by

$$a_\lambda = \tfrac{1}{12} \sum_{\mu=1}^{24} \delta p(t_\mu^*) \cos \lambda t_\mu^*,\ b_\lambda = \tfrac{1}{12} \sum_{\mu=1}^{24} \delta p(t_\mu^*) \sin \lambda t_\mu^*,\ \lambda = 1, 2, \ldots 11 \tag{2.4}$$

$$a_{12} = \tfrac{1}{24} \sum_{\mu=1}^{24} (-1)^\mu \, \delta p(t_\mu^*), \qquad b_{12} = 0$$

with $t_\mu^* = \dfrac{\pi}{12} \mu$ corresponding to μ^h local mean time.

The numbers in these formulas are true for the special, but most important analysis of 24 ordinates. A more general treatment of harmonic analysis can be found in many textbooks of applied mathematics.

Each term of the sum (2.2) can be represented as a vector in a harmonic dial (see Fig. 2). The quantity c_λ is the magnitude of the vector and ϵ_λ, counted positive from the b_λ axis, gives its direction in the $a_\lambda b_\lambda$ plane. The significance of such a term can be tested by comparing the length of its vector with the probable error which should not be larger than a third of the length of the vector. The probable error is obtained by application of the theory of errors in a plane to the dial points, the end points of the vectors of $c_\lambda \sin (\lambda t^* + \epsilon_\lambda)$, for instance when records of many years are available, and when c_λ and ϵ_λ are determined for each single year. In this case the number of the years would be equal to the number of the dial points which are distributed around the mean of all years. Bartels [11] showed how, from the theory of errors, a probable error ellipse follows which degenerates into a probable error circle if the dial points are randomly distributed. In this case, half the points should fall within the circle. The radius of this circle is the probable error of the vectors $c_\lambda \sin (\lambda t^* + \epsilon_\lambda)$. The probable error circle of the mean vector is then obtained by dividing this radius by $\sqrt{N}$ (N = number of dial points).

2.2 Planetary Representation of the Tidal Oscillations

When the analysis of δp has been carried out for a sufficiently large number of stations all over the world the harmonic coefficients a_λ and b_λ can be regarded as functions of colatitude ϑ and longitude ϕ. Because of the periodicity in ϕ which is given by the sphericity of the earth, a_λ and b_λ can be analyzed harmonically with respect to ϕ:

$$a_\lambda(\vartheta, \phi) = \sum_{\nu=0} (k_\lambda^\nu(\vartheta) \cos \nu\phi + l_\lambda^\nu(\vartheta) \sin \nu\phi), \qquad \lambda = 1, 2, \ldots 12 \tag{2.5}$$

$$b_\lambda(\vartheta, \phi) = \sum_{\nu=0} (q_\lambda^\nu(\vartheta) \cos \nu\phi + r_\lambda^\nu(\vartheta) \sin \nu\phi), \qquad \lambda = 1, 2, \ldots 12 \tag{2.6}$$

The upper limit of ν, which is merely an affix (as with s later)—not an index or exponent—depends on the number of the grid points chosen. The coefficients in (2.5) and (2.6) are still functions of ϑ. When a_λ and b_λ are substituted into (2.1) by means of (2.5) and (2.6), and the notation of Section 1 is used, we can write with the aid of some trigonometric formulas, for instance, for the λ^{-1}-diurnal solar pressure oscillation

$$S_\lambda(p) = a_\lambda(\vartheta, \phi) \cos \lambda t^* + b_\lambda(\vartheta, \phi) \sin \lambda t^* \tag{2.7}$$

$$= \tfrac{1}{2} \sum_{\nu=0} [(k_\lambda^\nu - r_\lambda^\nu) \cos (\lambda t^* + \nu\phi) + (l_\lambda^\nu + q_\lambda^\nu) \sin (\lambda t^* + \nu\phi)] \tag{2.8}$$

$$+ \tfrac{1}{2} \sum_{\nu=0} [(k_\lambda^\nu + r_\lambda^\nu) \cos (\lambda t^* - \nu\phi) - (l_\lambda^\nu - q_\lambda^\nu) \sin (\lambda t^* - \nu\phi)]$$

We see from (2.8) that S_λ consists of terms

$$A_\lambda^\nu(\vartheta) \cos (\lambda t^* + \nu\phi) + B_\lambda^\nu(\vartheta) \sin (\lambda t^* + \nu\phi)$$

where ν can be negative. Let us introduce universal time t' and eliminate local mean time t^* by

$$t^* = t' + \phi \tag{2.9}$$

Further, let us put $s = \lambda + \nu$. Then simple expressions of planetary meaning can be derived when A_λ^ν and B_λ^ν are represented by spherical harmonics of the order s so that an expression in terms of surface harmonics is obtained

$$A_\lambda^\nu = \sum_{m=s}^{\infty} \alpha_{\lambda,m}^s P_m^s(\vartheta); \qquad B_\lambda^\nu = \sum_{m=s}^{\infty} \beta_{\lambda,m}^s P_m^s(\vartheta) \tag{2.10}$$

The coefficients in (2.10) can be determined by the method of least squares. A more convenient procedure is attained by employing the orthogonality of the spherical harmonics. For this purpose (2.10) is multiplied by the weight factor $\sin \vartheta$ because of the change of the area of the earth's surface with $\sin \vartheta$.

Using the seminormalized associated Legendre functions according to Schmidt (tables were given, for instance, by Haurwitz and Craig [38]), we have

$$(2.11) \qquad \int_0^\pi P_m^s(\vartheta)\, P_{m'}^s(\vartheta) \sin\vartheta\, d\vartheta = \frac{2}{2m+1} \times \delta_{m,m'}^s$$

with

$$\delta_{m,m'}^s = \begin{cases} 0 \text{ if } m \neq m' \\ 1 \text{ if } m = m' \text{ and } s = 0 \\ 2 \text{ if } m = m' \text{ and } s = 1, 2, \ldots m \end{cases}$$

This kind of normalization, which is usual in geophysics, has the advantage that the coefficients, e.g., $\alpha_{\lambda,m}^s$ and $\beta_{\lambda,m}^s$ in (2.10) and (2.13) and $c_{\lambda,m}^s$ in (2.14), indicate at once the approximate orders of magnitude of the corresponding terms.

By means of (2.11) we obtain from (2.10)

$$(2.12) \qquad \alpha_{\lambda,m}^s = \frac{2m+1}{2\delta_{m,m}^s} \int_0^\pi A_\lambda^{\,\nu} P_m^s(\vartheta) \sin\vartheta\, d\vartheta$$

This procedure involves the minimum condition of the method of least squares with respect to $A_\lambda^{\,\nu} \sin\vartheta$. When the method of least squares is applied to $A_\lambda^{\,\nu}$, small differences from (2.12) result for the coefficients of those wave families which are mainly important at high latitudes. The integration in (2.12) must be carried out numerically. Analogous formulas are, of course, valid for $\beta_{\lambda,m}^s$.

With (2.10) the representation of a wave family is given by

$$(2.13) \qquad S_\lambda^s = \sum_{m=s}^{\infty} P_m^s(\vartheta)[\alpha_{\lambda,m}^s \cos(\lambda t' + s\phi) + \beta_{\lambda,m}^s \sin(\lambda t' + s\phi)]$$

or

$$(2.14) \qquad S_\lambda^s = \sum_{m=s}^{\infty} c_{\lambda,m}^s P_m^s(\vartheta) \sin(\lambda t' + s\phi + \epsilon_{\lambda,m}^s)$$

The numerical results of the analysis quoted in Section 2.4, are given according to (2.14).

For comparison with the theoretical results the spherical harmonics must be replaced by Hough's functions $\Theta_{\lambda,n}^s(\vartheta)$. According to Section 5, the following relationship holds

$$(2.15) \qquad P_m^s(\vartheta) = \sum_{n=s}^{\infty} \gamma_{\lambda,m}^{s,n} \Theta_{\lambda,n}^s(\vartheta)$$

When P_m^s is substituted into (2.13) by means of (2.15), we obtain finally a series of wave types $S_{\lambda,n}^s$ with

(2.16) $$S_{\lambda,n}^{s} = K_{\lambda,n}^{s} \Theta_{\lambda,n}^{s}(\vartheta) \sin (\lambda t' + s\phi + \kappa_{\lambda,n}^{s})$$

where

(2.17) $$K_{\lambda,n}^{s} = \Big\{\Big(\sum_{m=s}^{\infty} \gamma_{\lambda,m}^{s,n} \alpha_{\lambda,m}^{s}\Big)^2 + \Big(\sum_{m=s}^{\infty} \gamma_{\lambda,m}^{s,n} \beta_{\lambda,m}^{s}\Big)^2\Big\}^{1/2}$$

(2.18) $$\tan \kappa_{\lambda,n}^{s} = \sum_{m=s}^{\infty} \gamma_{\lambda,m}^{s,n} \alpha_{\lambda,m}^{s} \Big/ \sum_{m=s}^{\infty} \gamma_{\lambda,m}^{s,n} \beta_{\lambda,m}^{s}$$

The same formulism can, of course, be applied also to the lunar pressure oscillation and to the temperature variation. When the temperature variation at the earth's surface is studied, the simplifying assumption is mostly used that the amplitude of the variation disappears over the ocean. This assumption involves the harmonic analysis of the land-water distribution which was carried out by Kertz [39] for this purpose.

2.3 Treatment of Seasonal Variations of the Tidal Oscillations

Some of the pressure and temperature variations show characteristic seasonal changes. They are usually investigated by arranging the observational data with respect to the months and determining a_λ and b_λ in (2.1) for each month. Thus twelve points in the $a_\lambda b_\lambda$-harmonic dial are obtained which illustrate the seasonal variation of the analyzed λ^{-1}-diurnal variation. Following Siebert [40], it is possible to continue the analysis and represent these variations in a manner analogous to the partial tides of the gravitational tidal potential. The season may be denoted by η which is varying through 2π or 360° within a tropical year. Each month should have 30.437 mean solar days, in order to comply with the requirements of equidistant ordinates. However, civil months were employed hitherto for most of the analyses. When a_λ and b_λ which are now functions of η, are known for $\eta_\nu = \pi\nu/6$, $\nu = 1, 2, \ldots 12$, they can be represented by

(2.19) $$a_\lambda(\eta) = k_{\lambda,0} + \sum_{\mu=1}^{6} (k_{\lambda,\mu} \cos \mu\eta + l_{\lambda,\mu} \sin \mu\eta), \qquad \lambda = 1, 2, \ldots 12$$

(2.20) $$b_\lambda(\eta) = q_{\lambda,0} + \sum_{\mu=1}^{6} (q_{\lambda,\mu} \cos \mu\eta + r_{\lambda,\mu} \sin \mu\eta), \qquad \lambda = 1, 2, \ldots 12$$

The coefficients in (2.19) and (2.20) must be again determined according to the precepts of harmonic analysis. The coefficients $k_{\lambda,0}$ and $q_{\lambda,0}$ are annual mean values by which the mean vector, representing the λ^{-1}-diurnal variation, is given. The additional terms of (2.19) and (2.20) can be illustrated by six ellipses in the harmonic dial due to the period of $24/\lambda$ hr. The center of the ellipses is the end point of the mean vector $c_{\lambda,0} \sin (\lambda t^* + \epsilon_{\lambda,0})$. Each ellipse corresponds to an annual change with a period of μ^{-1} year, $\mu = 1, 2, \ldots 6$.

We can describe the ellipses by sine terms as follows: Using an xy system as presented in Fig. 2 (right-handside) the λ, μ terms in (2.19) and (2.20) can be written:

$$y = k_{\lambda,\mu} \cos \mu\eta + l_{\lambda,\mu} \sin \mu\eta \tag{2.21}$$

$$x = q_{\lambda,\mu} \cos \mu\eta + r_{\lambda,\mu} \sin \mu\eta \tag{2.22}$$

This is the parametric representation of an ellipse. On the other hand, the analytical expression of an ellipse in the harmonic dial must be derived. Although only sine waves with the same periods can be added vectorially in the same harmonic dial, it is possible to represent also the expression :

$$c_0 \sin (t^* + \epsilon_0) + u \sin (t^* + \eta + \alpha); \quad c_0, u, \epsilon_0, \alpha = \text{const.}$$

We consider both waves to have the period t^*. Then, the second wave has the periodically variable phase $(\eta + \alpha)$. Because the amplitude u is constant, the

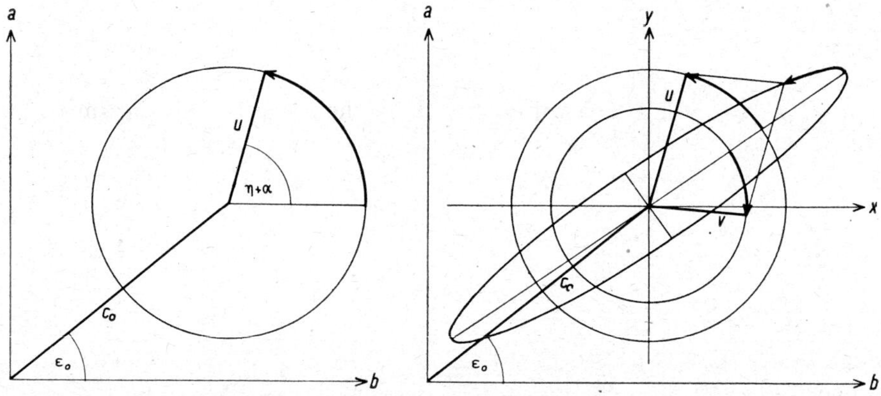

FIG. 2. Representation of two sine waves with different frequencies in the same harmonic dial (left-handside). On the representation of an elliptic seasonal variation in a λ^{-1}–diurnal harmonic dial (right-handside). (After Siebert [40].)

second vector with the length u describes a circle around the first, a constant vector with the length c_0 (Fig. 2, left-handside). It is now possible to construct an ellipse by the addition of two vectors, whose end points move with the same angular velocity but opposite rotational sense on the peripheries of two concentric circles. The radii of the circles are determined by the magnitudes of the vectors (Fig. 2, right-handside). Thus it must be possible to represent the ellipse given by (2.21) and (2.22), in the harmonic dial by the expression

$$u_{\lambda,\mu} \sin (\lambda t^* + \mu\eta + \alpha_{\lambda,\mu}) + v_{\lambda,\mu} \sin (\lambda t^* - \mu\eta + \beta_{\lambda,\mu}) \tag{2.23}$$

Resolving this vector in its x and y components, we find

$$y = u_{\lambda,\mu} \sin (\mu\eta + \alpha_{\lambda,\mu}) - v_{\lambda,\mu} \sin (\mu\eta - \beta_{\lambda,\mu}) \tag{2.24}$$

(2.25) $$x = u_{\lambda,\mu} \cos(\mu\eta + \alpha_{\lambda,\mu}) + v_{\lambda,\mu} \cos(-\mu\eta + \beta_{\lambda,\mu}).$$

The equations (2.24) and (2.25) are identical with (2.21) and (2.22). Hence, amplitudes and phases in (2.23) can be determined from the harmonic coefficients in (2.19) and (2.20). The relations are

(2.26) $$u_{\lambda,\mu} = \tfrac{1}{2}\sqrt{(k_{\lambda,\mu} - r_{\lambda,\mu})^2 + (q_{\lambda,\mu} + l_{\lambda,\mu})^2}$$

(2.27) $$v_{\lambda,\mu} = \tfrac{1}{2}\sqrt{(k_{\lambda,\mu} + r_{\lambda,\mu})^2 + (q_{\lambda,\mu} - l_{\lambda,\mu})^2}$$

(2.28) $$\tan \alpha_{\lambda,\mu} = \frac{k_{\lambda,\mu} - r_{\lambda,\mu}}{q_{\lambda,\mu} + l_{\lambda,\mu}}$$

(2.29) $$\tan \beta_{\lambda,\mu} = \frac{k_{\lambda,\mu} + r_{\lambda,\mu}}{q_{\lambda,\mu} - l_{\lambda,\mu}}$$

In order to obtain single-valued quantities from (2.26) to (2.29), we define $u_{\lambda,\mu}$ and $v_{\lambda,\mu}$ as always positive. Then the following schemes hold for the phases:

If	$k_{\lambda,\mu} - r_{\lambda,\mu}$	>0	>0	<0	<0
	$q_{\lambda,\mu} + l_{\lambda,\mu}$	>0	<0	<0	>0
then $\alpha_{\lambda,\mu}$ lies in quadrant		1	2	3	4

If	$k_{\lambda,\mu} + r_{\lambda,\mu}$	>0	>0	<0	<0
	$q_{\lambda,\mu} - l_{\lambda,\mu}$	>0	<0	<0	>0
then $\beta_{\lambda,\mu}$ lies in quadrant		1	2	3	4

The length of the semi-major axis of the ellipse is $(u_{\lambda,\mu} + v_{\lambda,\mu})$ and that of the semi-minor axis $|u_{\lambda,\mu} - v_{\lambda,\mu}|$ as can be seen from Fig. 2. The angle between the major axis and the positive x axis is $0.5(\alpha_{\lambda,\mu} + \beta_{\lambda,\mu})$.

Summarizing the results of this section we find that a λ^{-1}-diurnal variation whose amplitude and phase change with the season, can be represented in the following manner:

(2.30) $$\begin{aligned} c_\lambda(\eta) \sin(\lambda t^* + \epsilon_\lambda(\eta)) &= c_{\lambda,0} \sin(\lambda t^* + \epsilon_{\lambda,0}) \\ &+ \sum_{\mu=1}^{6} [u_{\lambda,\mu} \sin(\lambda t^* + \mu\eta + \alpha_{\lambda,\mu}) \\ &+ v_{\lambda,\mu} \sin(\lambda t^* - \mu\eta + \beta_{\lambda,\mu})] \end{aligned}$$

As for the determination of η, a proposal by Bartels [41] may be noticed according to which η should be assumed to be zero at December 21 for all geophysical elements. The most practical way of determining η is to begin the numbering of the months with January; that is, η is zero for January 16. In the final result a transformation to December 21 may be made.

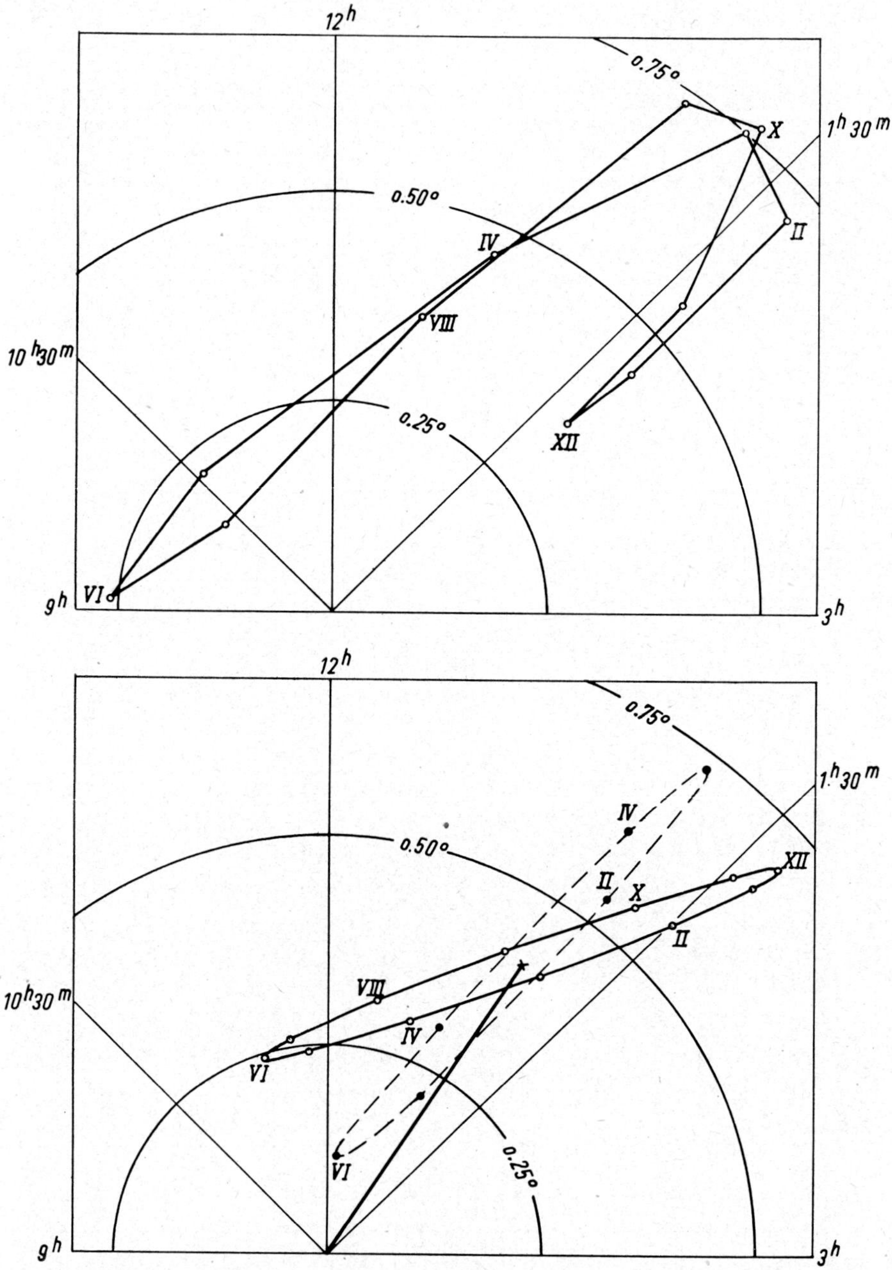

FIG. 3. The semidiurnal temperature variation in 50°–60° North: the seasonal variation (above) and the result of its analysis, given by the mean vector and ellipses with μ^{-1}–annual periods ($\mu = 1$ = annual ellipse (solid curve), $\mu = 2$ = semiannual ellipse (broken curve)), (below). (After Siebert [40]).

The formulism for the analysis of the seasonal change of the λ^{-1}-diurnal variations has been developed here with regard to the observational data of one station or such stations whose data can be combined. The result of such an analysis is given in Fig. 3 which shows the analyzed annual change of the semidiurnal temperature variation in 50°–60° North (above), the mean vector, the annual ellipse ($\mu = 1$), and the semiannual ellipse ($\mu = 2$) (below). The other ellipses are negligible. Data of the following stations were used: Oslo, Uppsala, Pawlowsk, Aberdeen, Potsdam, Irkutsk, Valentia, and Kew.

The formulism can also be extended to a planetary representation of the seasonal change, as was previously described for the mean vector only. However, such a development is not yet topical. The first step is the numerical determination of the amplitudes and phases in (2.30) for as many stations as possible. That has to be done in order to study the seasonal change of the λ^{-1}-diurnal variations appropriately.

2.4 *Numerical Results*

The best-known solar pressure oscillation is S_2. Hann [42] and Angot [43] were the first to determine the second harmonic coefficients (but also further harmonics) for a large number of stations all over the world. As early as 1890, Schmidt [44] pointed out that the distribution of the phase angle can be explained if S_2 is assumed to consist of two oscillations, one migrating with the sun around the earth (S_2^2 according to our notation) and the other an oscillation with maxima at the same universal time and only zonal dependence (S_2^0). An analytical representation of these oscillations was given by Simpson [45] in 1918 on the basis of the data of 214 stations:

$$S_2^2(p) = 1.25 \sin^3 \vartheta \sin (2t' + 2\phi + 154°) \text{ mb} \tag{2.31}$$

$$S_2^0(p) = 0.061(3 \cos^2 \vartheta - 1) \sin (2t' + 105°) \text{ mb} \tag{2.32}$$

The $\sin^3 \vartheta$ law was first noticed by Jaerisch [46].

Since that time new data have been obtained. Therefore a new study of S_2 was reasonable. It was carried out by Haurwitz [47] who used the data of 296 stations (see references quoted in [47]). Figure 4 shows, as a result, the observed distribution of the amplitude and Fig. 5 the observed distribution of the phase angle of S_2 in local solar time after Haurwitz. Kertz [26], who applied the formulism of Section 2.2 to Haurwitz' material, found a significant third wave family S_2^3, the amplitude of which is of the same magnitude as that of S_2^0. S_2^3 is a semidiurnal migrating wave with three maxima and three minima around the earth's circumference. The following analytical expressions are taken from Haurwitz [47] and Kertz [26]. Only the main terms are quoted:

$$S_2^2(p) = [1.23P_2^2(\vartheta) - 0.224P_4^2(\vartheta)] \sin (2t' + 2\phi + 158°) \text{ mb}; \tag{2.33}$$

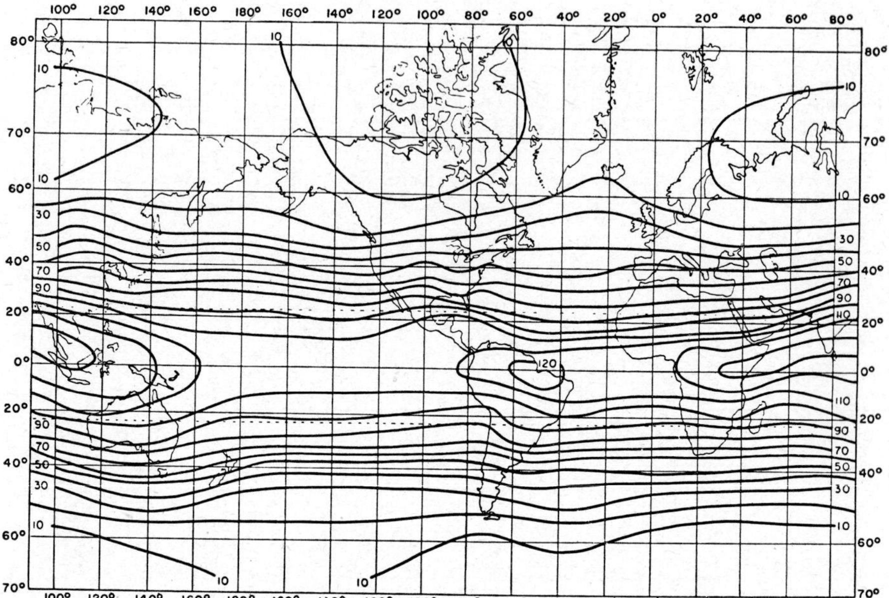

FIG. 4. Observed distribution of the amplitude of the semidiurnal solar pressure oscillation S_2 (unit 10^{-2} mb), (after Haurwitz [47]).

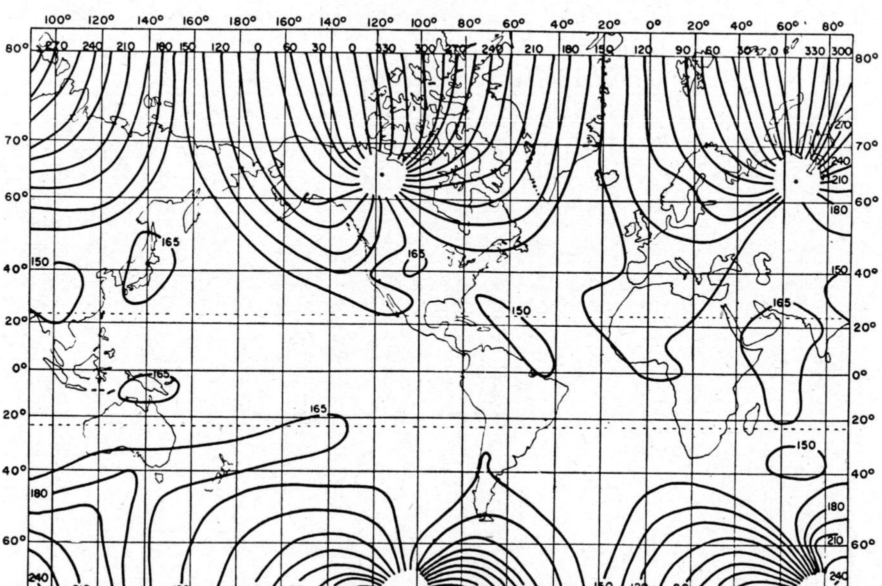

FIG. 5. Observed distribution of the phase constant of the semidiurnal solar pressure oscillation S_2 in local time (after Haurwitz [47]).

(2.34) $$S_2^0(p) = [7.18P_2^0(\vartheta)\sin(2t' + 135°) + 5.62P_3^0(\vartheta)\sin(2t' + 123°)]\ 10^{-2}\ \text{mb}$$

(2.35) $$S_2^3(p) = 10.70P_3^3(\vartheta)\sin(2t' + 3\phi + 88°)10^{-2}\ \text{mb}$$

The wind fields associated with $S_2^2(p)$ and $S_2^0(p)$ have been computed by Bartels [23] and Stolov [48] employing (2.31) and (2.32).

The seasonal variations of $S_2^2(p)$ and $S_2^0(p)$ were studied by Haurwitz and Sepúlveda [49] and planetary expressions were derived from data of 136 stations for the months of January, March, July, and September. A strong asymmetry in the latitudinal distribution of the amplitudes of S_2^2 was found for January, resulting in larger amplitudes in the northern hemisphere than in the southern hemisphere. Moreover, the amplitude of S_2^2 shows a summer minimum at all latitudes.

Spar [50] gave a detailed picture of the regional and seasonal variations of S_2 in the United States using the data of 100 stations. The influence of topography on the phase appears to be more striking than that on the amplitude. It agrees roughly with the theoretical result obtained by Kertz [13] concerning the influence of the Cordilleras on S_2^2.

The terdiurnal solar pressure oscillation also was investigated in detail by Hann [19]. Schmidt [51] derived, in a rough manner, planetary pressure waves from Hann's results. Computations having the same aim, were recently carried out more carefully by Siebert [27] again on the basis of Hann's analyses of more than 100 stations. The main features of the terdiurnal pressure oscillation are its annual variation which is exhibited as a change of phase by 180° from January to July (the months of maximum amplitudes) and again by 180° from July to January, and a nearly antisymmetric distribution of the amplitude with regard to the equator. For an exact description, therefore, the formulism of Section 2.3 should be used, but this requires laborious numerical calculations. An approximate description, which depicts the main features quite well [27], is based on the data of January and July only, from which annual means and the pure annual variation were derived. The results of this simplified method are:

(2.36) $$S_3^{3,*}(p) = [0.269P_4^3(\vartheta) - 0.014P_6^3(\vartheta)]\sin\left(3t' + 3\phi + \begin{matrix}175°\\355°\end{matrix}\right)\text{mb}\quad\begin{matrix}\text{July}\\\text{Jan.}\end{matrix}$$

(2.37) $$S_3^3(p) = [7.5P_3^3(\vartheta) + 4.4P_4^3(\vartheta) + 3.9P_5^3(\vartheta)]\sin(3t' + 3\phi + 10°)\ 10^{-2}\text{mb}$$

According to the definition of S_λ^s by (2.14) both wave families (2.36) and (2.37) must be denoted by S_3^3 because in both cases $\lambda = 3$ and $s = 3$.

A seasonal variation was not considered in (2.14) in order not to complicate the present classification more than necessary. Let us, therefore, denote wave families distinguished by annual variations only by using asterisks: $S_\lambda^{s,*}$. This is sufficient to attain an unequivocal classification in this article.

The six-hourly pressure oscillation is not as well known as the semidiurnal and terdiurnal oscillations. After Pramanik's investigation [20] in 1926, Kertz [26] recently investigated this component of atmospheric tides again and derived planetary expressions from the data of more than 60 stations. His results (see also [14]) are given in Table I as amplitudes and phases of the terms of S_4^4. In order to study the annual and even semiannual variations, the computations were carried out for the groups of the equinoctial and solstitial months.

TABLE I. Amplitudes $c_{4,m}^4$ and phases $\epsilon_{4,m}^4$ of the terms of S_4^4 [see (2.14)] of March, April, September, October = MASO; January, February, November, December = JFND; May, June, July, August = MJJA; and the Annual Mean.

	MASO		JFND		MJJA		Annual Mean	
m	$c_{4,m}^4$ 10^{-3}mb	$\epsilon_{4,m}^4$	$c_{4,m}^4$ 10^{-3}mb	$\epsilon_{4,m}^4$	$c_{4,m}^4$ 10^{-3}mb	$\epsilon_{4,m}^4$	$c_{4,m}^4$ 10^{-3}mb	$\epsilon_{4,m}^4$
4	17.4	173°	7.9	244°	19.1	180°	13.4	187°
5	27.9	204°	98.1	207°	24.2	359°	35.1	213°
6	34.7	336°	24.4	215°	26.8	222°	15.5	260°
7	13.9	252°	28.2	191°	22.8	331°	10.8	248°

The existence of the lunar air tide was first reliably demonstrated at tropical stations (St. Helena by Lefroy in 1842 and later by Smythe and Sabine [52], Singapore by Elliot [53] in 1852, and Batavia by Bergsma [54] in 1871). Outside the tropics the lunar pressure oscillation was first determined by Chapman [21] in 1918 for Greenwich, as already noted in Section 1. After many attempts of other workers had failed, Chapman succeeded by confining the observational material to those days on which the barometric range did not exceed 3.4 mb. For practical methods of the determination of the lunar air tide by means of harmonic analysis, see references [22, 55, 56, 57].

Meanwhile the data of 68 stations have been analyzed mainly by Chapman. Only the semidiurnal lunar variation L_2^2 was found significant. The amplitudes and phases due to this variation at the 68 stations are listed in Bartels *et al.* [58]. In reference [59] the results are discussed by Chapman and Westfold and the semidiurnal lunar tide is compared with the semidiurnal solar tide. Regional anomalies of amplitudes and phases are considered: "Both these tides decrease with increasing latitude, but not quite symmetrically relative to the equator; there is some indication that in northern latitudes the lunar tide decreases rather more rapidly than the solar. Both tides are abnormally great over east Africa; the solar tide is also large in India, and the lunar tide in the East Indies. The lunar tide is large relative to the solar tide along and near the east coast of Asia, and also in the East Indies; it is small relative to the solar tide over India and western North America. The phase difference between the solar and lunar tides is specially large along and near the east Asiatic coast." (Quotation from the abstract of [59].)

There is also an obvious annual change of the lunar air tide, which shows the strange feature that both amplitude and phase vary in the same sense in both hemispheres. An annual change, however, which depends on the season of the concerned hemisphere, would be expected from the variation of the generating gravitational lunar tidal force. A more detailed discussion and illustration of this behavior of the lunar tide are given by Chapman [29].

Siebert [60] derived from the data in Bartels *et al.* [58] a planetary representation of the semidiurnal lunar pressure oscillation. Assuming a symmetric distribution with regard to the equator, he found

$$L_2^2(p) = [7.1P_2^2(\vartheta) - 2.1P_4^2(\vartheta) + 0.7P_6^2(\vartheta)] \sin(2t_L' + 2\phi + 72°)10^{-2}\text{mb} \tag{2.38}$$

where t_L' denotes lunar universal time, that is Greenwich lunar mean time.

In addition to the pressure oscillations, a few wave families of semidiurnal and terdiurnal temperature variations shall be noted without discussion. They are the empirical basis for a treatment of thermal excitation by eddy conductivity. The qualitative comparison of the corresponding pressure and temperature terms already shows remarkable differences between the amplitude proportions of the pairs.

The following semidiurnal temperature waves were derived by Haurwitz and Möller [61]:

$$S_2^2(T) = [0.301P_2^2(\vartheta) + 0.368P_3^2(\vartheta)] \sin(2t' + 2\phi + 65°) + 0.113P_4^2(\vartheta) \sin(2t' + 2\phi + 68°)°\text{C} \tag{2.39}$$

$$S_2^0(T) = [0.076P_1^0(\vartheta) \sin(2t' + 194°) + 0.040P_2^0(\vartheta) \sin(2t' + 214°) + 0.112P_3^0(\vartheta) \sin(2t' - 1°) + 0.104P_4^0(\vartheta) \sin(2t' + 56°)]°\text{C} \tag{2.40}$$

The expressions of the terdiurnal temperature waves are taken from Siebert's paper [27]:

$$S_3^{3,*}(T) = [1.8P_3^3(\vartheta) + 7.5P_4^3(\vartheta) + 6.4P_5^3(\vartheta) + 5.7P_6^3(\vartheta) + 5.0P_7^3(\vartheta)] \sin(3t' + 3\phi + 45°/225°)10^{-2}\,°\text{C(July/Jan.)} \tag{2.41}$$

$$S_3^3(T) = [6.7P_3^3(\vartheta) + 3.0P_4^3(\vartheta) + 1.4P_6^3(\vartheta)] \sin(3t' + 3\phi + 12°)10^{-2}\,°\text{C} \tag{2.42}$$

3. Foundation of the Theory

3.1 Assumptions and Basic Equations

The conditions under which atmospheric tides occur in the earth's atmosphere are so numerous and complex in such a manner that a quantitative consideration of all these conditions in theory is impossible. Therefore, it is necessary to neglect less important influences in order to simplify the problem so far that it becomes amenable to mathematical treatment.

For this purpose the earth is assumed to be a sphere of radius a with smooth surface and to rotate with the uniform angular velocity $\vec{\omega}$. The change of the acceleration of gravity $\mathbf{g}$ with colatitude ϑ and height z is neglected. The error thus introduced is about 3 % at 100 km height. The vector $\mathbf{g}$ may also include the centrifugal acceleration and have radial direction; that is, the direction of $-z$. The undisturbed atmosphere is described by the distribution of static pressure p_0, density ρ_0, and temperature T_0. These quantities are assumed to depend on z only, not on colatitude ϑ and longitude ϕ. They are connected by the hydrostatic equation

$$\frac{dp_0}{dz} = -g\rho_0 \tag{3.1}$$

and by the equation for an ideal gas

$$p_0 = \frac{R}{M}\rho_0 T_0 = g\rho_0 H \tag{3.2}$$

where R is the universal gas constant, M the mean molecular weight of air, and H the *scale height*

$$H = \frac{R\,T_0}{M\,g} \tag{3.3}$$

The molecular weight M is treated as a constant. This is justified by observational evidence up to the lower boundary of the ionosphere. The humidity of the air is neglected.

If the meridional temperature variation is considered as it exists in the actual earth's atmosphere, the mathematical problem becomes extremely complicated. Estimates of the influence of a meridional temperature gradient have been given by some authors [11, 12, 23, 25] by varying the mean surface temperature within reasonable limits or by using Hough's [62] solution for tides in an ocean with the variable depth $(h + k \sin^2 \vartheta)$. Using approximation methods, Haurwitz [63] and Siebert [64] succeeded in describing the actual circumstances somewhat more realistically. Their studies, however, are confined to the standing semidiurnal oscillation S_2^0. The influence of the meridional temperature variation was found from all estimates to be negligibly small.

The equations of motion employed are the Eulerian equations of hydrodynamics referred to the rotating earth

$$\frac{D\mathbf{v}}{Dt} + 2\vec{\omega} \times \mathbf{v} = -\frac{1}{\rho}\operatorname{grad} p + \mathbf{g} - \operatorname{grad} \Omega \tag{3.4}$$

with

$$p = p_0 + \delta p \qquad \text{and} \qquad \rho = \rho_0 + \delta\rho \tag{3.5}$$

Because a gravitational tidal force is a potential force the acceleration due to it can be represented as the gradient of the tidal potential Ω. The vector $\mathbf{v}$ is the velocity of tidal winds. It has the components (u, v, w) positively directed southward, eastward, and vertically upward. Friction is neglected. This assumption has to be kept in mind when energy considerations are made. We also do not take into account the fact that the undisturbed atmosphere is not at rest relative to the earth. The effect of an undisturbed zonal wind with constant angular velocity upon the free periods of a homogeneous ocean and an autobarotropic atmosphere was investigated by Chiu [65]. The periods of the oscillations of interest in tidal theory are affected very little in this way. The tidal oscillations occur, at least in the lower atmosphere, with small amplitudes so that the variations of pressure δp, density $\delta \rho$, and temperature δT are small in comparison with p_0, ρ_0, and T_0. Also the tidal wind components (u, v, w) are small compared to $a\omega$. Therefore, the squares and products of these small quantities are neglected. Then we obtain from (3.4) when (3.1) and (3.5) are used

$$\frac{\partial \mathbf{v}}{\partial t} + 2\vec{\omega} \times \mathbf{v} = -\frac{1}{\rho_0}\operatorname{grad} \delta p + \frac{\delta \rho}{\rho_0}\mathbf{g} - \operatorname{grad} \Omega \tag{3.6}$$

The effect of the nonlinear terms even in the lower atmosphere was considered by Thrane [66, 67, 105] in an approximate manner by using expansions. The point of interest is the vertical transport of energy and its connection with a variation of phase with height, the so-called *tilt* of tidal waves. A dependence of the phase of S_2^2 on height was derived by Wagner [68] and Stapf [69] from observations in the Alps. They found a phase retardation of about one hour at approximately 2500-meter height and a retardation of nearly two hours at 4500-meter height compared to the surface phase. Bjerknes [70] interpreted these findings as a necessary world-wide property of S_2^2. If that is true, further difficulties arise in explaining the magnitude of the observed tilt on the basis of the resonance theory, as discussed by Haurwitz [71]. Chapman [9] on the other hand, pointed out that the tilt of S_2^2 does not appear, to the same extent, in all height observations. Hence, local effects might be responsible for some extreme cases. Preliminary analyses of the semidiurnal pressure oscillation at the 250-mb, 200-mb, and the 175-mb levels [72] exhibit an equal phase retardation by somewhat more than 1.5 hr compared to the surface phase.

Neither the extent of the tilt nor its physical cause are exactly known, but they might be explainable within the scope of a linearized theory; the use of which is allowed, according to estimates, up to the lower boundary of the ionosphere (see Section 6.2 with Fig. 6). Moreover, the vertical acceleration is omitted in (3.6) as was first done by Laplace, and the vertical component w of the velocity is assumed to be small in comparison with the horizontal components u and v. Finally, let us replace the distance r from the earth's center to a point in space at height z ($r = a + z$) by the earth's radius a if r appears as a factor of another variable. Under these conditions we obtain from the vector equation (3.6), introducing spherical polar coordinates:

$$\frac{\partial u}{\partial t} - 2\omega v \cos \vartheta = -\frac{1}{a}\frac{\partial}{\partial \vartheta}\left(\frac{\delta p}{\rho_0} + \Omega\right) \tag{3.7}$$

$$\frac{\partial v}{\partial t} + 2\omega u \cos \vartheta = -\frac{1}{a \sin \vartheta}\frac{\partial}{\partial \phi}\left(\frac{\delta p}{\rho_0} + \Omega\right) \tag{3.8}$$

$$\frac{\partial \delta p}{\partial z} = -g\delta\rho - \rho_0 \frac{\partial \Omega}{\partial z} \tag{3.9}$$

Furthermore, we add the equation of continuity

$$\frac{D\rho}{Dt} + \rho_0 \chi = 0 \tag{3.10}$$

in which we use for the divergence of velocity the simplified form

$$\operatorname{div} \mathbf{v} = \chi = \frac{1}{a \sin \vartheta}\frac{\partial}{\partial \vartheta}(u \sin \vartheta) + \frac{1}{a \sin \vartheta}\frac{\partial v}{\partial \phi} + \frac{\partial w}{\partial z} \tag{3.11}$$

Solberg [73], who calls this kind of treatment of tides *quasistatic*, showed that his *exact dynamical method*, employing the complete system of the linearized equations, leads to significant differences when it is applied to tidal oscillations with periods longer than 12 sidereal hours. His study is restricted to tides in a homogeneous ocean covering the rotating earth. For periods shorter than 12 sidereal hours, the results of the exact dynamical theory agree with those of the quasistatic theory of oceanic tides as developed by Hough [62] who advanced Laplace's tidal theory. Hylleraas [74] studied the behavior of tidal oscillations in a stable ocean but did not find an advantage of the exact method over the quasistatic method if a rapid density decrease with height exists. Thrane [66], finally, applied the exact method to atmospheric oscillations resembling tidal waves. He inferred from his computations that the quasistatic method may lead to erroneous results if applied to the semidiurnal oscillation even in a medium with great static stability. This is contrary to Hylleraas' results. Hence, the justification for neglecting the vertical acceleration remains in doubt.

The replacement of r by a in some terms of (3.7) to (3.11) as well as the assumption of a height-independent g made Mariani [75] investigate the influence of these simplifications. He considered all terms of first order in z/a. Even the solution for such a simple model atmosphere as given by the assumption of a constant scale height, involves Bessel functions. The eigenvalue of this model is insignificantly smaller than the value derived from the equations of this section for the corresponding model (a) of Section 4.1.

Summarizing the results of this discussion, we find rather strong justification for applying these simplified basic equations to tidal motion in the troposphere and stratosphere. Further attention, however, must be paid to the omission of vertical acceleration and friction. The range of validity of the equations with regard to the height is limited by the lower boundary of the ionosphere. Not only do some of the simplifications—such as the nonlinear terms in (3.4)—become inadmissible, but also the physical properties of the high atmosphere are quite different from those of the troposphere. At least the equations of hydromagnetics should be used. On the other hand, it has become usual to employ, for the sake of simplicity, model atmospheres with isothermal tops. These models extend to infinity. Therefore, a boundary condition is necessary for the behavior of the oscillations at infinity. It has to be emphasized that the use of models of this kind is inconsistent with the preceding assumptions and can lead to erroneous interpretations of the results. This inconsistency is ineffective only when two conditions are satisfied: The state and the properties of the high atmosphere do not appreciably affect the observed pressure variations at the earth's surface. And the top of the model atmosphere does not have a noticeable influence upon the pressure variations computed for the ground. This point is considered again in Section 6.2.

A tidal pressure variation is also a thermodynamic process which can be described by the first law of thermodynamics for an ideal gas

$$\delta Q = c_v dT + p\, d\left(\frac{1}{\rho}\right) \tag{3.12}$$

Positive δQ (which need not be an exact differential) denotes an infinitesimal amount of heat added per unit mass of air. The quantity c_v is the specific heat at constant volume. In the following c_p, the specific heat at constant pressure, is also used. Both quantities are assumed to be constant throughout the whole atmosphere.

If in (3.12) δQ is proportional to dT the changes of state occur polytropically and are determined by

$$\delta Q = -\Gamma\, dT;\ T \propto \rho^{\gamma'-1} \tag{3.13}$$

with

$$\gamma' = \frac{c_p + \Gamma}{c_v + \Gamma},\ 0 \leqq \Gamma \angle \infty$$

When $\Gamma = 0$, (3.13) becomes the important case of adiabatic changes of state:

$$\delta Q = 0;\ T \propto \rho^{\gamma-1} \tag{3.14}$$

with

$$\gamma = \frac{c_p}{c_v} = 1.40$$

Hence, the changes of state according to (3.13) or (3.14) are connected with temperature changes with the exception of isothermal changes of state, that is $\gamma' = 1$ in (3.13). If undamped oscillations are assumed, the variations of heat are reversible. The *secondary* temperature variations due to them must be distinguished from the *primary* temperature variations produced in the atmosphere by outside influences. Only the primary variations are due to the thermal tidal forces. The observed λ^{-1}-diurnal temperature variations consist of a superposition of primary and secondary variations.

When a tidal pressure variation is exclusively gravitationally excited, a primary temperature variation of the same period is absent and the secondary variation can be determined by harmonic analysis. This was done by Chapman [76], who analyzed the temperature variation due to the known semidiurnal lunar pressure variation at Batavia. The heating of the atmosphere by moonlight is quite negligible. In this way, Chapman found that these world-wide pressure changes take place adiabatically within the margin of accuracy of the determination.

This result may be generalized, to the extent that all tidal pressure changes are assumed to cause secondary temperature changes which follow the adiabatic law and do not produce a variation of δQ. If a pressure variation is generated thermally, the excitation may be described by a function J which gives the amount of heat absorbed or emitted by a unit mass of air per unit time. Only J contributes to δQ and we have

$$\delta Q = J dt \tag{3.15}$$

The relationship between J and the primary temperature variation is treated in Section 7.1.

When δQ is replaced in (3.12) by (3.15), γ according to (3.14) is introduced into (3.12) by use of the equation

$$R = M(c_p - c_v) \tag{3.16}$$

and the individual differential operator with respect to time is denoted by D/Dt [as in (3.4) and (3.10)], we obtain from (3.12)

$$\frac{R}{M(\gamma - 1)} \frac{DT}{Dt} = \frac{p}{\rho^2} \frac{D\rho}{Dt} + J \tag{3.17}$$

It is possible to eliminate T in (3.17) with the aid of the gas equation (3.2) which is also valid in the case of a disturbed atmosphere, that is, for p, ρ, T. Differentiating this equation we get

$$\frac{dp}{p} + \frac{d\rho}{\rho} = \frac{dT}{T} \tag{3.18}$$

When (3.3) and (3.5) are also used and again small quantities are neglected, two equivalent representations of the first law of thermodynamics are obtained from (3.17)

$$\frac{R}{M(\gamma-1)}\frac{DT}{Dt}=\frac{gH}{\rho_0}\frac{D\rho}{Dt}+J \tag{3.19}$$

and

$$\frac{Dp}{Dt}=\gamma gH\frac{D\rho}{Dt}+(\gamma-1)\rho_0 J \tag{3.20}$$

where the individual differential operator may be reduced to

$$\frac{Dp}{Dt}=w\frac{dp_0}{dz}+\frac{\partial\delta p}{\partial t} \tag{3.21}$$

and likewise for ρ and T.

3.2 Formal Development of the Theory

The equations of Section 3.1 render it possible, as Pekeris [15] showed, to derive formal solutions for the pressure and temperature variations and the components of velocity connected with a tidal oscillation. Because only periodical processes are considered, we can put

$$u,\ v,\ w,\ \delta p,\ \delta\rho,\ \delta T,\ \chi,\ \Omega,\ J,\ \propto e^{i\sigma t} \tag{3.22}$$

where σ is the angular frequency of the oscillation. Then (3.7) and (3.8) can be solved for u and v:

$$u=\frac{\sigma}{4a\omega^2(f^2-\cos^2\vartheta)}\left[i\frac{\partial}{\partial\vartheta}+\frac{\cot\vartheta}{f}\frac{\partial}{\partial\phi}\right]\left(\frac{\delta p}{\rho_0}+\Omega\right) \tag{3.23}$$

$$v=\frac{i\sigma}{4a\omega^2(f^2-\cos^2\vartheta)}\left[\frac{i}{f}\cos\vartheta\frac{\partial}{\partial\vartheta}+\frac{1}{\sin\vartheta}\frac{\partial}{\partial\phi}\right]\left(\frac{\delta p}{\rho_0}+\Omega\right) \tag{3.24}$$

with

$$f=\frac{\sigma}{2\omega} \tag{3.25}$$

Substituting the expressions (3.23) for u and (3.24) for v into (3.11), we find

$$\chi-\frac{\partial w}{\partial z}=\frac{i\sigma}{4a^2\omega^2}F\left(\frac{\delta p}{\rho_0}+\Omega\right) \tag{3.26}$$

where F denotes the following differential operator

$$F=\frac{1}{\sin\vartheta}\frac{\partial}{\partial\vartheta}\left(\frac{\sin\vartheta}{f^2-\cos^2\vartheta}\frac{\partial}{\partial\vartheta}\right)+\frac{1}{f^2-\cos^2\vartheta}\left[\frac{i}{f}\frac{f^2+\cos^2\vartheta}{f^2-\cos^2\vartheta}\frac{\partial}{\partial\phi}+\frac{1}{\sin^2\vartheta}\frac{\partial^2}{\partial\phi^2}\right] \tag{3.27}$$

If we substitute the equation of continuity (3.10) into (3.19), using (3.3) and (3.22), we obtain

$$i\sigma\delta T = \frac{\kappa M}{R}\left(\gamma J - \gamma g H\chi - \frac{wg}{\kappa}\frac{dH}{dz}\right) \tag{3.28}$$

where

$$\kappa = \frac{\gamma - 1}{\gamma} = \frac{c_p - c_v}{c_p} = \frac{2}{7} \tag{3.29}$$

for dry air. Substituting (3.10) into (3.20) and considering (3.1), (3.21), and (3.22) we have

$$i\sigma\delta p = wg\rho_0 - \gamma g\rho_0 H\chi + (\gamma - 1)\rho_0 J \tag{3.30}$$

If (3.30) and (3.2) are differentiated with respect to z, (3.1) used, and δp and $\delta\rho$ eliminated with the aid of (3.9) and (3.10), it follows that

$$\frac{\partial w}{\partial z} = \gamma H\frac{\partial\chi}{\partial z} - (\gamma - 1)\chi - \frac{\gamma - 1}{g\rho_0}\frac{\partial}{\partial z}(\rho_0 J) - \frac{i\sigma}{g}\frac{\partial\Omega}{\partial z} \tag{3.31}$$

We differentiate (3.26) and (3.31) with respect to z and eliminate $\partial^2 w/\partial z^2$ neglecting a term in $\partial^2\Omega/\partial z^2$, which is small in comparison with a term in $\partial\Omega/\partial z$. We further differentiate $\delta p/\rho_0$ with respect to z, eliminate $\partial w/\partial z$ by means of (3.31), combine (3.1) and (3.2) to obtain

$$\frac{1}{\rho_0}\frac{d\rho_0}{dz} = -\frac{1}{H}\left(1 + \frac{dH}{dz}\right) \tag{3.32}$$

and use these relations in order to derive a partial differential equation for χ. The result is

$$H\frac{\partial^2\chi}{\partial z^2} + \left(\frac{dH}{dz} - 1\right)\frac{\partial\chi}{\partial z} - \frac{\kappa}{g}\frac{\partial}{\partial z}\left[\frac{\partial J}{\partial z} - \left(1 + \frac{dH}{dz}\right)\frac{J}{H}\right] - \frac{g}{4a^2\omega^2}F\left\{\left(\kappa + \frac{dH}{dz}\right)\chi - \frac{\kappa}{gH}\left(1 + \frac{dH}{dz}\right)J\right\} = 0 \tag{3.33}$$

The equation (3.33) can be solved by the method of separation of variables. For this purpose we represent χ and J by series expansions in terms of the eigenfunctions $\Psi_n(\vartheta, \phi)$ of the operator F:

$$\left\{\begin{aligned} \chi &= \sum_n \chi_n(z)\Psi_n(\vartheta, \phi)e^{i\sigma t} \\ J &= \sum_n J_n(z)\Psi_n(\vartheta, \phi)e^{i\sigma t} \end{aligned}\right. \tag{3.34}$$

Substituting χ and J into (3.33) by (3.34) and denoting the constant of separation by $1/h_n$, we obtain differential equations which must be valid for each term of the series because the Ψ_n's form a system of orthogonal functions. These equations are

$$F\Psi_n + \frac{4a^2\omega^2}{gh_n}\Psi_n = 0 \tag{3.35}$$

and

$$H\frac{d^2\chi_n}{dz^2} + \left(\frac{dH}{dz} - 1\right)\frac{d\chi_n}{dz} + \left(\kappa + \frac{dH}{dz}\right)\frac{\chi_n}{h_n} = \frac{\kappa}{g}\left\{\left(1 + \frac{dH}{dz}\right)\frac{J_n}{Hh_n} - \frac{d}{dz}\left[\left(1 + \frac{dH}{dz}\right)\frac{J_n}{H}\right] + \frac{d^2J_n}{dz^2}\right\} \tag{3.36}$$

Now assume that u, v, w, δp, δT, Ω can also be represented by series expansions corresponding to (3.34). Then from (3.26), (3.31), (3.32), and (3.35) it follows that

$$\frac{\delta p_n}{\rho_0} + \Omega_n = \frac{\gamma g h_n}{i\sigma}\left[\left(H\frac{d}{dz} - 1\right)\left(\chi_n - \frac{\kappa J_n}{gH}\right)\right] \tag{3.37}$$

In (3.37) a term $h_n d\Omega_n/dz$ which is of the order of $h_n\Omega_n/a$ has been neglected in comparison with Ω_n. When δp in (3.30) is expanded and δp_n is substituted by (3.37), we find for w_n:

$$w_n = -\frac{i\sigma}{g}\Omega_n + \gamma\left[\left(Hh_n\frac{d}{dz} + H - h_n\right)\left(\chi_n - \frac{\kappa J_n}{gH}\right)\right] \tag{3.38}$$

With these results, u_n, v_n, and δT_n, according to (3.23), (3.24), and (3.28) can be represented as functions of χ_n, Ω_n, J_n and their derivatives. A formal simplification of these expressions yields the following transformation (gradually completed by Pekeris [15], Wilkes [17], and Siebert [60] and similarly Sawada [77] and White [78]):

$$x = \int_0^z \frac{d\zeta}{H(\zeta)} \tag{3.39}$$

$$y_n(x)e^{x/2} = \chi_n(z) - \frac{\kappa J_n(z)}{gH(z)} \tag{3.40}$$

With (3.39), the distribution of static pressure becomes from (3.1) and (3.2)

$$p_0(x) = p_0(0)e^{-x} \tag{3.41}$$

The differential equation (3.36) is transformed by (3.39) and (3.40) into

$$\frac{d^2y_n}{dx^2} - \frac{1}{4}\left[1 - \frac{4}{h_n}\left(\kappa H(x) + \frac{dH}{dx}\right)\right]y_n(x) = \frac{\kappa J_n(x)}{\gamma g h_n}e^{-x/2} \tag{3.42}$$

After elimination of ρ_0 by means of (3.2) the equations (3.37) and (3.38) can also be transformed from χ_n and z to y_n and x, employing (3.39) and (3.40).

Then the same transformation can be carried out for (3.23), (3.24), and (3.28), too. Thus we obtain the following expressions by which the coefficients of the series expansions are formally given:

$$u_n(x) = \frac{\gamma g h_n e^{\frac{1}{2}x}}{4a\omega^2(f^2 - \cos^2\vartheta)} \left(\frac{dy_n}{dx} - \frac{1}{2}y_n\right)\left(\frac{\partial}{\partial\vartheta} - \frac{i}{f}\cot\vartheta\frac{\partial}{\partial\phi}\right) \tag{3.43}$$

$$v_n(x) = \frac{i\gamma g h_n e^{\frac{1}{2}x}}{4a\omega^2(f^2 - \cos^2\vartheta)} \left(\frac{dy_n}{dx} - \frac{1}{2}y_n\right)\left(\frac{\cos\vartheta}{f}\frac{\partial}{\partial\vartheta} - \frac{i}{\sin\vartheta}\frac{\partial}{\partial\phi}\right) \tag{3.44}$$

$$w_n(x) = -\frac{i\sigma}{g}\Omega_n(x) + \gamma h_n e^{\frac{1}{2}x}\left[\frac{dy_n}{dx} + \left(\frac{H(x)}{h_n} - \frac{1}{2}\right)y_n\right] \tag{3.45}$$

$$\delta p_n(x) = \frac{p_0(0)}{H(x)}\left[-\frac{\Omega_n(x)}{g}e^{-x} + \frac{\gamma h_n}{i\sigma}e^{-\frac{1}{2}x}\left(\frac{dy_n}{dx} - \frac{1}{2}y_n\right)\right] \tag{3.46}$$

$$\delta T_n(x) = \frac{M}{R}\left\{\frac{\Omega_n(x)}{H(x)}\frac{dH}{dx} - \frac{\gamma g h_n}{i\sigma}e^{\frac{1}{2}x}\left[\frac{\kappa H(x)}{h_n} + \frac{1}{H(x)}\frac{dH}{dx}\left(\frac{d}{dx} + \frac{H(x)}{h_n} - \frac{1}{2}\right)\right]y_n + \frac{\kappa J_n(x)}{i\sigma}\right\} \tag{3.47}$$

If $f < 1$; that is, for oscillations with periods longer than 12 sidereal hours, the denominators in (3.27), (3.43), and (3.44) vanish for a certain ϑ value (critical colatitude). As Brillouin [79] showed, the numerators disappear for the same colatitude so that singularities do not appear.

The formal development of the theory is finished after (3.43) to (3.47) have been established. In order to treat a specific case, the functions H, Ω, and J must be known for this case, and the differential equations (3.35) and (3.42) must be solved under certain boundary conditions. Analytical solutions exist only if severely simplifying assumptions are employed. These solutions and their consequences are studied in the following sections. Because the differential equations are linear, we can separately treat gravitational and thermal excitations and also the different kinds of thermal excitation. Moreover, we can consider the effect of every term of an exciting function separately.

Some importance attaches to the boundary conditions which have to be imposed upon the solutions of (3.42). It is generally assumed that at the earth's surface the vertical component w of velocity must vanish. Then we have, according to (3.45),

$$\left[\frac{dy_n}{dx} + \left(\frac{H}{h_n} - \frac{1}{2}\right)y_n\right]_{x=0} = \frac{i\sigma}{\gamma g h_n}\Omega_n(0) \tag{3.48}$$

In the case that $w_n(0) \neq 0$, regular world-wide and periodic oscillations of the solid and fluid parts of the earth's surface, as a whole, would exist. They could be caused by gravitational tidal forces and would act as a third kind of tide-generating forces besides gravitational and thermal tidal forces. Such an influence of the elasticity of the earth's body upon the lunar pressure variation was considered in a former state of the resonance theory [11, 23]. Later this surmise was dropped.

A second boundary condition for (3.42), determining the behavior of the solution for $x \to \infty$, can be derived from the reasonable assumption that the kinetic energy per column of unit cross-section shall be finite:

$$\int_0^\infty \rho_0(x)\mathbf{v}^2(x)H(x)dx < \infty \tag{3.49}$$

When ρ_0 is eliminated by means of (3.1), (3.39), and (3.41), $\mathbf{v}^2$ replaced by its components (3.43), (3.44), (3.45), and H assumed to remain finite for $x \to \infty$, we find from (3.49) that as $x \to \infty$, y_n must vanish more rapidly than $x^{-\frac{1}{2}}$:

$$y_n = o(1/\sqrt{x}); \qquad \text{that is} \qquad \lim_{x\to\infty} [y_n(x) \,.\, \sqrt{x}] = 0 \tag{3.50}$$

However, this boundary condition is not applicable to wave types with small h_n values if model atmospheres with isothermal tops are employed. The possible reasons and consequences are discussed in Section 6.2.

When this formulism is used to treat oscillations of such a kind as observed in the earth's atmosphere, we are mainly interested in computing the pressure variations at ground level. These quantities are best known from observational data and are, therefore, well suited for testing the hypotheses of the theory. It follows from (3.46) and (3.48) that the amplitudes of the pressure variations on the ground can be represented in a simple manner which holds for gravitational as well as thermal generation:

$$\delta p_n(0) = \frac{i\gamma}{\sigma} p_0(0)y_n(0). \tag{3.51}$$

4. Free Oscillations

4.1 *Simple Model Atmospheres and Their Eigenvalues*

In order to study the conditions for free oscillations from the equations of Section 3.2, we assume that exciting forces are absent, that is

$$\Omega_n = J_n = 0 \tag{4.1}$$

Then the equations contain, besides some empirical constants, only the function H which has to be derived from observations. According to (3.3), H is determined by the vertical distribution of the temperature T_0. The actual distribution of T_0 in the earth's atmosphere, however, is complex to such a degree that H cannot be represented by elementary functions leading to analytical solutions. For the sake of simplicity we introduce model atmospheres which can be treated analytically. Let us consider four simple models:

(a) The isothermal atmosphere: Here T_0 is uniform throughout the whole atmosphere and, hence, we have

$$H(x) = H(0) = \text{const.} \tag{4.2}$$

(b) The atmosphere in convective equilibrium or, more general, the autobarotropic atmosphere: This model is determined by the assumption that atmospheric pressure and density are connected by a power law which is valid for the atmosphere in both the disturbed and undisturbed state. In the case of adiabatic changes of state the index in both relations is given by $\gamma = c_p/c_v$

$$\frac{p}{\rho^\gamma} = \frac{p_0}{\rho_0^\gamma} = \text{const.} \tag{4.3}$$

When p and ρ are substituted from (3.5), only linear terms in δp and $\delta\rho$ considered, and (3.2) used, we obtain

$$\delta p = \gamma g H \delta\rho \tag{4.4}$$

Equation (4.4) replaces the general relation (3.20). Combining (4.4) and (3.20) and considering that (3.20) must hold for $J = 0$ as well as for $J \neq 0$, we find that the autobarotropic model is not appropriate for the treatment of thermal excitation of tides.

The height dependence of H is obtained by introducing H into (4.3) with the aid of (3.2) and by eliminating ρ_0 through differentiation of the static part of (4.3) with respect to x, eliminating the constant, and employing (3.41). With the aid of (3.29) integration over x then yields

$$H(x) = H(0)e^{-\kappa x} \tag{4.5}$$

Thus the model of an undisturbed atmosphere in convective equilibrium is described. The autobarotropic models follow formally from (4.5) when κ is replaced by κ' which is defined analogous to κ. That means that according to (3.13) and (3.29) κ' may vary from $0 \leq \kappa' \leq \kappa$. It is, of course, possible to study thermal action for a model given by (4.5). Then, however, p and ρ no longer satisfy the power law (4.3). That is, if disturbed by thermal forces, the barotropic stratification of the undisturbed atmosphere ceases to exist. On the contrary, gravitational tidal forces do not remove autobarotropy.

(c) This model atmosphere has no simple characteristics. It is a generalization of (a) and (b) including both [15]:

$$H(x) = \Delta H e^{-\kappa x} + H(\infty) \tag{4.6}$$

with

$$\Delta H = H(0) - H(\infty) \geq 0 \tag{4.7}$$

The atmosphere (c) is the most general model known hitherto which leads to elementary solutions. It is, therefore, taken as a basis for some numerical calculations in this article (see Fig. 10 in Section 7.2).

(d) Atmospheres with linear temperature decrease with height: Assuming only stable stratifications, we can write for the temperature gradient

$$\frac{dT_0}{dz} = -\frac{gM}{R}\epsilon \text{ with } 0 \leqq \epsilon \leqq \kappa \tag{4.8}$$

Therefore, we have with (3.3)

$$H(z) = H(0) - \epsilon z \tag{4.9}$$

When this value of H is substituted into (3.39), we find the relationship between z and x in this special case and can transform (4.9) into

$$H(x) = H(0)e^{-\varepsilon x} \tag{4.10}$$

Free atmospheric oscillations consist of wave types, whose equivalent depths agree with the eigenvalue(s) of the atmosphere. The eigenvalues $\hat{h}_n$ which belong to one of our model atmospheres, result when (3.42) is solved for the given H function and the solution subjected to the boundary conditions (3.48) and (3.50), always assuming (4.1).

Let us apply this method to model (c). Replacing the H in (3.42) by (4.6) we obtain

$$\frac{d^2y_n}{dx^2} - \frac{1}{4}\left(1 - \frac{4\kappa H(\infty)}{\hat{h}_n}\right)y_n = 0 \tag{4.11}$$

When $\hat{h}_n > 4\kappa H(\infty)$, the solution of this differential equation of second order consists of two exponential functions, one decreasing, the other one increasing with height. Because of (3.50) only the function vanishing as $x \to \infty$ can be used. That is,

$$y_n(x) = A_n \exp\left\{-\frac{x}{2}\sqrt{1 - \frac{4\kappa H(\infty)}{\hat{h}_n}}\right\}, \tag{4.12}$$

$$A_n = \text{arbitrary constant}$$

With y_n given by (4.12) we find from (3.48) with $\Omega = 0$ that model (c) has only one eigenvalue:

$$\hat{h} = \frac{H^2(0)}{H(0) - \kappa H(\infty)} > 4\kappa H(\infty) \text{ if } \Delta H \geq 0 \tag{4.13}$$

Remembering that model (c) involves the models (a) and (b), we obtain from (4.13) with (3.29) the eigenvalue of an isothermal atmosphere when $H(\infty) = H(0)$:

$$\hat{h} = \gamma H(0) \tag{4.14}$$

and the eigenvalue of an autobarotropic atmosphere when $H(\infty) = 0$:

$$\hat{h} = H(0) \tag{4.15}$$

Model atmospheres leading to (4.15) were used by Laplace [1], Margules [4], and Lamb [7] in the beginning of the theoretical atmospheric tidal work. Since that time they have been the basis for many estimates and approximative results. Bartels [11, 23] demonstrated a complete analogy between tidal oscillations of an ocean and an autobarotropic atmosphere. Haurwitz [80] extended Hough's [62] investigations of the oscillations of a homogeneous ocean to oscillations of a two-layer autobarotropic atmosphere.

With $T_0(0) = 288°\text{K}$ as mean surface temperature, the eigenvalue of the corresponding isothermal atmosphere turns out to be $\hat{h} = 11.8$ km and that of the corresponding atmosphere in convective equilibrium (with adiabatic lapse rate) $\hat{h} = 8.44$ km. When the model (c) is employed in Sections 6.2 and 7.2 to obtain numerical results, $T_0(\infty) = 160°\text{K}$ is assumed in addition to $T_0(0) = 288°\text{K}$. The eigenvalue due to this model is $\hat{h} = 10.0$ km. These values have been chosen to be in agreement with the generally accepted eigenvalue of the earth's atmosphere derived with the aid of more complicated models. It is also obtainable, however, by the employment of a simple two-layer model which correctly represents only the troposphere and the lower stratosphere (see Fig. 7 in Section 6.2).

The eigenvalue of an atmosphere in which the temperature decreases with height at a uniform rate follows by replacing H in (3.42) by (4.10). Introducing the new independent variable

$$\xi = \frac{2}{\epsilon}\sqrt{(\kappa - \epsilon)\frac{H(0)}{\hat{h}_n}}\, e^{-\varepsilon x/2} \tag{4.16}$$

we obtain the differential equation of Bessel functions

$$\frac{d^2 y_n}{d\xi^2} + \frac{1}{\xi}\frac{dy_n}{d\xi} + \left(1 - \frac{1}{\epsilon^2 \xi^2}\right) y_n = 0 \tag{4.17}$$

Apart from the case when $\epsilon = 0$ (isothermal atmosphere), the solution of (4.17) is given either by Bessel functions of the order $1/\epsilon$ and $-1/\epsilon$ (if $1/\epsilon$ is not an integer) or by a Bessel function and a Neumann function of the order $1/\epsilon$ (if $1/\epsilon$ is an integer). By means of the boundary condition (3.50), it can be shown that in both cases only the Bessel function of the order $1/\epsilon$ satisfies this condition:

$$y_n = A_n J_{1/\varepsilon}(\xi), \qquad A_n = \text{arbitrary constant.} \tag{4.18}$$

We use a differential formula of Bessel functions

$$\frac{dJ_{1/\varepsilon}}{d\xi} = J_{-1+1/\varepsilon} - \frac{1}{\epsilon\xi} J_{1/\varepsilon} \tag{4.19}$$

for determining dy_n/dx with the aid of (4.16) and (4.18). Let us denote the ξ value in (4.16) by ξ_n when the surface value ($x = 0$) is concerned. Then we find from (4.16)

$$\hat{h}_n = 4\frac{\kappa - \epsilon}{\epsilon^2}\frac{H(0)}{\xi_n{}^2} \tag{4.20}$$

and from the second boundary condition (3.48) with $\Omega = 0$

$$\xi_n J_{1/\varepsilon}(\xi_n) = 2\left(\frac{\kappa}{\epsilon} - 1\right)J_{-1+1/\varepsilon}(\xi_n) \tag{4.21}$$

The transcendental equation (4.21) has, in general, an infinite number of roots ξ_n for a given ϵ. To each root belongs an eigenvalue $\hat{h}_n$ according to (4.20). Hence, an atmosphere with uniform lapse rate has, in general, an infinite number of eigenvalues.

This result was first given, in a different way, by Lamb [6] when he studied the velocities of long horizontal waves in such atmospheres. It was later applied by Taylor [8], who determined the equivalent depths of these models when subjected to tidal oscillations. The most successful continuation was Pekeris' [15] development of five-layer models which resemble the earth's atmosphere (see Fig. 6, left-handside, in Section 6.2) and have two eigenvalues in agreement with the modern version of the resonance theory (see Section 1). In each layer the temperature was assumed to be uniform or to have a positive or negative uniform lapse rate. At the boundary between any two layers the continuity of pressure and velocity is required. In order to satisfy this condition, y_n and dy_n/dx must be continuous at the boundary. But the numerical treatment is rendered difficult by the appearance of different Bessel functions. Approximative solutions of (4.20) and (4.21) can be found in Siebert [60] and Solberg [81].

4.2. *Atmospheric Tsunamis*

The exact method of analyzing the observations of free atmospheric oscillations corresponds to the analysis of forced oscillations as described in Section 2. In other words, the oscillations should be resolved into wave types determined by the three parameters of frequency and geographical distribution. It is then possible to compute the equivalent depths of these wave types as shown in Section 5. Because the analyzed oscillation shall be a free oscillation of the earth's atmosphere, its equivalent depths would be identical with the eigenvalue(s) of the earth's atmosphere.

This method is most suitable for free oscillations which appear more or less periodically all over the world for a sufficiently long time. The only actual case of a free oscillation we have available, is the famous Krakatoa air wave caused by the great explosion of the volcano in 1883. This horizontal pressure wave, however, is better treated as a pulse generated by a point source and spreading out nearly concentrically around the source (atmospheric tsunami). Therefore, it is appropriate to seek a relationship between the velocity of propagation and the eigenvalue of the atmosphere. This was done by Taylor [12], who inferred from the analogy of long oceanic and atmospheric waves that

$$V = \sqrt{g\hat{h}} \tag{4.22}$$

is also valid in the case of long atmospheric waves. The quantity V is the velocity of propagation of the disturbance. Using Lamb's [6] results for the horizontal propagation of plane waves in a plane, nonrotating atmosphere, Taylor [8] later established (4.22) exactly for these conditions. Therefore, (4.22) is sometimes called *Taylor's theorem.*

It follows from Taylor's investigation that the validity of (4.22) is restricted to long waves, that is, to the limiting case of vanishing angular frequency σ and wavelength λ approaching infinity. However, the product $\sigma\lambda = 2\pi V$ must remain finite. On the sphere the limiting case $\lambda \to \infty$ becomes meaningless. Hence, the *long waves* in the sense of Taylor's theorem are restricted to a range of wavelengths limited by both a lower and upper boundary. In order to study this point, let us regard free oscillations of the Krakatoa type in a spherical atmosphere after Siebert [60]. In general, the same assumptions may be valid as in Section 3.1. In addition, we neglect the influence of the Coriolis force because the mean period of the oscillation was about 1.2 hr and is, therefore, smaller by about one order than the periods of the large tidal oscillations. Moreover, Taylor [12] showed on the basis of the data collected by Strachey [82], that deviations from a concentric spreading out of the Krakatoa wave can be explained mainly by the effect of world-wide wind circulations. Hence, the Coriolis force could not be of importance. Contrary to the treatment of the tides, the vertical acceleration is now considered. Then we have the hydrostatic equation for the undisturbed atmosphere,

$$\frac{dp_0}{dz} = -g\rho_0, \qquad g = \text{constant} \tag{4.23}$$

the linearized Eulerian equations,

$$\rho_0 \frac{\partial \mathbf{v}}{\partial t} + \operatorname{grad} p = \mathbf{g}\rho, \qquad \mathbf{g} \text{ has the direction of } -z, \tag{4.24}$$

the equation of continuity,

$$\frac{D\rho}{Dt} + \rho_0 \chi = 0, \qquad \chi = \operatorname{div} \mathbf{v} \tag{4.25}$$

and the first law of thermodynamics,

$$\frac{Dp}{Dt} = \gamma g H \frac{D\rho}{Dt} \qquad \text{[see (3.20) and (3.21)]} \tag{4.26}$$

With (3.5) it follows from (4.23) and (4.24) that

$$\rho_0 \frac{\partial \mathbf{v}}{\partial t} + \text{grad } \delta p = \mathbf{g} \delta \rho \tag{4.27}$$

Neglecting terms of higher order, we obtain from (4.25), (4.26), and (4.23)

$$\frac{\partial \delta p}{\partial t} = w \rho_0 g - \gamma p_0 \chi \tag{4.28}$$

If (4.27) is differentiated with respect to t and δp and $\delta \rho$ are eliminated by means of (4.28) and (4.25), we find with the aid of (4.23) and (3.2)

$$\frac{\partial^2 \mathbf{v}}{\partial t^2} = -g \text{ grad } w + \gamma g H \text{ grad } \chi + \mathbf{g}(\gamma - 1)\chi \tag{4.29}$$

Scalar multiplication of (4.29) by $\boldsymbol{\nabla}$ leads to

$$\frac{\partial^2 \chi}{\partial t^2} = -g \boldsymbol{\nabla}^2 w + \gamma g H \boldsymbol{\nabla}^2 \chi - \left[\gamma g \frac{dH}{dz} - (\gamma - 1) g\right] \frac{\partial \chi}{\partial z} \tag{4.30}$$

We introduce now spherical polar coordinates and write for the Laplacian operator

$$\boldsymbol{\nabla}^2 = \Lambda + \frac{\partial^2}{\partial z^2} \tag{4.31}$$

with

$$\Lambda = \frac{1}{a^2 \sin \vartheta} \frac{\partial}{\partial \vartheta}\left(\sin \vartheta \frac{\partial}{\partial \vartheta}\right) + \frac{1}{a^2 \sin^2 \vartheta} \frac{\partial^2}{\partial \phi^2} \tag{4.32}$$

where again r has been replaced by a. By differentiating (3.11) with respect to z, applying Λ to (3.11), and adding both equations we obtain with (4.31) and (4.32)

$$\boldsymbol{\nabla}^2 w = \frac{\partial \chi}{\partial z} - \frac{1}{a \sin \vartheta} \frac{\partial}{\partial \vartheta}[\sin \vartheta (\text{curl } \mathbf{v})_\phi] + \frac{1}{a \sin \vartheta} \frac{\partial}{\partial \phi} (\text{curl } \mathbf{v})_\vartheta \tag{4.33}$$

The suffixes ϑ and ϕ denote the ϑ and ϕ components of the vector curl $\mathbf{v}$. These components of the second derivative of $\mathbf{v}$ with respect to time can be found by vector multiplication of (4.29) by $\boldsymbol{\nabla}$. Substitution into (4.33) after (4.33) has been differentiated twice with respect to t, yields with (4.32)

$$\boldsymbol{\nabla}^2 \frac{\partial^2 w}{\partial t^2} = \frac{\partial^3 \chi}{\partial z \partial t^2} - g\left[\gamma \frac{dH}{dz} + (\gamma - 1)\right] \Lambda \chi \tag{4.34}$$

When (4.30) is differentiated twice with respect to t we can eliminate the w term in (4.34) and have with (3.29)

$$\frac{\partial^4 \chi}{\partial t^4} = \gamma g H \frac{\partial^4 \chi}{\partial z^2 \partial t^2} + \gamma g\left(\frac{dH}{dz} - 1\right)\frac{\partial^3 \chi}{\partial z \partial t^2} + \gamma g H \Lambda \frac{\partial^2 \chi}{\partial t^2} + \gamma g^2\left(\kappa + \frac{dH}{dz}\right)\Lambda\chi \tag{4.35}$$

This partial differential equation will be solved by the method of separation of variables. For this purpose we represent χ by a series expansion in terms of spherical surface harmonics $Y_n(\vartheta, \phi)$ which satisfy the equation

$$\Lambda Y_n + \frac{n(n+1)}{a^2} Y_n = 0 \tag{4.36}$$

Let us assume that pulses traveling with different velocities, belong to different eigenvalues. Regarding the propagation of only one pulse, we can put

$$\chi = \sum_n \chi_n(z) Y_n(\vartheta, \phi) e^{i\sigma_n t} \tag{4.37}$$

We must sum (4.37) over all frequencies because a pulse like the Krakatoa disturbance is not a completely harmonic oscillation. With (4.37) and (4.36) it follows from (4.35)

$$H\frac{d^2\chi_n}{dz^2} + \left(\frac{dH}{dz} - 1\right)\frac{d\chi_n}{dz} + \left[\frac{\sigma_n{}^2}{\gamma g} - \frac{n(n+1)}{a^2} H + \frac{n(n+1)g}{a^2\sigma_n{}^2}\left(\kappa + \frac{dH}{dz}\right)\right]\chi_n = 0 \tag{4.38}$$

The boundary condition on the ground $w_n(0) = 0$ can be expressed by means of (4.29), (4.30), (4.32), (4.36), and (4.37):

$$\left[H\frac{d\chi_n}{dz} + \left(\frac{n(n+1)}{a^2}\frac{gH}{\sigma_n{}^2} - 1\right)\chi_n\right]_{z=0} = 0 \tag{4.39}$$

The restriction to long waves is qualitatively given by limitation to small σ and small n values. Then the first two terms in the brackets of (4.38) can be neglected. The omission of these two terms determines the lower limit of long waves as is shown later. Comparing (4.38) without the negligible terms with the corresponding equation (3.36) in the theory of tides [assuming for free oscillations $J_n = 0$, $\sigma = \sigma_n$, and $h_n = \hat{h}$ in (3.36)] we find a complete agreement if

$$\hat{h} = \frac{a^2\sigma_n^2}{n(n+1)g} \qquad \text{[see also (5.7)]} \tag{4.40}$$

The formulism shall now be specialized for the idealized Krakatoa wave spreading out circularly around the source. Putting the ϑ axis through the source, we do not have any ϕ dependence and can replace (4.37) by

(4.41) $$\chi = \sum_n \chi_n(z) P_n(\vartheta) e^{i\sigma_n t}$$

When V_n denotes the phase velocity of the component $P_n(\vartheta)$, we have

(4.42) $$2\pi V_n = \sigma_n \lambda_n$$

The wavelength λ_n shall be determined from the distances of the roots of

(4.43) $$P_n(\vartheta) = 0$$

With increasing n the distance of the roots soon becomes more uniform and tends towards the limit

(4.44) $$\vartheta_{\nu+1} - \vartheta_\nu = 2\pi/(2n+1)$$

if ϑ_ν, $\nu = 1, 2, 3, \ldots n$, are the roots of (4.43). Only the most external roots ϑ_1 and ϑ_n have the distances $3\pi/2(2n+1)$ from the poles which are not roots. Hence, we assume that the wavelengths are empirically determined within $\vartheta_1 \leqq \vartheta \leqq \vartheta_n$. Then, considering (4.44) we can define the constant wavelength by

(4.45) $$\lambda_n = a(\vartheta_{\nu+2} - \vartheta_\nu) = 4\pi a/(2n+1)$$

From (4.40), (4.42), and (4.45) we obtain

(4.46) $$\hat{h} = \frac{(2n+1)^2}{4n(n+1)} \frac{V_n^2}{g}$$

The equation (4.46) replaces (4.22). Because V_n depends on λ_n according to (4.45) and (4.46), dispersion should be expected. However, (4.46) becomes (4.22) if $4n^2 \gg 1$. Then V_n is independent of n and, hence, also of λ_n. If we require an accuracy of 1 % from (4.22) compared to (4.46), we find from (4.46) that this condition is satisfied for $n \geq 5$. Thus, for the application of Taylor's theorem the analyzed pulse should not contain appreciable components of $n < 5$. In this manner we have found the upper limit of long waves within the meaning of (4.22). There is no sense, though, in applying (4.46) to long waves with $n < 5$ because we would then leave the range of the validity of (4.44), and also the neglected Coriolis force would become important.

The lower limit of, what can be called, long waves results from the omission of the first two terms in the bracket of (4.38). It is a noticeable fact that the second of these terms is the only one containing the temperature. Hence, such isothermal atmospheres may be visualized whose temperatures are adequately high that the other terms in the bracket can be neglected compared to the H term. Because this term does not appear in (3.36) we can conclude that it is due to $\partial w/\partial t$ which has been considered in this section, contrary to Section 3. However, we would need temperatures of $T_0 > 10^8$ °K in order for this term to become important in the theory of atmospheric tides.

Pekeris [83] investigated the influence of this term in the case of the propagation of plane waves in a plane, nonrotating atmosphere in which the temperature increases linearly with height.

Assuming actual temperatures, we have $H \sim \hat{h}$ and nearly half an adiabatic lapse rate in the troposphere. With the aid of (4.40) the influence of the disturbing terms in (4.38) turns out to be, under those conditions, smaller than 1 % for $n \leq 30$. Thus, we find with (4.45) from both limiting n values that Taylor's theorem (4.22) is a very good approximation if the mean wavelength $\bar{\lambda}_n$ of a pulse in the earth's atmosphere lies between

$$1300 \text{ km} < \bar{\lambda}_n < 7000 \text{ km} \tag{4.47}$$

When an influence of 10 % of the disturbing terms is permitted, the lower limit can be reduced to 400 km.

The observed Krakatoa wave had a mean wavelength within 1300 km $< \bar{\lambda} < 1400$ km. Hence, the main components lie around $n = 29$ and we are allowed to employ (4.22). The velocity of propagation of the pulse determined by Strachey [82] with the aid of barograms of tropical stations, was $V = 318.8$ meters/sec from which follows $\hat{h} = 10.4$ km. The velocity of the pulse along the line Krakatoa–England was $V = 314$ meters/sec, yielding the eigenvalue $\hat{h} = 10.1$ km. These results agree with the theoretically computed eigenvalue of about 10 km (see Section 4.1 and Fig. 7 in Section 6.2).

Pekeris [84] used his five-layer models which led him to the second eigenvalue of about 8 km, in order to study the spreading out of a supposed second pulse also produced by the Krakatoa eruption and due to that second eigenvalue. He estimated the total energy of the atmospheric disturbance to be of the order of 10^{24} ergs and found that the amplitude of the first pulse corresponding to $\hat{h} \sim 10$ km turns out to be larger by a factor varying from 2.4 to 2.9 than the amplitude of the second pulse corresponding to $\hat{h} \sim 8$ km. The indications of the second pulse on the barograms are doubtful. It should travel with a velocity of about 280 meters/sec.

The pressure wave caused by the fall of the great Siberian meteor in 1908 is frequently quoted as another example of a free atmospheric oscillation. The velocity of the propagation of the first pressure minimum was determined by Whipple [85] to be 318 meters/sec and by Solberg [81] to be 316 meters/sec. It followed a wave train of four oscillations with a mean period of about 140 sec and, hence, a mean wavelength of about 44 km. The pressure wave of the Siberian meteor is thus too short by at least one order of magnitude [see (4.47)] to treat it as a long atmospheric wave. Therefore, eigenvalues of the earth's atmosphere cannot be derived from it. This wave is obviously another kind of atmospheric oscillation whose theoretical treatment requires the complete equation (4.38) or the analogous equation for a plane wave in a plane atmosphere. On this basis Solberg [81] and later Pekeris [86] investigated pressure waves of that type. Calculating the group velocity which such a pulse should have in a two-layered atmosphere, Pekeris found 293 meters/second for a mean period of 2 min. It is noticeable in this connection that the barographic records of pressure oscillations due to H-bomb explosions are very similar to those of the pressure wave of the Siberian meteor and that the H-bomb waves were found to propagate with velocities of about 285 to 307 meters/sec (after Japanese observations [87]).

5. Laplace's Tidal Equation

A theoretical treatment of an atmospheric tidal oscillation requires, according to (3.34), series expansions of those quantities by which the oscillation can be described. The expansions have to be carried out in terms of the eigenfunctions Ψ_n of the operator F which satisfy the differential equation (3.35). This equation first appeared in Laplace's [1] theory of oceanic tides.

It is easy to derive (3.35) from the equations of Section 3.1 for the case of tides in a homogeneous ocean. For that purpose we have to assume uniform density, that is $\rho_0 =$ const. Variations $\delta\rho$ of density then signify variations of the quantity of the incompressible fluid per column of unit cross section and, hence, variations of the surface elevations usually denoted by $\zeta(\vartheta, \phi, t)$. Let $z = 0$ be the earth's surface, $z = h$ the surface of the undisturbed ocean, and $z = h + \zeta$ the surface of the disturbed ocean. Then we have

$$\delta\rho = \begin{cases} 0 & \text{if } z > h + \zeta \\ \rho_0 & \text{if } z < h + \zeta \end{cases} \tag{5.1}$$

and

$$\zeta = \frac{1}{\rho_0} \int_h^z \delta\rho dz', \qquad z \geq h + \zeta$$

Moreover, it is usual to neglect the variation of the disturbing potential Ω with respect to z. By integrating (3.9) over z, using (5.1), and putting $\delta p(\zeta) = 0$ (free surface), we obtain

$$\frac{\delta p}{\rho_0} = g\zeta \tag{5.2}$$

From (3.7) and (3.8) and with the aid of (5.2) we see that u and v do not depend on z. Hence, we can integrate (3.11) from $z = 0$ to $z = h + \zeta$. Considering that $\chi = \operatorname{div} \mathbf{v} = 0$ (because of the incompressibility of the fluid), that $w(0) = 0$, and that $w(\zeta) = \partial\zeta/\partial t = i\sigma\zeta$, we find from (3.11) when ζ is assumed to be small in comparison with h:

$$\frac{h}{a \sin \vartheta} \frac{\partial}{\partial \vartheta} (u \sin \vartheta) + \frac{h}{a \sin \vartheta} \frac{\partial v}{\partial \phi} + i\sigma\zeta = 0 \tag{5.3}$$

If we introduce the surface elevation $\bar{\zeta}$ of the equilibrium tide by

$$\bar{\zeta} = -\frac{\Omega}{g} \tag{5.4}$$

and use (3.23) and (3.24) with (5.2) and (5.4) to eliminate u and v in (5.3), we have

$$F(\zeta - \bar{\zeta}) + \frac{4a^2\omega^2}{gh} \zeta = 0 \tag{5.5}$$

where F is the differential operator defined by (3.27). Equation (5.5) is the fundamental equation of free and forced tidal oscillations of a homogeneous ocean covering the whole earth and having the uniform depth h. The effect of mutual attraction of the disturbed masses, which is appreciable in the case of a water ocean, is not considered in (5.5).

Comparing (3.35) with (5.5) we see that both equations become identical when (5.5) is restricted to free oceanic oscillations ($\bar{\zeta} = 0$) and ζ is expanded in series of Ψ_n. The constant h_n of separation in (3.35) replaces h in (5.5) and is, therefore, called *equivalent depth.*

Let us first consider the determination of Ψ_n in the case of a nonrotating earth. We divide (3.35) by ω^2, use (3.27) with (3.25), put $\omega = 0$, and find

$$\frac{1}{\sin \vartheta} \frac{\partial}{\partial \vartheta}\left(\sin \vartheta \frac{\partial \Psi_n}{\partial \vartheta}\right) + \frac{1}{\sin^2 \vartheta} \frac{\partial^2 \Psi_n}{\partial \phi^2} + \frac{a^2\sigma^2}{gh_n} \Psi_n = 0 \tag{5.6}$$

Equation (5.6) is well known as the differential equation of spherical surface harmonics. It follows from the theory of these functions that solutions of (5.6), finite over the entire sphere, can exist only if

$$h_n = \frac{a^2\sigma^2}{n(n+1)g}, \qquad n = 1, 2, 3, \ldots \tag{5.7}$$

This relation is of the same type as (4.40). However, (4.40) indicates that free harmonic oscillations with different frequencies σ_n belong to the same eigenvalue $\hat{h}$ of the earth's atmosphere. Equation (5.7), on the contrary, indicates that a forced oscillation of a given frequency σ consists of wave types with the equivalent depths h_n.

It may be noticed that in a mathematical sense the h_n's are the eigenvalues of the homogeneous differential equation (5.6), if $\omega = 0$, and the eigenvalues of the homogeneous differential equation (5.12) if $\omega \neq 0$ (with regard to certain boundary conditions). Yet they are not eigenvalues of the earth's atmosphere or model atmospheres, but are of use for the description of forced oscillations. Therefore, in order to avoid confusion, they are always called in this article equivalent depths.

It follows from the theory of spherical harmonics that $(2n+1)$ functions belong to each h_n value given by (5.7). These functions (wave types according to our terminology) are

$$\Psi_n(\vartheta, \phi) = P_n^s(\vartheta)e^{is\phi} \tag{5.8}$$

$$s = -n, -(n-1), \ldots -1, 0, +1, \ldots (n-1), n$$

The $P_n^s(\vartheta)$ are the associated Legendre functions as in Section 2. Hence, each wave type is given by the three parameters σ [or λ, see (5.11)], n, and s, in agreement with its definition in Section 2.2. Its nodal lines on the sphere consist of 2 s meridians and $(n-s)$ parallels of colatitude. A wave is traveling westwards if $s > 0$ and eastwards if $s < 0$. Standing (zonal) oscillations are given by $s = 0$. A peculiarity of the wave types determined for a nonrotating earth, is the nondependence of their equivalent depths on s [see (5.7)]. Besides these waves there is also a system of steady motions due to $\sigma = 0$.

Let us now consider the determination of Ψ_n in the case of the rotating earth. Then Ψ_n can no longer be represented by (5.8). However, when in Section 2.2 the planetary representation of an oscillation observed on the rotating earth was treated, the periodicity in ϕ rendered it possible to distinguish wave families by their frequencies and their periodicities in ϕ. Hence, it is reasonable to split again $\Psi_n(\vartheta, \phi)$ into a function of ϑ and a function periodic in ϕ. According to (2.16) the function of ϑ shall be denoted by $\Theta_{\lambda,n}^s(\vartheta)$. Instead of (5.8) we, therefore, put

$$\Psi_n(\vartheta, \phi) = \Theta_{\lambda,n}^s(\vartheta)e^{is\phi} \tag{5.9}$$

A complete (solar) wave family is thereby represented in the theory by

$$S_\lambda^s(p) = \sum_n \delta p_{\lambda,n}^s(x)\Theta_{\lambda,n}^s(\vartheta)e^{i(\sigma t + s\phi)} \tag{5.10}$$

Both σ and λ denote the frequency. The quantity σ is measured in $\sec^{-1}$ and λ in commensurable parts of the day. Both quantities are related by

$$\lambda = \frac{\sigma t_d}{2\pi} = \frac{\omega f t_d}{\pi} = 1, 2, 3, \ldots; \qquad \lambda t' = \sigma t \tag{5.11}$$

where t_d designates the length of a mean solar, lunar or sidereal day in seconds. Then t in (5.10) denotes universal time in seconds.

When Ψ_n from (5.9) is substituted into (3.35) we obtain with (3.27) and the usual abbreviation $\cos\vartheta = \mu$:

$$\frac{d}{d\mu}\left(\frac{1-\mu^2}{f^2-\mu^2}\frac{d\Theta_{\lambda,n}^s}{d\mu}\right) - \frac{1}{f^2-\mu^2}\left[\frac{s}{f}\frac{f^2+\mu^2}{f^2-\mu^2} + \frac{s^2}{1-\mu^2}\right]\Theta_{\lambda,n}^s + \frac{4a^2\omega^2}{gh_n}\Theta_{\lambda,n}^s = 0 \tag{5.12}$$

In order to integrate (5.12), special values of f and s are assumed (see Section 2). Then an infinite number of h_n values which are due to one and the same pair of f and s values, can be found by means of continued fractions. Thus h_n depends on f (that is, σ or λ), s, and n. One function $\Theta^s_{\lambda, n}$ belongs to each $h_n(\lambda, s)$. Because a phenomenon on a sphere is concerned, the representation of $\Theta^s_{\lambda, n}$ by series expansions in terms of associated Legendre functions is appropriate and has the advantage of more rapid convergence than the use of power series in μ. This integration of (5.12) by means of spherical harmonics, was first carried out by Hough [62] and, therefore, the $\Theta^s_{\lambda, n}(\vartheta)$ will be called *Hough's functions*. They form a complete system of orthogonal functions for each pair of f and s values, as can be proved from general principles.

In the special case of a zonal oscillation ($s = 0$) which has the period of half a sidereal day ($f = 1$ or $\sigma = 2\omega$) the solution of (5.12) can be given in a closed form (Solberg [73]), namely by

$$\Theta^0_{2,n} = A^0_{2,n} \sin\left(\frac{2a\omega}{\sqrt{gh_n}}\mu\right) + B^0_{2,n} \cos\left(\frac{2a\omega}{\sqrt{gh_n}}\mu\right), \tag{5.13}$$

$$t_d = 86{,}164.09 \text{ sec} \qquad [\text{see (5.11)}]$$

Equation (5.13) contains wave types which are symmetric or antisymmetric with regard to the equator according to whether

$$A^0_{2,n} = 0 \text{ or } B^0_{2,n} = 0.$$

In addition to the general condition that a solution of (5.12) must be finite and single valued over the entire sphere, we use the boundary condition that the north–south component of velocity shall vanish at the poles. That means, that $u_n\Theta^s_{\lambda,n} = 0$ for $\mu = \pm 1$. Using (3.43) and applying this condition to (5.13), we find the spectrum of the h_n:

$$h_n = \frac{16a^2\omega^2}{g\pi^2 n^2}, \qquad n = 1, 2, 3, \ldots \tag{5.14}$$

In (5.14) the equivalent depths with even n values belong to symmetric and the equivalent depths with odd n values to antisymmetric wave types.

The results (5.13) and (5.14) are of some practical importance because the observed semidiurnal standing solar wave family S^0_2 is determined by $s = 0$ and $f = 0.99727$, which is only slightly different from $f = 1$. Haurwitz [47, 61], therefore, employs (5.13) with (5.14) for the representation of S^0_2.

In general, (5.12) is solved by putting

$$\Theta^s_{\lambda, n}(\mu) = \sum_\nu C^{s, \nu}_{\lambda, n} P^s_\nu(\mu) \tag{5.15}$$

We do not consider the oscillations of Hough's second class which would be converted into steady motions ($\sigma = 0$) if the rotation of the globe were to vanish. In the case of the rotating globe they consist of slowly westward migrating waves ($\sigma \leq \omega$); a steady motion is still possible, but only if $s = 0$.

The method of solving (5.12) by means of (5.15) is described in detail in Hough's papers [62]. Only the results best suited for computing the oscillations of Hough's first class are given as follows. A notation slightly different from that of Hough is used in order to avoid ambiguity of the representation:

$$M^{s,n}_{\lambda,n} = \frac{\mathrm{a}^s_{n-1}|}{|N^s_{\lambda,n-1}} - \frac{\mathrm{a}^s_{n-2}|}{|M^{s,n-2}_{\lambda,n}} - \frac{\mathrm{a}^s_{n-3}|}{|N^s_{\lambda,n-3}} - \cdots + \frac{\mathrm{a}^s_{n}|}{|N^s_{\lambda,n+1}} - \frac{\mathrm{a}^s_{n+1}|}{|M^{s,n+2}_{\lambda,n}} - \frac{\mathrm{a}^s_{n+2}|}{|N^s_{\lambda,n+3}} - \cdots \tag{5.16}$$

with

$$\mathrm{a}^s_\nu = \frac{\nu^2(\nu+2)^2(\nu-s+1)(\nu+s+1)}{(2\nu+1)(2\nu+3)} \quad \text{if } \nu \geq s,$$

$$N^s_{\lambda,\nu} = \nu(\nu+1) - \frac{s}{f}$$

$$M^{s,\nu}_{\lambda,n} = f^2 N^s_{\lambda,\nu} - \nu^2(\nu+1)^2 \frac{gh_n(s,\lambda)}{4a^2\omega^2}$$

It is appropriate to our problem to use the continued fraction (5.16) in such a way that s and f are given and the h_n values due to them are determined from (5.16). The number of terms necessary depends on the quality of the convergence and the accuracy wanted. Symmetric wave types are obtained when $(n - s)$ is an even integer and antisymmetric ones when $(n - s)$ is odd. When h_n is known, the determination of the coefficients in (5.15) is possible by means of the following recurrence formula

$$C^{s,\nu+2}_{\lambda,n} = \sqrt{(\nu-s+1)(\nu-s+2)(\nu+s+1)(\nu+s+2)} \times \frac{(\nu+2)^2(\nu+3)^2}{(2\nu+1)(2\nu+3)} \frac{C^{s,\nu}_{\lambda,n}}{E^{s,\nu}_{\lambda,n} F^{s,\nu+1}_{\lambda,n}} \tag{5.17}$$

with

$$E^{s,\nu}_{\lambda,n} = \frac{\mathrm{a}^s_\nu}{M^{s,\nu}_{\lambda,n} - F^{s,\nu-1}_{\lambda,n}}; \qquad F^{s,\nu}_{\lambda,n} = \frac{\mathrm{a}^s_\nu}{N^s_{\lambda,\nu} - E^{s,\nu-1}_{\lambda,n}}$$

and

$$C^{s,\nu}_{\lambda,n} = F^{s,\nu}_{\lambda,n} = E^{s,\nu}_{\lambda,n} = 0 \text{ for } \begin{cases} n \text{ and } \nu \angle s \text{ if } s = 0 \\ n \text{ and } \nu \angle s \text{ if } s = 1, 2, 3, \ldots \nu \end{cases}$$

When $f > 0$ and $s > 0$ we have waves traveling westward. The simplest way of treating waves which travel eastward is to use negative f values in the preceding formulism.

The associated Legendre functions in (5.15) are assumed to be normalized according to Schmidt [see (2.11)] and, therefore, the recurrence formula

(5.17) has been rewritten with regard to this normalization. Moreover, we must also normalize the Hough's functions because one of the coefficients to be computed from (5.17) remains indeterminable. A usual normalization would be given by

$$\int_{-1}^{+1} [\Theta^s_{\lambda,n}(\mu)]^2 d\mu = \int_0^{\pi} [\Theta^s_{\lambda,n}(\vartheta)]^2 \sin\vartheta \, d\vartheta = 1 \tag{5.18}$$

However, when (5.18) is applied in practice, the accuracy of the normalization would depend on the number of terms considered in (5.15); and in the case of slow convergence it would be necessary to compute a high number of coefficients. Therefore, another condition was chosen in order to normalize the $\Theta^s_{\lambda,n}$ in Table II. This condition is

$$C^{s,n}_{\lambda,n} = 1 \tag{5.19}$$

or

$$\int_{-1}^{+1} \Theta^s_{\lambda,n}(\mu) P^s_n(\mu) d\mu = \frac{2}{2n+1} \delta^s_{n,n}$$

The quantity $\delta^s_{n,n}$ has been defined by (2.11). The coefficient $C^{s,n}_{\lambda,n}$ is distinguished in that it is the largest one in (5.15) for most (but not all) of Hough's functions required. Moreover, the term $C^{s,n}_{\lambda,n} P^s_n(\mu)$ is the only one in (5.15) which does not disappear when the earth's rotation is assumed to vanish.

Besides the representation of Hough's functions by means of associated Legendre functions, we also have to know the expansions of Legendre functions in series of Hough's functions in order to give the observational results a form appropriate to a theoretical discussion [see (2.16)]. Multiplying (2.15) by $\Theta^s_{\lambda,n}$ and using the orthogonality of Hough's functions and spherical harmonics we obtain with (5.15) and (2.11) for the coefficients of the expansion (2.15):

$$\gamma^{s,n}_{\lambda,m} = \frac{C^{s,m}_{\lambda,n}}{(2m+1)\sum_{\nu}(C^{s,\nu}_{\lambda,n})^2/(2\nu+1)} \tag{5.20}$$

The series expansions of some important Hough's functions and the reversed expansions of associated Legendre functions, which were computed by Siebert [60] on the basis of the preceding formulas are given in Table II. Graphic representations of h_n as function of f or $2\pi/\sigma$ for different values of n and s can be found in the papers of Wilkes [17] (Fig. 13) and Kertz [14] (Fig. 5).

TABLE II. Hough's functions represented in terms of seminormalized associated Legendre functions and expansions of associated Legendre functions in series of Hough's functions for different values of σ, s, and n after [60]. The values of the equivalent depths and the coefficients of the expansions based on the terrestrial constant,

$$gh_n/4a^2\omega^2 = 0.011349\ h_n\ [\text{km}]$$

Solar wave types, $t_d = 86{,}400.00$ sec

Hough's functions of semidiurnal migrating solar wave types:

$s = 2$, f $= 0.99727$, $\sigma = 1.4544 \,.\, 10^{-4}$ sec^{-1}

$n = 2$: $h_2 = 7.85$ km

$$\Theta^2_{2,2}(\vartheta) = P^2_2 - 0.339\ P^2_4 + 0.041\ P^2_6 - 2\,.\,10^{-3}\ P^2_8 + - \cdots$$

$n = 4$: $h_4 = 2.11$ km

$$\Theta^2_{2,4}(\vartheta) = 0.202\ P^2_2 + P^2_4 - 0.819\ P^2_6 + 0.24\ P^2_8 - 0.04\ P^2_{10} + - \cdots$$

$n = 6$: $h_6 = 0.957$ km

$$\Theta^2_{2,6}(\vartheta) = 0.13P^2_2 + 0.755P^2_4 + P^2_6 - 1.72P^6_8 + 0.88P^2_{10} - 0.2P^2_{12} + - \cdots$$

Expansions of P^2_m in series of $\Theta^2_{2,n}$:

$$P^2_2(\vartheta) = 0.939\Theta^2_{2,2} + 0.231\Theta^2_{2,4} + 0.0703\Theta^2_{2,6} + \cdots$$
$$P^2_4(\vartheta) = -\ 0.177\Theta^2_{2,2} + 0.638\Theta^2_{2,4} + 0.235\Theta^2_{2,6} + \cdots$$

Hough's functions of semidiurnal standing solar wave types:

$s = 0$, $f = 0.99727$, $\sigma = 1.4544 \,.\, 10^{-4}$ sec^{-1}

$n = 2$: $h_2 = 8.85$ km

$$\Theta^0_{2,2}(\vartheta) = P_2 - 0.386P_4 + 0.044P_6 - 2\,.\,10^{-3}P_8 + - \cdots$$

$n = 4$: $h_4 = 2.21$ km

$$\Theta^0_{2,4}(\vartheta) = 0.229P_2 + P_4 - 0.846P_6 + 0.24P_8 - 0.04P_{10} + - \cdots$$

$n = 6$: $h_6 = 0.981$ km

$$\Theta^2_{2,6}(\vartheta) = 0.15P_2 + 0.767P_4 + P_6 - 1.75P_8 + 0.89P_{10} - 0.2P_{12} + - \cdots$$

Expansions of P^0_m in series of $\Theta^0_{2,n}$:

$$P_2(\vartheta) = 0.923\Theta^0_{2,2} + 0.254\Theta^0_{2,4} + 0.0797\Theta^0_{2,6} + \cdots$$
$$P_4(\vartheta) = -\ 0.198\Theta^0_{2,2} + 0.617\Theta^0_{2,4} + 0.232\Theta^0_{2,6} + \cdots$$

Hough's functions of terdiurnal migrating solar wave types:

$s = 3$, $f = 1.4959$, $\sigma = 2.1817 \,.\, 10^{-4}$ sec^{-1}

$n = 3$: $h_3 = 12.89$ km

$$\Theta^3_{3,3}(\vartheta) = P^3_3 - 0.105P^3_5 + 5.2\,.\,10^{-3}P^3_7 - 10^{-4}P^3_9 + - \cdots$$

$n = 4$: $h_4 = 7.66$ km

$$\Theta^3_{3,4}(\vartheta) = P^3_4 - 0.164P^3_6 + 0.012P^3_8 - 5\,.\,10^{-4}P^3_{10} + - \cdots$$

$n = 5$: $h_5 = 5.09$ km

$$\Theta^3_{3,5}(\vartheta) = 0.0675P^3_3 + P^3_5 - 0.216P^3_7 + 0.020P^3_9 - 10^{-3}P^3_{11} + - \cdots$$

$n = 6$: $h_6 = 3.62$ km

$$\Theta^3_{3,6}(\vartheta) = 0.115P^3_4 + P^3_6 - 0.265P^3_8 + 0.030P^3_{10} - 2\,.\,10^{-3}P^3_{12} + - \cdots$$

$n = 7$: $h_7 = 2.71$ km

$$\Theta^3_{3,7}(\vartheta) = 8.3\,.\,10^{-3}P^3_3 + 0.161P^3_5 + P^3_7 - 0.312P^3_9 + 0.042P^3_{11} - + \cdots$$

Expansions of P^3_m in series of $\Theta^3_{3,n}$:

$$P^3_3(\vartheta) = 0.993\Theta^3_{3,3} + 0.102\Theta^3_{3,5} + 0.0160\Theta^3_{3,7} + \cdots$$
$$P^3_4(\vartheta) = 0.982\Theta^3_{3,4} + 0.155\Theta^3_{3,6} + \cdots$$
$$P^3_5(\vartheta) = -\ 0.0665\Theta^3_{3,3} + 0.960\Theta^3_{3,5} + 0.197\Theta^3_{3,7} + \cdots$$
$$P^3_6(\vartheta) = -\ 0.112\Theta^3_{3,4} + 0.932\Theta^3_{3,6} + \cdots$$

TABLE II—*continued*

Lunar wave types, $t_d = 89{,}428.33$ sec

Hough's functions of semidiurnal migrating lunar wave types:

$s = 2, f = 0.96350, \sigma = 1.4052 \,.\, 10^{-4}\ \mathrm{sec}^{-1}$

$n = 2$: $h_2 = 7.07$ km

$$\Theta^2_{2,2}(\vartheta) = P^2_2 - 0.375P^2_4 + 0.050P^2_6 - 3\,.\,10^{-3}P^2_8 + - \ldots$$

$n = 4$: $h_4 = 1.85$ km

$$\Theta^2_{2,4}(\vartheta) = 0.227P^2_2 + P^2_4 - 0.951P^2_6 + 0.32P^2_8 - 0.06P^2_{10} + - \ldots$$

$n = 6$: $h_6 = 0.825$ km

$$\Theta^2_{2,6}(\vartheta) = 0.19P^2_2 + 1.03P^2_4 + P^2_6 - 2.42P^2_8 + 1.5P^2_{10} - 0.5P^2_{12} + - \ldots$$

Expansions of P^2_m in series of $\Theta^2_{2,n}$:

$$P^2_2(\vartheta) = 0.927\Theta^2_{2,2} + 0.230\Theta^2_{2,4} + 0.0584\Theta^2_{2,6} + \ldots$$
$$P^2_4(\vartheta) = -0.193\Theta^2_{2,2} + 0.564\Theta^2_{2,4} + 0.174\Theta^2_{2,6} + \ldots$$

6. GRAVITATIONAL EXCITATION OF ATMOSPHERIC TIDES

6.1. Gravitational Tidal Forces

When an astronomical body is moving under the action of the gravitational forces of other astronomical bodies, the orbit of the moving body is determined by the orbit of its center of mass; that is, the whole body can be replaced by a mass point having the mass of the body and the position of the body's center of mass. Considering the finite extent of the body two new effects of the gravitational forces appear. If the body is not a sphere with a homogeneous or concentric distribution of mass, the forces produce a torque which causes a precession of the axis of rotation in the case of a rotating body.

The second effect is the appearance of tidal forces. Under the assumption that the body is rigid, the forces of inertia due to the motion along the orbit (i.e., revolution without rotation) are uniform for each mass element of the body. They have the same magnitude and opposite direction as the gravitational forces acting on the body's center of mass. These gravitational forces, however, are not uniform throughout the body, so that outside the center of mass differences appear between gravitational forces and forces of inertia. These differences are the tidal forces. They are potential forces and depend: (*1*) on the mass of the disturbing astronomical body, as the moon or sun; (*2*) on the distance of the disturbing body from the points where the tidal forces act, say, on the earth's surface; (*3*) on the distance from these points to the center of mass of the moving body, say, the earth; and (*4*) on the zenith distance of the disturbing body when it is observed from the earth's center. In order to study the daily variation of the tidal forces at a fixed point on the rotating earth, the zenith distance is eliminated with the aid of the declination and the hour-angle of the disturbing body, and the colatitude of the point of observation.

For a detailed investigation it is more convenient to analyze the tidal potential than the tidal forces. Such a detailed harmonic analysis was carried out by Doodson [88]. He succeeded in developing a consistent representation of the tidal potential by a sum of sine and cosine terms with constant frequencies and amplitudes. These depend only on the distance of the center of the earth and the geocentric colatitude. The latter is nearly identical with the geographical colatitude. A clear treatment of this topic was recently given by Bartels [89]. The following particulars were derived from the data of his paper (where V corresponds to $-\Omega$ in this article). A brief introduction is also contained in Melchior's [90] contribution to Vol. 4 of this series.

Let us denote the local solar time by t^*, local lunar time by t_L^*, mean longitude of the moon by μ (period = 1 tropical month), mean longitude of the sun by η (period = 1 tropical year), and mean longitude of the perigee by ψ (period = 8.847 years). Only the diurnal and semidiurnal main terms of the tidal potential $\Omega(0)$ are considered. The coefficients are exactly valid on the surface of a spherical earth with the radius of 6371.221 km. The dependence on the colatitude ϑ is again described by seminormalized associated Legendre functions. The terms of Ω are denoted by the customary symbols for this topic. Diurnal terms:

$$O_1 = -11{,}405 P_2^1(\vartheta) \sin (t_L^* - \mu) \frac{\text{cm}^2}{\text{sec}^2}, \qquad \text{(diurnal lunar)} \tag{6.1}$$

$$P_1 = -5312 P_2^1(\vartheta) \sin (t^* - \eta) \frac{\text{cm}^2}{\text{sec}^2} \qquad \text{(diurnal solar)} \tag{6.2}$$

The consideration of μ and η also leads to two terms, a lunar one and a solar one, which have the same period because $t_L + \mu = t + \eta$ = mean sidereal time. Both terms are combined into

$$K_{1m} + K_{1s} = +16{,}053\, P_2^1(\vartheta) \sin (t^* + \eta) \frac{\text{cm}^2}{\text{sec}^2}, \qquad \text{(diurnal luni-solar)} \tag{6.3}$$

Semidiurnal terms:

$$N_2 = -5261\, P_2^2(\vartheta) \cos (2t_L^* - \eta + \psi) \frac{\text{cm}^2}{\text{sec}^2}, \qquad \text{(large lunar elliptic)} \tag{6.4}$$

$$M_2 = -27{,}480\, P_2^2(\vartheta) \cos 2t_L^* \frac{\text{cm}^2}{\text{sec}^2}, \qquad \text{(semidiurnal lunar)} \tag{6.5}$$

$$S_2 = -12{,}818\, P_2^2(\vartheta) \cos 2t^* \frac{\text{cm}^2}{\text{sec}^2}, \qquad \text{(semidiurnal solar)} \tag{6.6}$$

It may be noticed that S_2 according to (6.6) includes a very small lunar variation term with the period of half a solar day. Analogous to (6.3) there are again two terms, a lunar one and a solar one, having the same period which in this case is half a sidereal day. The sum of both terms is

$$K_{2m} + K_{2s} = -3482\, P_2^2(\vartheta) \cos\,[2(t^* + \eta)]\, \frac{\text{cm}^2}{\text{sec}^2}, \tag{6.7}$$

(semidiurnal luni-solar)

The terms of higher orders are quite insignificant.

Terms of the observed tidal pressure variation which correspond respectively to the preceding terms, are the semidiurnal lunar and solar variations $L_2^2(p)$ and $S_2^2(p)$ (see Section 2.4). No diurnal lunar pressure variation could be detected as yet, and even the contribution of (6.6) in generating $S_2^2(p)$ besides the thermal tidal forces is very doubtful. In the following, we therefore consider only (6.5) and (6.6).

It is convenient for the treatment of gravitational tidal oscillations of the atmosphere to introduce the *atmospheric equilibrium tide* defined by the analogy with the surface elevation $\bar{\zeta}$ of the oceanic equilibrium tide. The relationship between pressure variation δp and surface elevation ζ has already been given by (5.2). We need only replace ζ in (5.2) by $\bar{\zeta}$ according to (5.4). Denoting the atmospheric equilibrium tide by $\overline{\delta p}$ we have

$$\overline{\delta p} = -\rho_0 \Omega \tag{6.8}$$

The quantity $\overline{\delta p}$ plays the double role of a gravitational tide-generating function and of a unit of resonance magnification. When universal time t' or t [see (5.11)] is introduced, P_2^2 replaced by the corresponding series of Hough's functions from Table II, and the complex form of the periodic factor used, we obtain with $\rho_0(0) = 1.226 \,.\, 10^{-3}$ gm/cm³ from (6.5) and (6.6): The semidiurnal lunar equilibrium tide of the atmosphere

$$\overline{\delta p_2^2(0)} = 3.369\, P_2^2(\vartheta) \sin\,(2t_L' + 2\phi + 90°) \,.\, 10^{-2}\ \text{mb} \tag{6.9}$$

$$= (3.12\Theta_{2,2}^2 + 0.77\Theta_{2,4}^2 + 0.20\Theta_{2,6}^2 + \ldots)e^{i(\sigma t_L + 2\phi)}\ 10^{-2}\ \text{mb} \tag{6.10}$$

with $\sigma = 1.4052 \,.\, 10^{-4}\ \text{sec}^{-1}$;

and the semidiurnal solar equilibrium tide of the atmosphere

$$\overline{\delta p_2^2(0)} = 1.571\, P_2^2(\vartheta) \sin\,(2t' + 2\phi + 90°) \,.\, 10^{-2}\ \text{mb} \tag{6.11}$$

$$= (1.48\Theta_{2,2}^2 + 0.36\Theta_{2,4}^2 + 0.11\Theta_{2,6}^2 + \ldots)e^{i(\sigma t + 2\phi)}\ 10^{-2}\ \text{mb} \tag{6.12}$$

with $\sigma = 1.4544 \,.\, 10^{-4}\ \text{sec}^{-1}$.

It can be seen from (6.9) and (6.11) that the times of the maxima correspond in both cases to phase angles of 90° in the semidiurnal harmonic dials. Therefore, they agree with the times of upper and lower transits.

6.2. Gravitationally Generated Oscillations

In order to illustrate the treatment of these kinds of oscillations, we determine the tidal pressure oscillations at the ground due to $\Omega_n \neq 0$ and $J_n = 0$. As a model we employ atmosphere (c) of Section 4.1. The important differential equation (3.42) is the same as in the case of free oscillations. Hence, we can at once use the solution (4.12) in which we replace $\hat{h}_n$ by h_n. This procedure results from (4.11) when the boundary condition (3.50) is considered. Substituting $y_n(0)$ from (4.12) into the second boundary condition (3.48), we have an equation for determining the integration constant A_n:

$$A_n = \frac{2i\sigma\Omega_n(0)}{\gamma g[2H(0) - h_n(1 + \beta_n)]} \tag{6.13}$$

with

$$\beta_n = +\sqrt{1 - 4\kappa\frac{H(\infty)}{h_n}} \qquad \text{if } h_n \geq 4\kappa H(\infty) \tag{6.14}$$

With the known A_n we substitute $y_n(0)$ from (4.12) into (3.51). Introducing the atmospheric equilibrium tide (6.8), we obtain with (3.2):

$$\delta p_n(0) = \frac{2H(0)}{2H(0) - h_n(1 + \beta_n)} \cdot \overline{\delta p_n(0)} \tag{6.15}$$

When the equilibrium tide is used as the unit of resonance magnification M_n, the latter is defined by

$$M_n = \left|\frac{\delta p_n(0)}{\overline{\delta p_n(0)}}\right| \tag{6.16}$$

Hence, the resonance curve of the model 4(c) is given by

$$M_n = \left|\frac{2H(0)}{2H(0) - h_n(1 + \beta_n)}\right| \tag{6.17}$$

This result also includes the special resonance curves due to isothermal and autobarotropic atmospheres. In the case of an isothermal model, (6.17) remains formally valid. We have only to replace $H(\infty)$ in (6.14) by $H(0)$ and to use this modified form of β_n. In the case of an autobarotropic model we have to put $H(\infty) = 0$ (see Section 4) and obtain from (6.17) with (6.14):

$$M_n = \left|\frac{H(0)}{H(0) - h_n}\right| \tag{6.18}$$

The resonance function (6.18) is a well-known relation and is often used for estimates of M_n. However, the sharpness of the resonance given by (6.18) is not as great as that following from other model atmospheres, the structures of which more closely resemble that of the actual earth's atmosphere.

Since the exponent of the solution (4.12) which we used must be real, we find from (6.14) that (6.15) and also (6.17) are valid only if

$$h_n \geq 4\kappa H(\infty) \tag{6.19}$$

or with (3.3) and (3.29):

$$h_n \geq 0.0335\ T_0(\infty) \quad (h_n \text{ in km, } T_0 \text{ in °K}) \tag{6.20}$$

This condition means, formally in regard to the boundary condition (3.49), that wave types with equivalent depths which do not satisfy (6.20), should not appear. It is important that more complicated models which may resemble the earth's atmosphere to a great extent, also involve (6.19) and (6.20), if these models have isothermal tops with $T_0(\infty)$. Hence, if an ionospheric temperature of only $T_0(\infty) = 270°\text{K}$ is assumed, wave types with $h_n < 9$ km should be forbidden. However, nearly all appreciable wave types which are observed in the earth's atmosphere, have equivalent depths smaller than 9 km contrary to the reasonable boundary condition (3.49). This caused some confusion. The reason for the appearance of a condition such as (6.20) is probably not physical, but rather a result of the simplifications with which the theoretical treatment has been infected. By some of the assumptions the range of validity of the basic equations has been restricted to heights which are small in comparison with the earth's radius (see Section 3.1). Above all, the neglected nonlinear terms become increasingly important at heights above 100 km as estimated by Pekeris [91, 107] and shown in Fig. 6. Since the isothermal model extends to infinity, (3.49) is integrated for an unbounded interval. This procedure implies an infinitely large range within which the formulas used for $\mathbf{v}^2$ are not valid. The influence of the simplifications apparently increases with decreasing h_n so that the integral (3.49) formally diverges when h_n becomes sufficiently small. The contradictions disappear when an autobarotropic top is assumed which involves a finite height of the whole model atmosphere (with the exception of $\kappa' = 0$). However, both kinds of models, those which extend to infinity and those which have a finite upper limit, are not free from artificial assumptions such as the temperature at infinity or the height of the upper limit.

In practice models with isothermal tops are used as much as ever. Only the boundary condition (3.49) is replaced by other assumptions. Thus the application of a condition introduced by Wilkes [17] was used more and more. This condition is based on the assumption that energy is put into the atmosphere by the tidal forces. This process is assumed to occur mainly in the lower atmosphere where the air density is high. From there the energy spreads out into the upper atmosphere where it is lost through damping effects. Then,

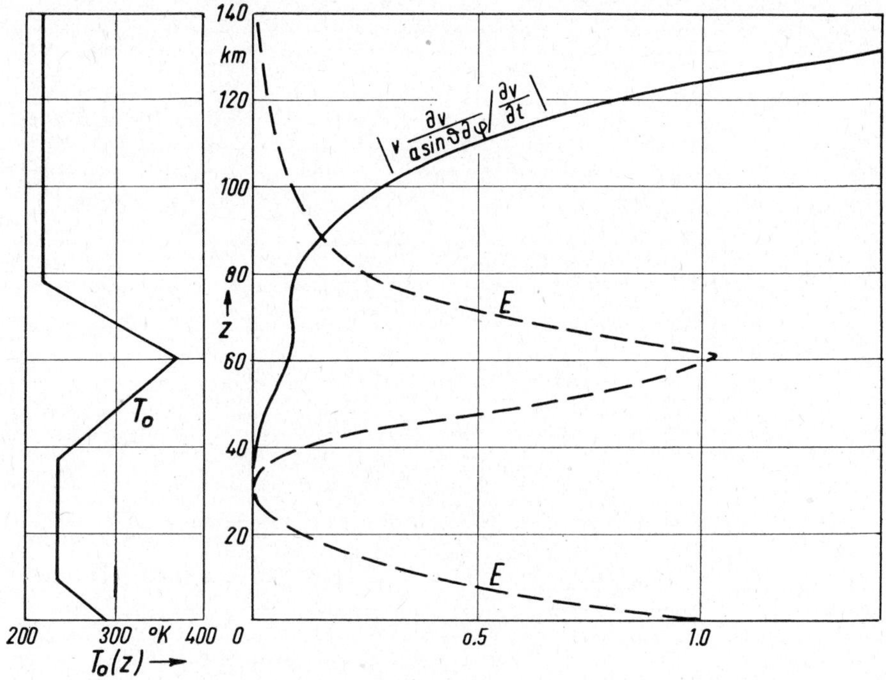

FIG. 6. Ratio of the neglected term $\left(\dfrac{v\,\partial v}{a \sin\vartheta\,\partial\phi}\right)$ to the term $(\partial v/\partial t)$, which is retained in the equations of atmospheric oscillations. Curve T_0 is the assumed temperature distribution, E the wave energy density (after Pekeris [91]).

on the average, a flow of energy in vertically upward direction occurs in connection with the tidal motion. This average flow is given by

$$W = \overline{\text{Re}\,(\delta p)\,.\,\text{Re}\,(w)}^{t} = \tfrac{1}{2}\,\text{Re}\,(\delta p\,.\,w^*) \tag{6.21}$$

Re = real part of, * = complex conjugate of.
The flow of energy per unit cross section, due to a certain wave type follows from (6.21) with (3.45), (3.46), and (5.9). If x is sufficiently large, we obtain

$$W_n = \frac{\gamma^2 h_n p_0(0)}{2\sigma}\left[\Theta^s_{\lambda,n}(\vartheta)\right]^2 \text{Im}\left(y_n^* \frac{dy_n}{dx}\right) \tag{6.22}$$

Im = imaginary part of.
If the energy is flowing upward, as assumed, W_n is positive and the new boundary condition is

$$\operatorname{Im}\left(y_n^* \frac{dy_n}{dx}\right) > 0 \tag{6.23}$$

Combining (6.23) with (3.42) Weekes and Wilkes [16] try to interpret the resonance properties of the earth's atmosphere by means of layers where the upward propagation of the energy is blocked or hindered. These layers are distinguished by low temperature and/or negative temperature gradients. Each barrier gives rise to resonance, a total barrier to strong resonance and a partial barrier to weaker resonance.

When (6.23) is applied to model (c) of Section 4.1, one of the possible solutions of (3.42) for the range $h_n \leq 4\kappa H(\infty)$ is

$$y_n(x) = A_n e^{i\tilde{\beta}_n x/2} \tag{6.24}$$

with

$$\tilde{\beta}_n = +\sqrt{4\kappa \frac{H(\infty)}{h_n} - 1} \qquad \text{if } h_n \leq 4\kappa H(\infty). \tag{6.25}$$

In the same way as before we can derive, with (6.24) and (6.25), the pressure variation at ground level for the h_n range now considered:

$$\delta p_n(0) = \frac{\gamma[2H(0) - h_n(1 + i\tilde{\beta}_n)]}{2(\gamma H(0) - h_n)} \overline{\delta p_n(0)} \tag{6.26}$$

The relation (6.26) is also true for a completely isothermal model when $H(\infty)$ in (6.25) is replaced by $H(0)$ and $\tilde{\beta}_n$ is employed in this modified manner.

The striking feature of (6.26), in contrast to (6.15), is the appearance of an imaginary part of $\delta p_n(0)$, which depends on h_n and the temperature profile. As a consequence, arbitrary phase differences according to the temperature profile assumed, are possible between exciting forces and disturbance, whereas no angles other than 0° and 180° can be found from (6.15). This consequence of (6.23) and (6.24) was used by Sawada [92] for an explanation of the phase retardation of the semidiurnal lunar oscillation L_2^2 and also of the peculiar seasonal change of this phase. For this purpose he studied many different temperature profiles and found an appreciable influence of the mesospheric temperature on the surface pressure oscillation. His results are more satisfactory for cool ozonospheres (262 to 271°K) than warm ozonospheres.

However, the boundary condition (6.23) is not free of inconsistencies considering the scope of the whole theory; for the assumption that on the average energy is flowing upward, is not compatible with the model described by the basic equations. In this model the tides are treated as a frictionless periodic motion in the field of the potential force: $-\operatorname{grad} \Omega$. Hence, the theorem of conservation of mechanical energy is valid. Since the equations do not contain other energy-generating terms, no vertical average flow of energy can appear if permanent oscillations are to exist. An equivalent consideration is also true for thermal excitation with respect to the manner in which thermal excitation is treated in the theory (see Section 7). Hence, the boundary condition appropriate to the model must require that the flow of energy vanishes on the average, even if such a flow is carried by the actual tidal oscillations in the earth's atmosphere. A more general treatment of tides in which this effect is to be considered, must also consider the influence of frictional terms.

Contradictions immediately result from the application of (6.23), if for the sake of simplicity, again the model 4(c) is regarded. The only possible solution (4.12) (with h_n instead of $\hat{h}_n$) for the range $h_n \geqslant 4\kappa H(\infty)$ requires the absence of a vertical flow of energy. A different behavior is shown by the wave types of the range $h_n \leqslant 4\kappa H(\infty)$ when (6.23) is applied. The limit separating the two ranges is determined by the temperature at infinity. Hence, $T_0(\infty)$ essentially affects the oscillations, even at ground level. This is an obvious contradiction to the conclusion at the end of Section 3.1 that the top of the model atmosphere should not affect the computed surface pressure oscillation. Moreover, we can see this break in the resonance curve. This curve is not smooth (the first derivative dM_n/dh_n is not continuous) for the critical h_n value where both ranges meet.

Therefore, Siebert [60] uses the boundary condition that the vertical flow of energy must also vanish for $h_n \leqslant 4\kappa H(\infty)$. Then we have, according to (6.22)

$$\text{Im}\left(y_n^* \frac{dy_n}{dx}\right) = 0 \tag{6.27}$$

The variety of solutions which satisfy (6.27), can be restricted by requiring that $\delta p_n(x)$ and the components of the velocity be continuously differentiable functions for the critical h_n value. This requirement implies a smooth resonance curve. The solution obtained for model 4(c) under these assumptions is

$$y_n(x) = A_n[(1 - i)e^{i\tilde{\beta}_n x/2} + (1 + i)e^{-i\tilde{\beta}_n x/2}] \tag{6.28}$$

or

$$y_n(x) = 2A_n(\cos \tfrac{1}{2}\tilde{\beta}_n x + \sin \tfrac{1}{2}\tilde{\beta}_n x) \tag{6.29}$$

When (6.29) is used to compute the surface pressure variation in the previously described manner, we find an expression which agrees with (6.15) after replacing β_n in (6.15) by $-\tilde{\beta}_n$. As a consequence of the last boundary condition we can represent $\delta p_n(0)$ within the whole range $0 \leq h_n < \infty$ by the expression

$$\delta p_n(0) = \frac{2H(0)}{2H(0) - h_n(1 + b_n)} \cdot \overline{\delta p_n(0)} \tag{6.30}$$

with

$$b_n = \begin{cases} \beta_n = +\sqrt{1 - 4\kappa \dfrac{H(\infty)}{h_n}} & \text{if } h_n \geq 4\kappa H(\infty) \\ -\tilde{\beta}_n = -\sqrt{4\kappa \dfrac{H(\infty)}{h_n} - 1} & \text{if } h_n \leq 4\kappa H(\infty) \end{cases} \tag{6.31}$$

According to (6.30) $\delta p_n(0)$ and $\overline{\delta p_n(0)}$ have either equal or opposite phase angles, but no angles between these extremes.

This model may be used for the numerical calculation of the pressure waves generated by (6.10) and (6.12). For this purpose we put, as in Section 4.2, $T_0(0) = 288°\text{K}$ and $T_0(\infty) = 160°\text{K}$. With these values the only eigenvalue of the model is found to be $\hat{h} = 10$ km. Then we obtain with the aid of (6.30) and (6.31) and the equivalent depths of Table II: The theoretical semidiurnal lunar wave family

$$L_2^2(p) = (8.33\Theta_{2,2}^2 + 0.74\Theta_{2,4}^2 + \ldots) \sin(2t_L' + 2\phi + 90°) \,.\, 10^{-2} \text{ mb} \tag{6.32}$$

and the theoretical semidiurnal solar wave family

$$S_2^2(p) = (5.43\Theta_{2,2}^2 + 0.35\Theta_{2,4}^2 + \ldots) \sin(2t' + 2\phi + 90°) \,.\, 10^{-2} \text{ mb} \tag{6.33}$$

For comparison we repeat the observed wave families (2.33) and (2.38) after introducing Hough's functions from Table II:

$$L_2^2(p) = (6.99\Theta_{2,2}^2 + 0.43\Theta_{2,4}^2 + \ldots) \sin(2t_L' + 2\phi + 72°) \,.\, 10^{-2} \text{ mb} \tag{6.34}$$

$$S_2^2(p) = (119\Theta_{2,2}^2 + 14\Theta_{2,4}^2 + \ldots) \sin(2t' + 2\phi + 158°) \,.\, 10^{-2} \text{ mb} \tag{6.35}$$

The agreement between (6.32) and (6.34) is quite satisfactory. The difference of the phases was briefly discussed already in Section 1. The resonance magnification of the main lunar wave type $L_{2,2}^2$ turns out to be 2.7. The disagreement between (6.33) and (6.35), on the other hand, is quite striking. The resonance magnification of the main solar wave type $S_{2,2}^2$ is 3.7, whereas we would need an eightyfold amplification to explain $S_{2,2}^2$ as a purely gravitationally generated pressure oscillation. Hence, if it is true that $\hat{h} \sim 10$ km is the only eigenvalue of the earth's atmosphere, the contribution of the gravitational tidal forces to the generation of S_2^2 is quite negligible and thermal excitation must be the only cause.

It is not difficult to compute the height dependence of δp_n on the basis of model 4(c) when (3.46) is employed. Since this model is very simple and does not contain the temperature maximum around 50 km height, the application of the results to the observations is uncertain. Thus the computed phase does

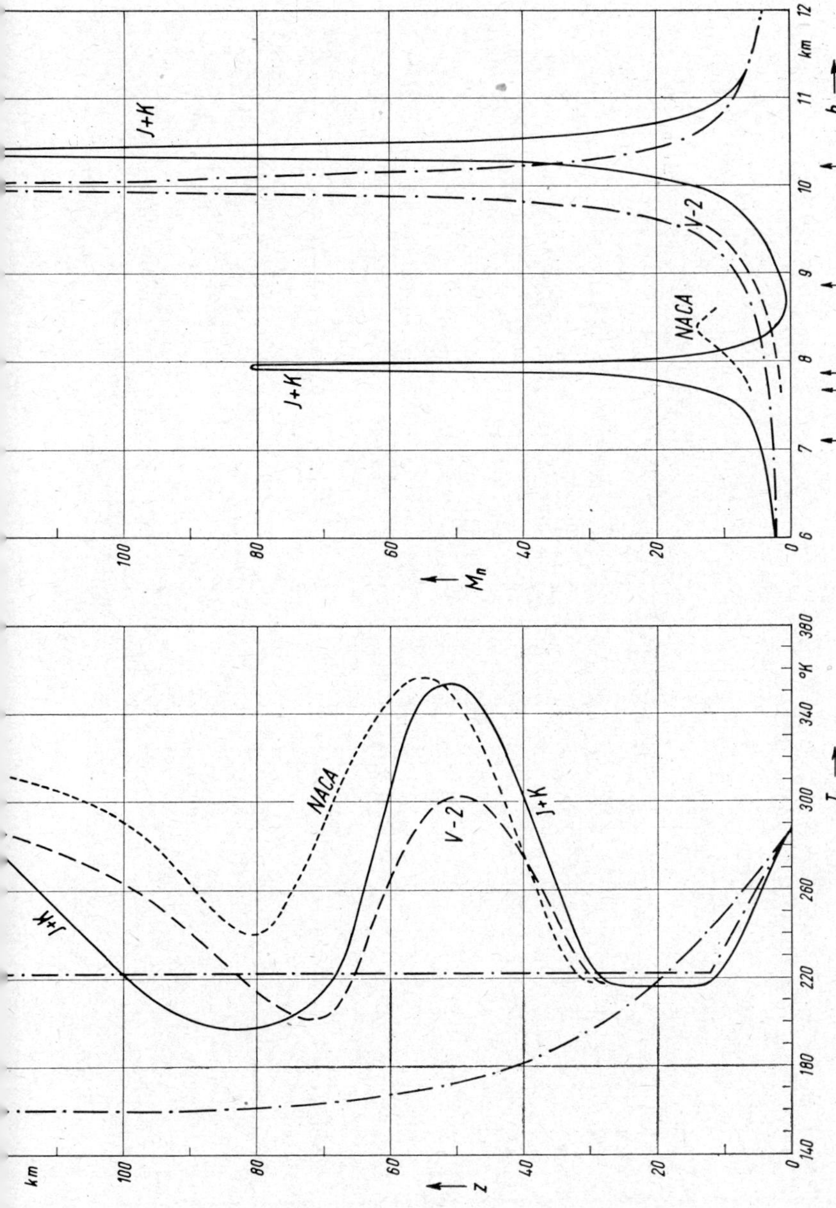

FIG. 7. Temperature profiles of the earth's atmosphere and some model atmospheres (left-handside) and the resonance curves (or pieces of resonance curves) due to these atmospheres in the case of gravitational excitation (right-handside) (after Jacchia and Kopal [18]). The profile of model 4(c) and the two-layer profile leading to the same resonance curve, have been added, and the positions of some important wave types are indicated by their symbols $S^s_{\lambda,n}$, and $L^s_{\lambda,n}$.

not show a reversal with height which has been deduced from the lunar magnetic variations with the aid of the dynamo theory. This reversal was also ascertained by direct observations of the ionospheric E layer at equatorial stations. On the other hand, Appleton and Weekes [32] found at Cambridge, England, that the lunar tides of the E layer are in phase with the surface pressure oscillation. The lunar tides of the F_2 layer were found to be out of phase for all stations considered.

The resonance magnifications of gravitationally caused tides, computed by Jacchia and Kopal [18] for more complicated and more realistic models than ours, 4(c), are shown in Fig. 7 (right-handside). The corresponding temperature profiles can be seen in the left diagram of Fig. 7. The resonance curves were computed by use of the boundary condition (6.23) which also affects the amplitudes, but not to such a degree as the phases. The resonance curve of model 4(c) with its temperature profile (4.6) has been added. This curve was obtained by use of the boundary condition (6.27), (6.28) and is nearly identical with that of the two-layer profile also drawn in the left diagram of Fig. 7.

The figure clearly exhibits the high mesospheric temperature required for the appearance of the eigenvalue $\hat{h}_2 \sim 8$ km. It shows, moreover, the sensitivity of this second resonance maximum to changes of the mesospheric temperature and its nonappearance for temperatures as they were derived from observations in the ozonosphere.

The equivalent depths of some important wave types are indicated by the symbols of the wave types. Although most of them are thermally excited, the resonance curves due to gravitational generation can still be used under conditions discussed in Section 7.2.

7. Thermal Excitation of Atmospheric Tides

7.1. Thermal Tidal Forces

The rotation of the earth in the radiation field of the sun causes periodic heating processes in the earth's atmosphere. These processes are very complex, and it is impossible to describe them quantitatively as exactly as the gravitational tidal forces. In developing the theory of atmospheric tides in Section 3, thermal action has formally been considered by introducing the function J of thermal excitation. In this section we have to determine J for special periodic heating processes.

Instead of J let us consider the temperature variation connected with J. Because we are interested only in small periodic deviations of the temperature from the mean values at each height, the customary temperature can be used instead of the potential temperature. Therefore, we have

$$J = c_p \frac{\partial \tau}{\partial t} \tag{7.1}$$

The quantity τ is, of course, a periodic function as J. Expanding both quantities according to (3.34) and eliminating c_p by means of (3.16) and (3.29) we find

$$J_n(z) = \frac{i\sigma R}{\kappa M} \tau_n(z) \tag{7.2}$$

We have already discussed in Section 3.1 that the observed temperature variation δT consists of a primary variation due to the excitation process and a secondary variation due to the pressure variation resulting from the primary temperature variation. The latter is identical with τ according to (7.1). Let us denote the secondary temperature variation by $\tilde{\tau}$. Then the amplitudes of the corresponding terms of the expansions are related by

$$\delta T_n(z) = \tau_n(z) + \tilde{\tau}_n(z) \tag{7.3}$$

These amplitudes can be complex quantities. When τ_n is known, $\tilde{\tau}_n$ can be calculated for a given model atmosphere from (3.47) with (7.2) and (7.3). We are mostly interested in the ground level value $\tilde{\tau}_n(0)$. It is obtained by eliminating χ from (3.28) and (3.30) and putting $w(0) = 0$ (according to the boundary condition assumed to be valid on the ground). We further use (7.2), (7.3), and (3.2) after replacing δp, δT, and J by series expansions such as (3.34). The result is

$$\tilde{\tau}_n(0) = \kappa \frac{T_0(0)}{p_0(0)} \delta p_n(0) \tag{7.4}$$

This well-known relation does not remain completely valid if $z > 0$.

Now we consider two special periodic heating processes: The first is the heat transfer by turbulent mass exchange, called *eddy conductivity*. Detailed treatments can be found in most of the meteorological textbooks. For the sake of simplicity we use a constant *austausch* coefficient and a constant coefficient K of eddy conductivity. These are, of course, extensively simplifying assumptions, and special attention should be paid to the daily variation of K. Again we neglect the difference between potential temperature and customary temperature. The reason is the same as given in the case of (7.1). Under these assumptions J can be expressed by

$$J = c_p K \frac{\partial^2 \tau}{\partial z^2} \tag{7.5}$$

When J in (7.5) is eliminated by means of (7.1), the amplitude τ_n of each term of τ is governed by

$$\frac{d^2 \tau_n}{dz^2} - \frac{i\sigma}{K} \tau_n = 0 \tag{7.6}$$

The solution of this differential equation consists of two exponential functions. In order to have agreement with the observations the exponential function increasing with increasing z, must be excluded. Hence, the appropriate solution of (7.6) can be written

$$\tau_n(z) = \tau_n(0)e^{-kz/H(0)} \tag{7.7}$$

with

$$k = H(0)\sqrt{\frac{i\sigma}{K}} = H(0)\sqrt{\frac{\sigma}{K}}\, e^{\pi i/4} \tag{7.8}$$

Observations lead to a mean value of K of about 10^4 cm^2/sec. Regarding the semidiurnal solar temperature wave ($\sigma = 1.4544 \,.\, 10^{-4}$ sec^{-1}) we find from (7.8) that $|k| \sim 100$ and from (7.7) that $|\tau_n(z)| = 0.1\,|\tau_n(0)|$ for a height as low as $z = 270$ meters.

Since the thermal excitation function must be substituted into (3.42), we have to transform (7.7) from z to x according to (3.39). Because of the large negative exponent in (7.7) an approximation is sufficient. For this purpose the troposphere is represented by a model with the constant lapse rate: $-6°$/km. That is, we use (4.9) with $\epsilon = 0.176$. Substituting H from (4.9) into (3.39) we find

$$x = -\frac{1}{\epsilon}\log\left(1 - \frac{\epsilon}{H(0)}z\right) = \frac{z}{H(0)} + \frac{\epsilon z^2}{2H^2(0)} + \ldots \tag{7.9}$$

For $z = 1$ km the second term of the expansion (7.9) is still smaller by a factor of 100 than the first term. At this height, however, $\tau_n(z)$ is negligibly small and quite ineffective in generating tides. Hence, we can replace $z/H(0)$ in (7.7) by x without introducing a noticeable error. Then it follows from (7.2) and (7.7) that

$$J_n(x) = \frac{i\sigma R}{\kappa M}\tau_n(0)e^{-kx} \tag{7.10}$$

Thus the thermal excitation function J has been determined and $J_n(x)$ in the differential equation (3.42) can be replaced by the simple analytic expression (7.10).

For the numerical treatment of atmospheric tides caused in this way, $\tau_n(0)$ must be numerically known. It can be easily computed when the corresponding temperature and pressure amplitudes, $\delta T_n(0)$ and $\delta p_n(0)$, are known from observations. Therefore, some important temperature waves have been given in Section 2.4. With (7.3) and (7.4) $\tau_n(0)$ is found when the addition of the amplitudes is carried out vectorially in the harmonic dial and when Hough's functions are employed.

Because the investigation of temperature waves based on observational data has not been carried out to the same extent as that of pressure waves, Kertz [39] (see also [14]) theoretically computed temperature waves on the earth's surface by harmonic analysis of cos ζ (ζ = zenith angle of the sun). The analysis was made for different latitudes and seasons; the heat transfer from the earth's surface into the atmosphere was described by turbulent mass exchange and that into the ground by conduction. Kertz also considered the influence of the land-water distribution. His results are given in terms of spherical harmonics.

The second special heating process considered is the direct absorption of insolation by water vapor in the troposphere [93]. The quantity B of energy absorbed by U cm of precipitable water under circumstances as they are true at ground level can be represented by the empirical Mügge-Möller formula [94]:

$$B = B_0(U \sec \zeta)^{0.3} \cos \zeta \text{ with } B_0 = 0.172 \text{ cal min}^{-1} \text{ gm}^{-0.3} \text{ cm}^{-1.4} \tag{7.11}$$

which was derived on the basis of Fowle's measurements. The quantity ζ is again the zenith angle of the sun. Since the absorption depends on pressure and takes place in a beam at different heights in the earth's atmosphere, the influence of pressure must still be considered in (7.11). It is quite sufficient for the accuracy needed to employ a linear pressure correction [95] and to neglect completely the temperature correction. We also neglect the correction which is obtained by introducing the relative air mass instead of sec ζ. The pressure correction is determined by defining the optical thickness U by

$$U = \int_{z_0}^{z} \rho_w(z') \frac{p_0(z')}{p_0(0)} dz' \tag{7.12}$$

where ρ_w is the density of water vapor. Using U according to (7.12) we obtain from (7.11) the energy absorbed within a slab with the lower boundary z_0 (level of observation) and the upper boundary z, when the sun's position is ζ. The cosine of ζ can be expressed with the aid of spherical trigonometry by

$$\cos \zeta = \cos \vartheta \sin \delta + \sin \vartheta \cos \delta \cos t^+ \geq 0 \tag{7.13}$$

with δ = declination of the sun and t^+ = solar local time beginning with $t^+ = 0$ at noon. The times of sunrise and sunset are given by $\cos \zeta = 0$. The relation (7.13) and, hence, also (7.11) hold only in daytime. At night we have to put $B = 0$.

With regard to the mechanism of thermal tidal action we have to know the absorbed energy per unit mass. We find it from (7.11) with (7.12):

$$\frac{1}{\rho_0} \frac{\partial B}{\partial z} = 0.3\, B_0 \frac{\rho_w p_0}{\rho_0 p_0(0)} U^{-0.7} \cos^{0.7} \zeta \tag{7.14}$$

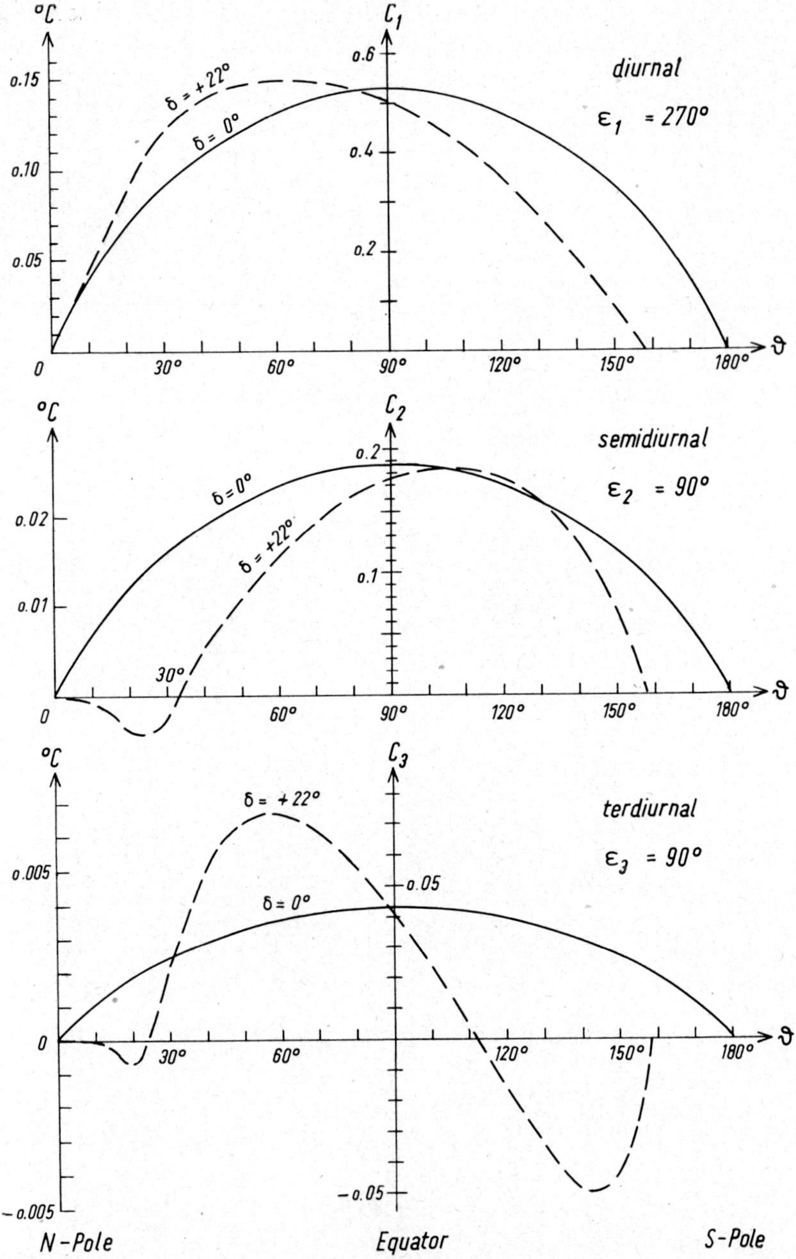

FIG. 8. Colatitude distribution of the first three Fourier coefficients C_λ of daily variation of absorption of insolation by water vapor in the troposphere, determined by analyzing the Mügge–Möller absorption law for the sun's declinations $\delta = 0°$ and $\delta = +22°$. The distribution for $\delta = -22°$ is obtained by interchanging the hemispheres. The phase constants ϵ_λ have been noted [cf. (7.17)]. The left-hand temperature scale would be true if the specific humidity on the earth's surface were uniformly 10 gm/kg.

When the harmonic analysis of $\cos^{0.7}\zeta \geq 0$ is carried out we obtain for certain ϑ and δ values the daily mean absorption and the λ^{-1}-diurnal variations of the daily absorption. It is assumed that an interdiurnal change of the daily mean temperature does not appear and, therefore, the mean daily emission is equal to the mean daily absorption. Moreover, we assume that the emission occurs uniformly during the whole day. Then the periodic variations of this heating process are due to the absorption only and found by harmonic analysis of (7.14):

$$\sum_{\lambda=1}^{\infty} C_\lambda^+ \cos \lambda t^+ = \begin{cases} \cos^{0.7}\zeta & \text{during day} \\ 0 & \text{at night} \end{cases} \tag{7.15}$$

The Fourier coefficients in (7.15) are determined by

$$C_\lambda^+ = \frac{1}{\pi}\int_{-t_0}^{+t_0} (\cos\vartheta \sin\delta + \sin\vartheta \cos\delta \cos t^+)^{0.7} \cos \lambda t^+ dt^+ \tag{7.16}$$

where $+ t_0$ is the time of sunset which must be computed from (7.13). The integration in (7.16) must be carried out numerically. This was done for $\delta = 0°$ and $\delta = \pm 22°$ [as a mean of June and July $(+ 22°)$ and December and January $(- 22°)$] and $\vartheta = 0°, 10°, 20°, \ldots 180°$. Introducing local mean time t^*, beginning at midnight, by means of $t^* = t^+ + \pi$ and changing the cosine terms into sine terms, we can write (7.14) by

$$J = 0.3\, B_0 \frac{\rho_w p_0}{\rho_0 p_0(0)} U^{-0.7} \sum_{\lambda=1}^{\infty} C_\lambda(\vartheta) \sin(\lambda t^* + \epsilon_\lambda) \tag{7.17}$$

By ignoring the daily mean absorption we can replace $\rho_0^{-1}\partial B/\partial z$ in (7.14) by the function J, the sum of the primary periodic heat variations per unit mass. The Fourier coefficients C_1, C_2, and C_3 are shown in Fig. 8 as functions of the colatitude ϑ for $\delta = 0°$ and $\delta = + 22°$. The distribution for $\delta = - 22°$ is obtained by interchanging the hemispheres (replacing $\vartheta°$ by $180° - \vartheta°$). The phases ϵ_λ have been noted. On the left-handside of the figure a temperature scale in degrees Celsius has been added which would be true if the specific humidity at the earth's surface would be uniformly $q = 10$ gm/kg.

The next step is to describe the dependence of ρ_w and U on colatitude and height. A possible dependence on longitude is later discussed. Let us denote the specific humidity at the earth's surface by $q_0(\vartheta)$. Its distribution for $\delta = 0°$ and $\delta = + 22°$ is shown in Fig. 9. These smoothed curves were derived from compilations of observations [96]. To describe the height dependence of q we try the formula

$$q(z, \vartheta) = q_0(\vartheta) \cdot (p_0(z)/p_0(0))^\alpha \tag{7.18}$$

This power law was tested by comparison with observed water vapor distributions [97]. The result is that (7.18) fits the observations to a sufficient degree of accuracy up to the 500-mb level. The value of α was determined to lie between 2 and 3. For the upper troposphere α increases to about 4. This range of variation does not greatly affect the dependence of J on the height in

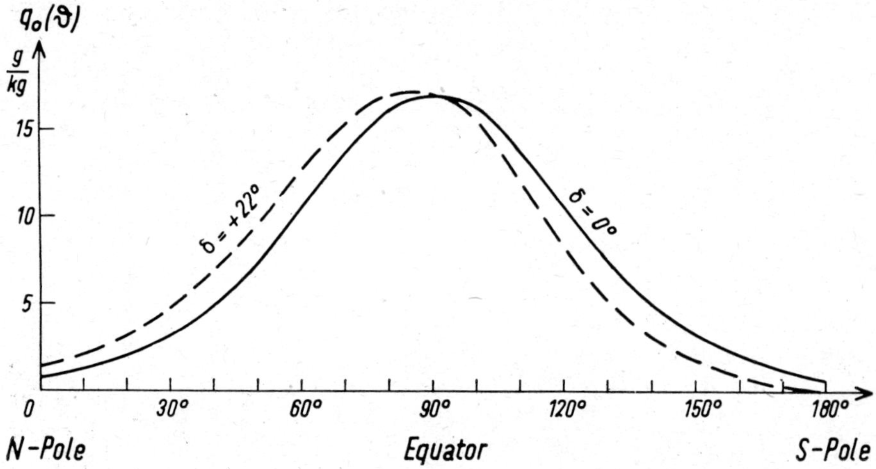

FIG. 9. Smoothed colatitude distribution of the specific humidity q_0 at the earth's surface for the sun's declinations $\delta = 0°$ and $\delta = +22°$. The distribution for $\delta = -22°$ is obtained by interchanging the hemispheres.

(7.20). We put $\alpha = 2.44$ to obtain a simple power law in (7.20). Then it follows from (7.18) and from the definition of q (when the difference between specific humidity and mixing ratio is neglected) that

$$\rho_w(z, \vartheta) = q_0(\vartheta) \cdot \rho_0(z) \, (p_0(z)/p_0(0))^{2.44} \tag{7.19}$$

By means of (3.1) and (7.19), the integration of (7.12) can be carried out. We let $z \to \infty$ and $p_0(\infty) = 0$, assuming for the sake of simplicity that (7.19) holds for the whole atmosphere. Thus ρ_w and U can be eliminated with the aid of (7.19), and (7.17) becomes (with z instead of z_0):

$$J = 0.852 \, B_0 g^{0.7} \, p_0^{-0.7}(0) q_0^{0.3}(\vartheta) \, (p_0(z)/p_0(0))^{1/3} \sum_{\lambda=1}^{\infty} C_\lambda(\vartheta) \sin (\lambda t^* + \epsilon_\lambda) \tag{7.20}$$

We see at once from (7.20) that the decrease of the amplitudes with height occurs very slowly.

The last step is to represent $q_0^{0.3}(\vartheta) \cdot C_\lambda(\vartheta)$ by associated Legendre functions of the order $s = \lambda$. Moreover, we can introduce the variable x by means of

(3.41) and replace each $J^s_{\lambda,m}$ term by the temperature amplitude $\tau^s_{\lambda,m}$ according to (7.2). Hence, we can represent a temperature wave family due to absorption of insolation by

$$S^s_\lambda(T) = \sum_{m=s}^{\infty} \tau^s_{\lambda,m}(0)e^{-x/3}\, P^s_m(\vartheta) \sin(\lambda t' + s\phi + \epsilon^s_{\lambda,m}) \tag{7.21}$$

where now $\tau^s_{\lambda,m}(0)$ and $\epsilon^s_{\lambda,m}$ may be determined. The amplitudes and phases of the terms of the important wave families $S^1_1(T)$, $S^2_2(T)$, and $S^3_3(T)$ are given in Table III. At the earth's surface these temperature waves are smaller by about one order of magnitude than the corresponding waves due to eddy conductivity.

TABLE III. Amplitudes $\tau^s_{\lambda,m}(0)$ in degrees Celsius and phases $\epsilon^s_{\lambda,m}$ of the terms of some important wave families $S^s_\lambda(T)$ of temperature [see (7.21)] caused by direct absorption of insolation by water vapor for $\delta = 0°$ and $\delta = \pm 22°$.

	$\tau^s_{\lambda,m}(0)$°C		$\epsilon^s_{\lambda,m}$		
	$\delta = 0°$	$\delta = \pm 22°$	$\delta = 0°$	$\delta = +22°$	$\delta = -22°$
m	$\lambda = s = 1$		$\lambda = s = 1$		
1	0.158	0.148	180°	180°	180°
2	—	0.049	—	180°	0°
3	0.020	0.022	0°	0°	0°
4	—	0.004	—	180°	0°
5	0.002	0.002	180°	0°	0°
	$\lambda = s = 2$		$\lambda = s = 2$		
2	0.036	0.030	0°	0°	0°
3	—	0.005	—	180°	0°
4	0.003	0.004	0°	180°	180°
5	—	0.001	—	180°	0°
6	0.002	0.003	0°	180°	180°
	$\lambda = s = 3$		$\lambda = s = 3$		
3	$6.4 \cdot 10^{-3}$	$5.5 \cdot 10^{-3}$	0°	0°	0°
4	—	$6.5 \cdot 10^{-3}$	—	0°	180°
5	$1.3 \cdot 10^{-3}$	$0.4 \cdot 10^{-3}$	0°	180°	180°
6	—	$1.3 \cdot 10^{-3}$	—	0°	180°
7	$0.7 \cdot 10^{-3}$	$0.7 \cdot 10^{-3}$	0°	180°	180°

The computations have been carried out without considering the influence of the land-water distribution. Such an influence can arise when the water-vapor density above the ocean areas is larger than above the land areas of the same latitude. Since the difference of both densities is probably not great and since, moreover, the specific humidity appears in the theory through $q_0^{0.3}$, the temperature waves computed before are hardly affected by this difference. On the other hand, this difference gives rise to new temperature waves which are not immediately involved in the formula (7.11) and which might be the cause of pressure waves such as the semidiurnal zonal wave family $S_2^0(p)$. In order to calculate these temperature waves, the land-water distribution is represented by

$$G(\vartheta, \phi) = \sum_{\nu=0}^{\infty} [a_\nu(\vartheta) \cos \nu\phi + b_\nu(\vartheta) \sin \nu\phi] = \begin{cases} 1 \text{ for land} \\ 0 \text{ for oceans} \end{cases} \tag{7.22}$$

The coefficients a_ν and b_ν can be found in Kertz [39]. Denoting the function of thermal excitation by J_L for land areas and by J_W for ocean areas we have with (7.22) the general representation

$$J = J_L G + J_W(1 - G) \tag{7.23}$$

For J_L and J_W we can use (7.20) with different q_L and q_W instead of q_0. From this formulism planetary temperature waves of very different types can be derived by means of spherical harmonics and trigonometric theorems. Because reliable data of q_L and q_W are not available, definite numerical results cannot be given. It is, however, possible to determine the phases by using the plausible assumptions that

$$q_L = cq_W \sim cq_0, \qquad c = \text{const}, \qquad 0 < c < 1 \tag{7.24}$$

With (7.24) the phases of $S_2^0(T)$ were computed and used to find the phases of $S_{2,2}^0(T)$ and the theoretical $S_{2,2}^0(p)$ in Table IV.

When the amplitudes and phases in (7.21) are known, Hough's functions can be introduced. The amplitudes of the new terms (wave types) arising in this way, may be denoted by $\tau_n(0)$ (omitting λ and s). Because the dependence of these amplitudes on height is given by exp $(- x/3)$, the amplitude $J_n(x)$ in the differential equation (3.42) is known. It can be represented by (7.10), the same expression as for thermal excitation by eddy conductivity. In the case of the excitation by direct absorption of solar radiation, however, we have

$$k = \tfrac{1}{3} \tag{7.25}$$

instead of k according to (7.8) for eddy conductivity.

The two examples of this section show how thermal tidal forces can be derived from observations.

7.2. Thermally Generated Oscillations

Analogous to Section 6.2 we determine the tidal pressure oscillations at the ground due to $J_n \neq 0$ and $\Omega_n = 0$ on the basis of the model atmosphere (c) of Section 4.1.

Before carrying out this compution let us consider an idea of Wilkes [10] and Haurwitz and Möller [61] which is to reduce the problem of thermally generated oscillations to that of gravitationally generated oscillations. Assuming a purely thermal air tide, we closely follow Haurwitz by putting

$$\delta p = P + \rho_0 \Phi \tag{7.26}$$

with

$$\Phi = \frac{\gamma - 1}{i\sigma} J \tag{7.27}$$

Substituting δp into (3.7) to (3.9), where $\Omega = 0$, by means of (7.26) and (7.27), we obtain

$$\frac{\partial u}{\partial t} - 2\omega v \cos\vartheta = -\frac{1}{a}\frac{\partial}{\partial\vartheta}\left(\frac{P}{\rho_0} + \Phi\right) \tag{7.28}$$

$$\frac{\partial v}{\partial t} + 2\omega u \cos\vartheta = -\frac{1}{a\sin\vartheta}\frac{\partial}{\partial\phi}\left(\frac{P}{\rho_0} + \Phi\right) \tag{7.29}$$

$$\frac{\partial P}{\partial z} = -g\delta\rho - \frac{\partial}{\partial z}(\rho_0\Phi) \tag{7.30}$$

The equation of continuity (3.10) remains unchanged. Finally we have to consider the first law of thermodynamics. Using the representation (3.30), we obtain with (7.26) and (7.27)

$$i\sigma P = wg\rho_0 - \gamma g\rho_0 H\chi \tag{7.31}$$

The equations (7.28), (7.29), and (7.31) show a complete analogy to the corresponding equations for a purely gravitational air tide ($J = 0$, $\Omega \neq 0$), if in the latter δp is replaced by P and Ω by Φ. The remaining equation (7.30) differs from the corresponding equation (3.9) because ρ_0 is under the differentiation sign. However, the additional term thus arising in (7.30) may become small. It is

$$\frac{\partial}{\partial z}(\rho_0\Phi) = \rho_0\Phi\left(\frac{1}{\Phi}\frac{\partial\Phi}{\partial z} + \frac{1}{\rho_0}\frac{\partial\rho_0}{\partial z}\right) \tag{7.32}$$

According to (3.32) we find with (3.2) for a troposphere with the lapse rate: $-6°$/km:

$$\left|\frac{1}{\rho_0}\frac{\partial\rho_0}{\partial z}\right| \sim \frac{1.2}{H(0)}$$

On the other hand, (7.27) and (7.10), with $x = z/H(0)$ and $|k| \sim 100$ (for semidiurnal oscillations), lead to the estimate

$$\left|\frac{1}{\Phi}\frac{\partial \Phi}{\partial z}\right| = \left|\frac{1}{J}\frac{\partial J}{\partial z}\right| = \left|\frac{k}{H(0)}\right| \sim \frac{100}{H(0)}$$

Hence, the second term in the parentheses of (7.32) can be neglected in comparison with the first. If then (7.32) is substituted into (7.30), this equation also agrees with its gravitational counterpart. Yet we see from the estimates that this analogy holds only for thermal excitation by eddy conductivity; for the same simplification is not allowed in the case of thermal excitation by absorption of insolation because in this case $k = 1/3$.

It is, however, not possible to use (7.26) and (7.27) in order to derive special results, say $\delta p_0(0)$, for thermal excitation from the corresponding results for gravitational excitation. This discrepancy is caused by convenient, but not necessary simplifications in the formal development of the theory of gravitational tides. The term $h_n d\Omega_n/dz$, for instance was neglected in Section 3.2 in comparison with Ω_n, whereas in general $|h_n dJ_n/dz| \gg |J_n|$ for thermal excitation by eddy conductivity.

The general treatment of thermally generated oscillations requires the solution of the inhomogeneous differential equation (3.42). With (4.6), (7.10), and (3.3) this differential equation becomes

$$\text{(7.33)} \qquad \frac{d^2y_n}{dx^2} - \frac{1}{4}\left(1 - 4\kappa\frac{H(\infty)}{h_n}\right)y_n(x) = \frac{i\sigma H(0)}{\gamma h_n T_0(0)}\cdot \tau_n(0)e^{-(1+2k)x/2}$$

The particular solution $\tilde{y}_0$ of the inhomogeneous equation is

$$\text{(7.34)} \qquad \tilde{y}_n(x) = \frac{i\sigma H(0)}{\gamma T_0(0)}\cdot\frac{\tau_n(0)e^{-(1+2k)x/2}}{(1+k)kh_n + \kappa H(\infty)}$$

The solution of the homogeneous equation has been discussed in Section 6.2. Again the problem of the boundary condition for $x \to \infty$ arises. If $h_n \geq 4\kappa H(\infty)$, there is no doubt about using the exponential solution (4.12). If $h_n \leq 4\kappa H(\infty)$, one might think that the the condition of an upward flow of energy is appropriate because the atmosphere is supplied with heat from the sun. Employing, therefore, (6.24) as a consequence of (6.23), we find again that the phase angle between exciting force and disturbance strongly depends on the temperature profile assumed (see, e.g., the paper of Sen and White [98]). However, because $\overline{J}^t = 0$, no energy is on the average put into the atmosphere by such periodic heating processes as have been described in Section 7.1. Hence, we use for $h_n \leq 4\kappa H(\infty)$ the solution (6.29) in the case of thermally generated oscillations too.

After adding $\tilde{y}_n$ according to (7.34) to the solution of the homogeneous equation we determine the arbitrary constant A_n by means of the boundary condition (3.48). We further find the pressure amplitude at ground level from (3.51):

$$(7.35)\quad \delta p_n(0) = -\frac{p_0(0)H(0)h_n(1+2k-b_n)\tau_n(0)}{T_0(0)[2H(0)-h_n(1+b_n)][(1+k)kh_n+\kappa H(\infty)]}$$

with b_n according to (6.31).

The result (7.35) holds for both thermal excitation by eddy conductivity and by absorption of insolation. Since in the former case $|k| \gg 1$ we can simplify (7.35) for this kind of excitation. Let us assume, moreover, $h_n \gtrsim H(\infty)$, a condition which is satisfied by the equivalent depths of all wave types of interest. Then we obtain from (7.35) with (7.8):

$$(7.36)\qquad \delta p_n(0) = \frac{2p_0(0)\tau_n(0)}{T_0(0)[2H(0)-h_n(1+b_n)]}\sqrt{\frac{K}{\sigma}}\,e^{3\pi i/4}$$

From (7.36) we immediately find the expressions for $\delta p_n(0)$ due to the isothermal and the former autobarotropic model atmospheres (see Section 4.1). In the case of the isothermal model, $H(\infty)$ in b_n [see (6.31)] must be replaced by $H(0)$, and in the case of the former autobarotropic model we have to put $b_n = 1$. These special expressions were first derived by Chapman [9] when he studied the thermal generation of the semidiurnal solar oscillation by eddy conductivity. The main feature of these formulas is the appearance of the two possible phase differences of 135° and 315° between the corresponding pressure and primary temperature variations.

Comparing (7.36) with (6.30) we find with (6.8) and (3.2) the following relations between a gravitationally generated variation (Gr) and a corresponding variation (Th) thermally excited by eddy conductivity:

$$(7.37)\qquad [\delta p_n(0)]_{\mathrm{Th}} = \frac{g\tau_n(0)}{T_0(0)\Omega_n(0)}\sqrt{\frac{K}{\sigma}}\,e^{-\pi i/4}\,[\delta p_n(0)]_{\mathrm{Gr}}$$

Because the unit of the gravitational resonance magnification is proportional to $\Omega_n(0)$, we need only define the unit of thermal resonance magnification proportional to $\tau_n(0)/\sqrt{\sigma}$. Then the resonance magnifications of both kinds of excitations are proportional to each other. This result has been derived on the basis of the special model atmosphere 4(c). In the same way that we found (6.30) and (7.36), we can show that (7.37) is also valid for an arbitrary atmosphere. We only have to assume that the solution of the homogeneous differential equation (already subject to the boundary condition for $x \to \infty$) can be represented by $y_n(x) = A_n\mathring{y}_n(x)$ and that $|d\mathring{y}_n/\mathring{y}_n dx|_{x=0}$ is negligible in comparison with $|d\tilde{y}_n/\tilde{y}_n dx|_{x=0}$. This condition is generally satisfied in the case of thermal excitation by eddy conductivity. Hence, if h_n is not too small resonance curves due to gravitational generation of atmospheric tides also hold for this kind of thermal excitation. Because of this analogy, some important thermally generated wave types have been admitted in Fig. 7 in Section 6.2 in addition to the gravitationally caused wave types.

In order to compute the amplitudes of the pressure variations due to thermal excitation by absorption of insolation, we must use (7.35). This has been done for the main wave types of the four most important solar wave families. For the primary temperature variations, the data of Table III were used after transforming them from terms of spherical harmonics into terms of Hough's functions by means of Table II. The results of this theoretical work are given in Table IV. The amplitudes and phases of the corresponding wave types known from observation according to Section 2.4, have been added for comparison. The unit of the resonance magnification has been defined in such a manner that the amplification of $S_{2,2}^{2}$ becomes formally equal for both gravitational and thermal excitations.

TABLE IV. Data of the four most important solar wave types. The amplitudes and phases of the temperature are due to the absorption of insolation by water vapor. By use of these temperature data the amplitudes and phases of the pressure have been computed on the basis of the model atmosphere 4(c). The observed amplitudes and phases of pressure have been taken from Section 2.4 after introducing Hough's functions.

Wave type	$S_{2,2}^{2}$	$S_{2,2}^{\circ}$	$S_{3,3}^{3}$	$S_{3,4}^{3,*}$	Unit
Equivalent depth	7.85	8.85	12.89	7.66	km
Resonance magnification	3.7	6.7	2.6	3.4	—
Temperature amplitude	$3.11 \cdot 10^{-2}$	$C_{2,2}^{\circ}$	$5.88 \cdot 10^{-3}$	$6.23 \cdot 10^{-3}$	°C
Temperature phase	0°	335°	0°	July 0° Jan. 180°	—
Pressure amplitude computed	0.360	$21C_{2,2}^{\circ}$	$4.89 \cdot 10^{-2}$	$6.66 \cdot 10^{-2}$	mb
Pressure phase computed	180°	155°	0°	July 180° Jan. 0°	—
Pressure amplitude observed	1.19	$6.98 \cdot 10^{-2}$	$7.19 \cdot 10^{-2}$	$26.6 \cdot 10^{-2}$	mb
Pressure phase observed	158°	137°	10°	July 175° Jan. 355°	—

Some details of Table IV have already been discussed in Section 1, especially the retardation of the observed phases of the semidiurnal wave types, and possible reasons for the differences between observed and computed amplitudes. As was explained in Section 7.1, it is at present not possible to compute the pressure amplitude of $S_{2,2}^{\circ}$. In addition to the main wave types given in Table IV, higher terms of these wave families were also computed. The results show good, poor, or no agreement between theoretical and observational data. Discrepancies in the higher terms, however, must be expected because of the numerous simplifications introduced in the theory.

Independent of the question whether or not the explanation here given for the appearance of atmospheric tides, is true, we can immediately see from Table IV that the heating process due to absorption of incoming radiation by water vapor is the most effective tide-generating source hitherto known. Comparing the pressure amplitude of $S_{2,2}^2$ in Table IV with the gravitationally generated amplitude of the same wave type according to (6.33) we find that the former one is nearly seven times larger. The same ratio of about ten to one holds when gravitational generation is replaced by thermal excitation by eddy conductivity as can be estimated with the aid of (7.37). Hence, all solar waves are thermally generated.

Without strong resonance of $S_{2,2}^2$ and without an appreciable influence of the gravitational forces on $S_{2,2}^2$ Holmberg's [99] explanation of the present angular velocity of the earth's rotation becomes questionable. According to this explanation which was first suggested by Kelvin [2], the gravitational attraction of the sun exerts an accelerating couple on the semidiurnal solar pressure oscillation because its maxima appear before noon and midnight. This torque is supposed to accelerate the earth by the action of the frictional drag exerted on the earth's surface and to be in equilibrium with the retarding torque due to the tidal friction of the oceans. However, the thermal tidal forces act similarly to gravitational tidal forces as can be seen from Haurwitz' analogy. If, therefore, magnitude and positions of the maxima of $S_2^2(p)$ are determined by thermal tidal forces and a couple of the gravitational tidal forces acts on $S_2^2(p)$, a restoring torque, set up by the thermal tidal forces, arises at once when the positions of the maxima change. It can be estimated that this restoring torque becomes adequately large to compensate for the accelerating torque of the gravitational forces. Hence, it is not necessary to assume that surface friction keeps the maxima in their positions before noon and midnight, and it is, therefore, doubtful that the earth is accelerated in this way.

Contrary to the case of pressure waves generated by eddy conductivity, we cannot apply the resonance curves for gravitational generation to pressure waves excited by direct absorption of solar radiation. A resonance curve for this kind of generation is shown in Fig. 10 (right-handside). It has been computed on the basis of (7.35). The temperature profile of this model is shown in the left diagram of Fig. 10. The relationship between the real height z and the modified height x [see (3.39)] is given in the middle diagram. The positions of the four wave types of Table IV and of four others have been indicated under the abscissa of the right-hand diagram. Among these, the main wave type $S_{1,1}^1$ of the migrating diurnal solar pressure wave is found (according to its equivalent depth of $h_1 = 0.63$ km [100]) on an unexpectedly vanishing branch of the resonance curve. This decrease of the magnifications strengthens the conjecture stated in Section 1, that the diurnal pressure wave

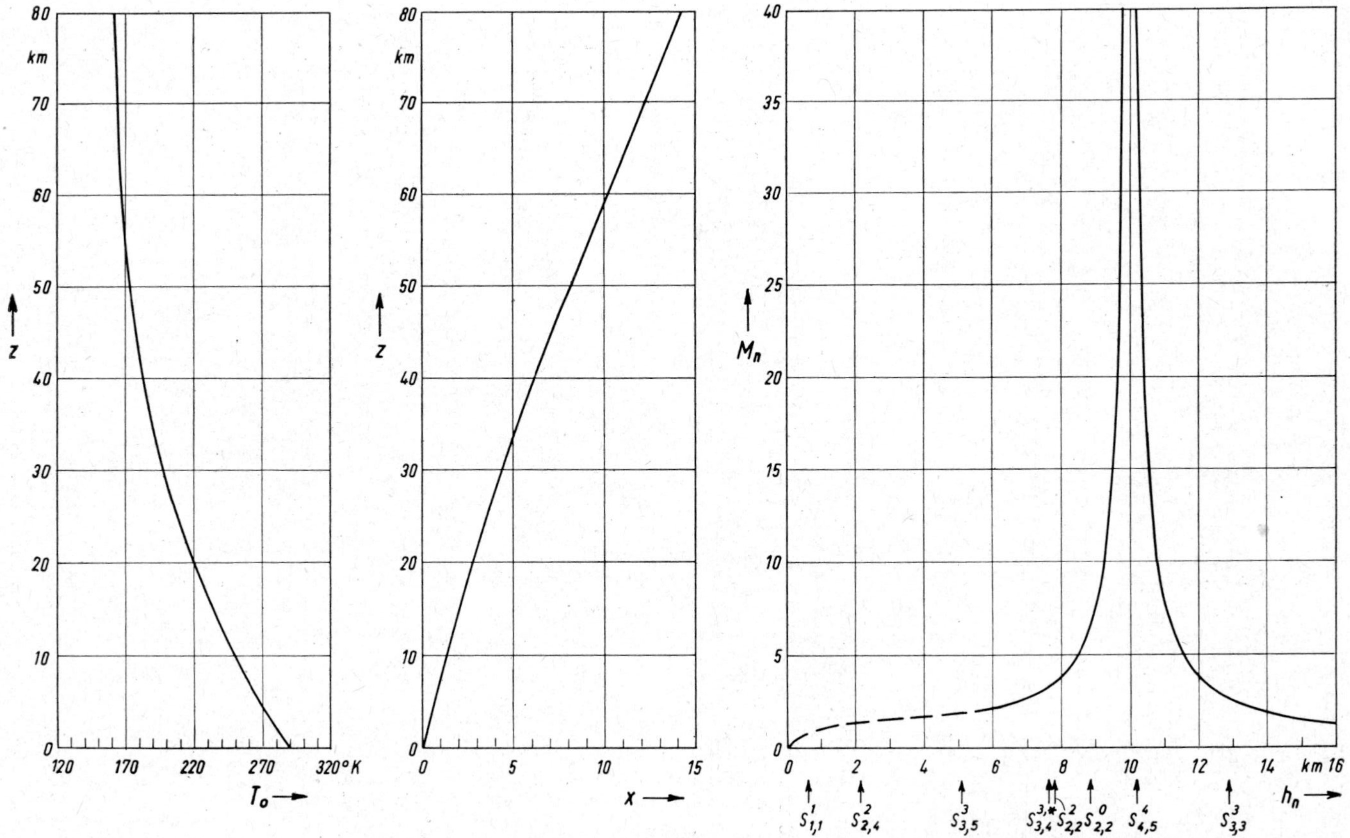

FIG. 10. Temperature profile of model 4(c) (left-handside), the relationship between the real height z and the modified height x (middle) and the resonance curve due to excitation by direct absorption of insolation by water vapor (right-handside), computed on the basis of model 4(c). The dotted part of the curve denotes the region where $h_n < 4\kappa H(\infty)$. The positions of some important wave types have been indicated by their symbols $S^s_{\lambda,n}$.

is suppressed in the earth's atmosphere. The wave type $S_{4,5}^4$ of the migrating 6-hourly solar pressure wave has been considered because it is the simplest wave type close to the resonance maximum belonging to $\hat{h} \sim 10$ km.

Let us finally consider the generation of atmospheric tides by an arbitrary thermal excitation function $J(x)$ which may be given by the arbitrary amplitudes $\tau_n(x)$ of the primary temperature wave. The excitation may occur within the layer $x_1 \leq x \leq x_2$. Again employing the model atmosphere 4(c), we find that the differential equation (3.42) becomes with (7.2):

$$\frac{d^2y_n}{dx^2} - \frac{1}{4}\left(1 - 4\kappa\frac{H(\infty)}{h_n}\right)y_n(x) = \frac{i\sigma H(0)}{\gamma h_n T_0(0)}\tau_n(x) \neq 0 \qquad \text{if } x_1 \leq x \leq x_2 \tag{7.38}$$

Because a particular solution of the homogeneous equation is known, we can solve the inhomogeneous equation by the method of variation of parameters. The double integral arising in this way, can be reduced to a single integral by integration by parts. For the complete solution of the homogeneous equation and the boundary condition for $x \to \infty$, one may notice that there is no difference between (7.38) and (7.33). After determining the arbitrary constant A_n by using the second boundary condition (3.48), the pressure amplitude at ground level is obtained from (3.51):

$$\delta p_n(0) = -\frac{2p_0(0)\,H(0)}{T_0(0)[2H(0) - h_n(1 + b_n)]}\int_{x_1}^{x_2}\tau_n(x)\mathring{y}_n(x)e^{-x/2}dx \tag{7.39}$$

with

$$\mathring{y}_n(x) = \begin{cases} e^{-\beta_n x/2} & \text{if } h_n \geq 4\kappa H(\infty) \\ \cos\tfrac{1}{2}\tilde{\beta}_n x + \sin\tfrac{1}{2}\tilde{\beta}_n x & \text{if } h_n \leq 4\kappa H(\infty) \end{cases} \tag{7.40}$$

and b_n, β_n, $\tilde{\beta}_n$ according to (6.31).

By means of the results from Johnson's [101] and Pressman's [102] studies of the diurnal temperature changes in the ozonosphere, the semidiurnal pressure oscillation on the earth's surface due to this periodic heating process was estimated from (7.39) to be

$$S_{2,2}^2(p) \sim 0.1\ \Theta_{2,2}^2(\vartheta)\sin(2t' + 2\phi + 180°)\text{ mb} \tag{7.41}$$

The amplitude obtained in (7.41) exceeds those of $S_{2,2}^2(p)$ computed for gravitational generation and thermal excitation by eddy conductivity. This remarkable result suggests that the influence of the ozonosphere as a tide-generating source must be considered in studies of the height distribution of $S_{2,2}^2(p)$ as well as in investigations of other solar wave types. Such studies however, require a better knowledge of the mesospheric temperature variations and the use of a more realistic model atmosphere.

From the treatment of this problem and from many other problems discussed in the preceding sections we see that many results are due to approximations and estimates and that a completely satisfactory explanation of atmospheric tides does not yet exist. The present incomplete status of the theory and of the observational information may act as a stimulant for further work.

List of Symbols

This list does not contain such unimportant symbols as the many harmonic coefficients and coefficients of series expansions found in Section 2.

Some quantities on this list appear in the text with lower and/or upper suffixes denoting amplitudes of series expansions of these quantities in terms of spherical harmonics or Hough's functions. Those quantities are listed without suffixes.

The number of symbols needed and the attempt to avoid unusual notations occasionally required that the same symbol has two or three meanings, but never different meanings in the same section.

A_n	Integration constant
a	Mean radius of the earth = 6371 km.
B	Energy of radiation absorbed by water vapor [see (7.11)]
B_0	Constant of the Mügge–Möller formula (7.11)
b_n	Defined by (6.31)
c_p	Specific heat of the air at constant pressure
c_v	Specific heat of the air at constant volume
F	Differential operator defined by (3.27)
f	Defined by (3.25)
G	Function of land–water distribution defined by (7.22)
$\mathbf{g}$	Mean acceleration of gravity; $g = \|\mathbf{g}\| = 979.8$ cm/sec^2
$H(z)$	Scale height defined by (3.3)
$H(0)$	Surface value of the scale height = 8.436 km
$h_n(\lambda,s)$	Equivalent depth of a wave type
$\hat{h}$	Eigenvalue of the earth's atmosphere or model atmosphere
i	Imaginary unit = $\sqrt{-1}$
J	Function of thermal excitation defined by (3.15)
$J_{1/\epsilon}(\xi)$	Bessel function of the order $1/\epsilon$
K	Coefficient of eddy conductivity
k	Defined by (7.8) for eddy conductivity and given by (7.25) for absorption of insolation
L_λ	λ^{-1}–diurnal lunar oscillation
L_λ^s	Lunar wave family
$L_{\lambda,n}^s$	Lunar wave type
M	Mean molecular weight of air = 28.97 gm/mol
M_n	Resonance magnification of a wave type
m	Integer
n	Integer
P	Defined by (7.26)
P_m^s	Seminormalized associated Legendre function [see (2,11)]
$p_0(z)$	Static pressure

$p_0(0)$ Surface value of static pressure = 1013 mb
δp Pressure variation
$\overline{\delta p}$ Atmospheric equilibrium tide defined by (6.8)
δQ Heat added per unit mass of air [see (3.12)]
q Specific humidity in the troposphere
$q_0(\vartheta)$ Surface distribution of specific humidity
R Universal gas constant = $8.314 \cdot 10^7$ ergs/mol degree
r Distance from the earth's center = $a + z$
S_λ λ^{-1}–diurnal solar oscillation of pressure (p) or temperature (T)
S_λ^s Solar wave family
$S_{\lambda,n}^s$ Solar wave type
$S_\lambda^{s,*}$ Solar wave family distinguished by an annual variation
$S_{\lambda,n}^{s,*}$ Solar wave type distinguished by an annual variation
s Integer indicating the periodicity in ϕ
$T_0(z)$ Undisturbed temperature in °K
$T_0(0)$ Surface value of the undisturbed temperature = 288°K
δT Temperature variation
t^+ Solar local time (in radians) beginning at noon
t^* Solar local time (in radians) beginning at midnight = $t^+ + \pi$
t' Solar universal time (in radians) = $t^* - \phi$
t Solar universal time (in seconds)
$t_L{}^*$ Lunar local time (in radians)
$t_L{}'$ Lunar universal time (in radians) = $t_L{}^* - \phi$
t_L Lunar universal time (in seconds)
t_d Length of a mean solar day (86, 400 sec), a mean lunar day (89, 428.33 sec) or a mean sidereal day (86, 164.09 sec)
U Precipitable water (in cm) or optical thickness
u Southward component of wind velocity (north wind)
V Magnitude of velocity of an atmospheric tsunami
$\mathbf{v}$ Tidal wind velocity = (u, v, w)
v Eastward component of wind velocity (west wind)
w Vertically upward component of wind velocity
x Modified height defined by (3.39)
$Y_n(\vartheta, \phi)$ Spherical surface harmonic
$y_n(x)$ Defined by (3.40), solution of the differential equation (3.42)
$\tilde{y}_n(x)$ Particular solution of the inhomogeneous differential equation (3.42)
z Height over the earth's surface
$\beta_n, \tilde{\beta}_n$ Defined by (6.31)
$\gamma_{\lambda,m}^{s,n}$ Coefficients of the expansion of P_m^s in terms of $\Theta_{\lambda,n}^s$
γ Ratio of the specific heats, adiabat index defined by (3.14)
γ' Polytrop index defined by (3.13)
δ Declination of the sun (in Section 7.1)
$\delta_{m,m'}^s$ Defined by (2.11)
ϵ Scale height gradient defined by (4.8)
ϵ_λ, $\epsilon_{\lambda,m}^s$ Phase constants
ζ Zenith angle of the sun (in Section 7.1)
ζ Surface elevation of oceanic tides (in Sections 5 and 6.1)
$\bar{\zeta}$ Oceanic equilibrium tide (in Sections 5 and 6.1)
η Mean longitude of the sun

$\Theta^s_{\lambda,n}(\vartheta)$ Hough's function, solution of the differential equation (5.12)
ϑ Geographical colatitude counted from the north pole.
κ Defined by (3.29)
κ' Defined analogous to κ when γ in (3.29) is replaced by γ'
Λ Differential operator defined by (4.32)
λ Integer, commensurable parts of the day
λ Wavelength of an atmospheric tsunami (in Section 4.2)
μ Integer (in Section 2)
μ $\cos\vartheta$ (in Section 5)
μ Mean longitude of the moon (in Section 6.1)
ν Integer
ξ Defined by (4.16)
$\rho_0(z)$ Undisturbed air density
$\rho_0(0)$ Surface value of the undisturbed air density $= 1.226.10^{-3}$ gm/cm^3
$\delta\rho$ Air density variation
ρ_{W} Density of water vapor
σ Angular frequency of an oscillation (in sec^{-1})
τ Primary temperature variation given by (7.1)
$\tilde{\tau}$ Secondary temparature variation $= \delta T - \tau$
Φ Defined by (7.27)
ϕ Geographical longitude, positive toward east
χ div **v** given by (3.11)
$\Psi_n(\vartheta, \phi)$ Solution of the differential equation (3.35)
ψ Mean longitude of the perigee
Ω Gravitational tidal potential
$\vec{\boldsymbol{\omega}}$ Angular velocity of the earth's rotation; $\omega = |\vec{\boldsymbol{\omega}}| = 7.2921 \,.\, 10^{-5}$ sec^{-1}
$\boldsymbol{\nabla}$ Vector operator *nabla*
$\boldsymbol{\nabla}^2$ Laplacian operator
D/Dt Individual differential operator with respect to time
C_λ, C^+_λ, $C^{s,\nu}_{\lambda,n}$, c_λ, $c^s_{\lambda,m}$ Fourier coefficients and coefficients of other series
$p = p_0(z) + \delta p(z, \vartheta, \phi, t)$ Total pressure
$T = T_0(z) + \delta T(z, \vartheta, \phi, t)$ Total temperature
$\rho = \rho_0(z) + \delta\rho\,(z, \vartheta, \phi, t)$ Total air density

References

1. Laplace, P. S. (1799, 1825). "Méchanique céleste." Paris.
2. Lord Kelvin (Thomson, W.) (1882). On the thermodynamic acceleration of the earth's rotation. *Proc. Roy. Soc. Edinburgh* **11**, 396–405.
3. Lord Rayleigh (Strutt, J. W.) (1890). On the vibrations of an atmosphere. *Phil. Mag.* (5) **29**, 173–180.
4. Margules, M. (1890). Über die Schwingungen periodisch erwärmter Luft. *Sitzber. Akad. Wiss. Wien Abt. IIa* **99**, 204–227; (1892). Luftbewegungen in einer rotierenden Sphäroidschale. *Sitzber. Akad. Wiss. Wien Abt. IIa* **101**, 597–626; (1893). *Sitzber. Akad. Wiss. Wien Abt. IIa* **102**, 11–56, 1369–1421.
5. Trabert, W. (1903). Die Theorie der täglichen Luftdruckschwankung von Margules und die tägliche Oszillation der Luftmassen. *Meteorol. Z.* **20**, 481–501, 544–563.
6. Lamb, H. (1910). On the atmospheric oscillations. *Proc. Roy. Soc.* (*London*) **A84**, 551–572.

7. Lamb, H. (1916). "Hydrodynamics," 4th ed. Cambridge Univ. Press, London; (1945). 1st American ed. Dover, New York,
8. Taylor, G. J. (1936). The oscillations of the atmosphere. *Proc. Roy. Soc.* (*London*) **A156**, 318–326.
9. Chapman, S. (1924). The semidiurnal oscillation of the atmosphere. *Quart. J. Roy. Meteorol. Soc.* **50**, 165–195.
10. Wilkes, M. V. (1951). The thermal excitation of atmospheric oscillations. *Proc. Roy. Soc.* (*London*) **A207**, 358–371.
11. Bartels, J. (1927). Über die atmosphärischen Gezeiten. *Abh. Preuss. Meteorol. Inst.* **8**, No. 9.
12. Taylor, G. I. (1929–1930). Waves and tides in the atmosphere. *Proc. Roy. Soc.* (*London*) **A126**, 169–183.
13. Kertz, W. (1951). Theorie der gezeitenartigen Schwingungen als Eigenwertproblem. *Ann. Meteorol.* **4**, Suppl. 1.
14. Kertz, W. (1957). Atmosphärische Gezeiten. *In* "Handbuch der Physik" (S. Flügge, ed.), Vol. 48, pp. 928–981. Springer, Berlin.
15. Pekeris, C. L. (1937). Atmospheric oscillations. *Proc. Roy. Soc.* (*London*) **A158**, 650–671.
16. Weekes, K., and Wilkes, M. V. (1947). Atmospheric oscillations and the resonance theory. *Proc. Roy. Soc.* (*London*) **A192**, 80–99.
17. Wilkes, M. V. (1949). "Oscillations of the Earth's Atmosphere." Cambridge Univ. Press, London and New York.
18. Jacchia, L. G., and Kopal, Z. (1952). Atmospheric oscillations and the temperature profile of the upper atmosphere. *J. Meteorol.* **9**, 13–23.
19. Hann, J. v. (1918). Untersuchungen über die tägliche Oszillation des Barometers. Die dritteltägige Luftdruckschwankung. *Denkschr. Akad. Wiss. Wien Math.-nat. Kl.* **95**, 1–64.
20. Pramanik, S. K. (1926). The six-hourly variations of atmospheric pressure and temperature. *Mem. Roy. Meteorol. Soc.* (*London*) **1**, 35–57.
21. Chapman, S. (1918). The lunar atmospheric tide at Greenwich, 1854–1917. *Quart. J. Roy. Meteorol. Soc.* **44**, 271–280.
22. Bartels, J. (1938). Berechnung der lunaren atmosphärischen Gezeiten aus Terminablesungen am Barometer. *Gerl. Beitr. Geophys.* **54**, 56–75.
23. Bartels, J. (1928). Gezeitenschwingungen der Atmosphäre. *In* "Wien-Harms, Handbuch der Experimentalphysik." (W. Wien, F. Harms, ed.), Vol. 25, Pt. 1, pp. 161–210. Akad. Verlagsgesellschaft, Leipzig.
24. Bartels, J. (1939). Sonnen- und mondentägige Luftdruckschwankungen. *In* "Hann-Süring, Lehrbuch der Meteorologie." (R. Süring, ed.), 5th ed. Vol. 1, pp. 276–306. W. Keller, Leipzig.
25. Chapman, S., Pramanik, S. K., and Topping, J. (1931). The world-wide oscillations of the atmosphere. *Gerl. Beitr. Geophys.* **33**, 246–260.
26. Kertz, W. (1959). Partialwellen in den halb- und vierteltägigen gezeitenartigen Schwingungen der Erdatmosphäre. *Arch. Meteorol. Geophys. Biokl.* **A11**, 48–63.
27. Siebert, M. (1957). A planetary representation of the terdiurnal variations of atmospheric pressure and temperature. *Sci. Rept. No.* **4**, *Project* 429, *N.Y. Univ., Dept. Meteorol. Oceanogr.*
28. Whipple, F. J. W. (1918). A note on the propagation of the semidiurnal pressure wave. *Quart. J. Roy. Meteorol. Soc.* **44**, 20–22.
29. Chapman, S. (1951). Atmospheric tides and oscillations. *In* "Compendium of Meteorology," (T. F. Malone, ed.), pp. 510–530. Am. Meteorol. Soc., Boston, Massachusetts.

30. Siebert, M. (1954). Zur Theorie der thermischen Erregung gezeitenartiger Schwingungen der Erdatmosphäre. *Naturwissenschaften* **41**, 446; (1956). Über die gezeitenartigen Schwingungen der Erdatmosphäre. *Ber. Deut. Wetterd*, **4**, No. 22, 65–71, 87–88.

31. Johnson, D. H. (1955). Tidal oscillations of the lower stratosphere. *Quart. J. Roy. Meteorol. Soc.* **81**, 1–8.

32. Appleton, E. V., and Weekes, K. (1939). On the lunar tides in the upper atmosphere. *Proc. Roy. Soc.* (*London*) **A171**, 171–187.

33. Matsushita, S. (1949). On the semidiurnal lunar tide (M_2) in region F. *J. Geomag. Geoelectricity* (*Japan*) **1**, 17–21; (1953) Lunar tidal variations in the sporadic-E region. *Rept. Ionos. Res. Japan* **7**, 45–52.

34. Bartels, J. (1950). Ebbe und Flut in der Ionosphäre. *Ber. Deut. Wetterd. U.S. Zone* No. 12, 30–33.

35. Duperier, A. (1946). A lunar effect on cosmic rays. *Nature* **157**, 296.

36. (1889). *Jahrb. Centr. Anstalt Meteorol. Erdmagn.* **25**, Jahrgang 1888, Wien.

37. Bartels, J. (1923). Zur Berechnung der täglichen Luftdruckschwankung. *Ann. Hydrog.* **51**, 153–160; (1923). Der lokale Anteil in der täglichen Luftdruckschwankung. *Beitr. Phys. frei. Atmos.* **11**, 51–60.

38. Haurwitz, B., and Craig, R. A. (1952). Atmospheric flow patterns and their representation by spherical surface harmonics. *Geophys. Res. Pap. No.* **14**, AFCRC.

39. Kertz, W. (1956). Die thermische Erregungsquelle der atmosphärischen Gezeiten. *Nachr. Akad. Wiss. Göttingen Math.-phys. Kl.* No. 6, 145–166.

40. Siebert, M. (1956). Analyse des Jahresganges der 1/n tägigen Variationen des Luftdruckes und der Temperatur. *Nachr. Akad. Wiss. Göttingen Math.-phys. Kl.* No. 6, 127–144.

41. Bartels, J. (1954). A table of daily integers, 1902–1952, seasonal, solar, lunar, and geomagnetic. *Sci. Rept. No.* **2**, *AF* 19 (604)–503, *Geophys. Inst. Univ. Alaska.*

42. Hann, J. v. (1886). Bemerkungen zur täglichen Oszillation des Barometers. *Sitzber. Akad. Wiss. Wien Abt. II* **93**, 981–994; (1889). Untersuchungen über die tägliche Oszillation des Barometers. *Denkschr. Akad. Wiss. Wien Math. -nat. Kl.* **55**, 49–121; (1892). *Denkschr. Akad. Wiss. Wien Math.-nat. Kl.* **59**, 297–356; (1898). Weitere Beiträge zu den Grundlagen für eine Theorie der täglichen Oszillation des Barometers. *Sitzber. Akad. Wiss. Wien Abt. IIa* **107**, 63–159; (1918). Die jährliche Periode der halbtägigen Luftdruckschwankung. *Sitzber. Akad. Wiss. Wien Abt. IIa* **127**, 263–365; (1919). Die ganztägige Luftdruckschwankung. *Sitzber. Akad. Wiss. Wien Abt. IIa* **128**, 379–506; and numerous notes in *Meteorol. Z.* **1–38**; (see also [19]).

43. Angot, A. (1887). Étude sur la marche diurne du baromètre. *Ann. Bur. Cen. Météorol. France Mem.*

44. Schmidt, A. (1890). Über die doppelte tägliche Oszillation des Barometers. *Meteorol. Z.* **7**, 182–185.

45. Simpson, G. C. (1918). The twelve-hourly barometer oscillation. *Quart. J. Roy. Meteorol. Soc.* **44**, 1–19.

46. Jaerisch, P. (1907). Zur Theorie der Luftdruckschwankungen auf Grund der hydrodynamischen Gleichungen in sphärischen Koordinaten. *Meteorol. Z.* **24**, 481–498.

47. Haurwitz, B. (1956). The geographical distribution of the solar semidiurnal pressure oscillation. *Meteorol. Pap.* **2**, (5).

48. Stolov, H. L. (1954). Tidal wind fields in the atmosphere. *J. Meteorol.* **12**, 117–140.

49. Haurwitz, B., and Sepúlveda, G. M. (1957). Geographical distribution of the semidiurnal pressure oscillation at different seasons. *J. Meteorol. Soc. Japan 75th*

Anniversary Vol. pp. 149–155; (1957). The geographical distribution and seasonal variation of the semidiurnal pressure oscillation in high latitudes. *Arch. Meteorol. Geophys. Biokl.* **A10**, 29–42.

50. Spar, J. (1952). Characteristics of the semidiurnal pressure waves in the United States. *Bull. Am. Meteorol. Soc.* **33**, 438–441.
51. Schmidt, A. (1919). Zur dritteltägigen Luftdruckschwankung. *Meteorol. Z.* **36**, 29.
52. Sabine, E. (1847). On the lunar atmospheric tide at St. Helena. *Phil. Trans. Roy. Soc. (London)* **137**, 45–50.
53. Elliot, C. M. (1852). On the lunar atmospheric tide at Singapore. *Phil. Trans. Roy. Soc. (London)* **142**, 125–129.
54. Bergsma, P. A. (1871). Lunar atmospheric tide. *Obsns. Magn. Meteorol. Obs. Batavia* **1**, 19–25.
55. Chapman, S., and Miller, J. C. P. (1940). The statistical determination of lunar daily variations in geomagnetic and meteorological elements. *Monthly Not. Roy. Astr. Soc. Geophys. Suppl.* **4**, 649–669.
56. Tschu, K. K. (1949). On the practical determination of lunar and luni-solar daily variations in certain geophysical data. *Australian J. Sci. Res.* **A2**, 1–24.
57. Chapman, S. (1952). The calculation of the probable error of determinations of lunar daily harmonic component variations in geophysical data: A correction. *Australian J. Sci. Res.* **A5**, 218–222.
58. Bartels, J., Chapman, S., and Kertz, W. (1952). Gezeitenartige Schwingungen der Atmosphäre. *In* Landolt-Börnstein "Zahlenwerte und Funktionen aus Physik, Chemie, Astronomie, Geophysik und Technik," (J. Bartels, P. ten Bruggencate, ed.) Vol. 3 (Astronomie und Geophysik), pp. 674–685, Springer, Berlin.
59. Chapman, S., and Westfold, K. C. (1956). A comparison of the annual mean solar and lunar atmospheric tides in barometric pressure, as regards their worldwide distribution of amplitude and phase. *J. Atmos. Terr. Phys.* **8**, 1–23.
60. Siebert, M. (1955). Zur Theorie der freien und erzwungenen gezeitenartigen Schwingungen der Erdatmosphäre. Dissertation, Univ. Göttingen.
61. Haurwitz, B., and Möller, F. (1955). The semidiurnal air-temperature variation and the solar air tide. *Arch. Meteorol. Geophys. Biokl.* **A8**, 332–350.
62. Hough, S. S. (1897). On the application of harmonic analysis to the dynamical theory of tides. *Phil. Trans. Roy. Soc. (London)* **A189**, 201–257; (1898). *Phil. Trans. Roy. Soc. (London)* **A191**, 139–185.
63. Haurwitz, B. (1957). Atmospheric oscillations and meridional temperature gradient. *Beitr. Phys. Atmos.* **30**, 46–54.
64. Siebert, M. (1957). Tidal oscillations in an atmosphere with meridional temperature gradient. *Sci. Rept. No.* **3**, *Project* 429, *N.Y. Univ. Dept. Meteorol. Oceanogr.*
65. Chiu, W. -C. (1953). On the oscillations of the atmosphere. *Arch. Meteorol. Geophys. Biokl.* **A5**, 280–303.
66. Thrane, P. (1951). Some hydrodynamical properties of simple atmospheric oscillations. *Geofys. Publik.* **18**, 1–36.
67. Thrane, P. (1955). The tilt of a 12-hourly atmospheric tidal wave. *Quart. J. Roy. Meteorol. Soc.* **81**, 9–17.
68. Wagner, A. (1932). Der tägliche Luftdruck- und Temperaturgang in der freien Atmosphäre und in Gebirgstälern. *Gerl. Beitr. Geophys.* **37**, 315-344.
69. Stapf, H. (1934). Der tägliche Luftdruck- und Temperaturgang in der freien Atmosphäre in seiner Abhängigkeit von den Jahreszeiten. Dissertation, Univ. Berlin.
70. Bjerkness, J. (1948). Atmospheric tides. *J. Mar. Res.* **7**, 154–162.
71. Haurwitz, B. (1952). Zur Resonanztheorie der halbtägigen Luftdruckschwankung. *Ber. Deut. Wetterd. U.S. Zone* No. 38, 12–16.

72. Haurwitz, B. (1947). Harmonic analysis of the diurnal variations of pressure and temperature aloft in the Eastern Caribbean. *Bull. Am. Meteorol. Soc.* **28,** 319–323; (1958). The semidiurnal pressure oscillation in the stratosphere. *Sci. Rept. No.* **7,** *Project* **429,** *N.Y. Univ., Dept. Meteorol. Oceanogr.*
73. Solberg, H. (1936). Über die freien Schwingungen einer homogenen Flüssigkeitsschicht auf der rotierenden Erde I. *Astrophys. Norveg.* **1,** 237–340.
74. Hylleraas, E. A. (1939). Über die Schwingungen eines stabil geschichteten, durch Meridiane begrenzten Meeres I. *Astrophys. Norveg.* **3,** 139–164.
75. Mariani, F. (1957). Sulla teoria delle maree atmosferiche gravitazionali. *Ann. Geofis.* **10,** 211–220.
76. Chapman, S. (1932). The lunar diurnal variation of atmospheric temperature at Batavia, 1866–1928. *Proc. Roy. Soc.* (*London*) **A137,** 1–24.
77. Sawada, R. (1954–1955). On the role of the thermal excitation in the atmospheric tides. *Geophys. Mag.* **26,** 267–281.
78. White, M. L. (1956). Gravitational and thermal oscillations in the earth's upper atmosphere. *J. Geophys. Res.* **61,** 489–499.
79. Brillouin, M. (1932). Les latitudes critiques. *Compt. rend.* **194,** 801–804.
80. Haurwitz, B. (1937). The oscillations of the atmosphere. *Gerl. Beitr. Geophys.* **51,** 195–233.
81. Solberg, H. (1936). Schwingungen und Wellenbewegungen in einer Atmosphäre mit nach oben abnehmender Temperatur. *Astrophys. Norveg.* **2,** 123–172.
82. Strachey (1888). The eruption of Krakatoa. *Rept. Krakatoa Com. Roy. Soc.*
83. Pekeris, C. L. (1950). Free oscillations of an atmosphere in which the temperature increases linearly with height. *Natl. Adv. Com. Aeronaut. Tech. Notes* **2209.**
84. Pekeris, C. L. (1939). The propagation of a pulse in the atmosphere. *Proc. Roy. Soc.* (*London*) **A171,** 434–449.
85. Whipple, F. J. W. (1930). The Great Siberian Meteor and the waves, seismic and aerial, which it produced. *Quart. J. Roy. Meteorol. Soc.* **56,** 287–303.
86. Pekeris, C. L. (1948). The propagation of a pulse in the atmosphere, Part II. *Phys. Rev.* **73,** 145–154.
87. Observation Division, Central Meteorological Observatory (1954–1955). Preliminary report on atmospheric pressure oscillation due to the H-bomb explosion. *Geophys. Mag.* **26,** 215–223.
88. Doodson, A. T. (1922). The harmonic development of the tide-generating potential. *Proc. Roy. Soc.* (*London*) **A100,** 305–329.
89. Bartels, J. (1957). Gezeitenkräfte. *In* "Handbuch der Physik" (S. Flügge, ed.), Vol. 48, pp. 734–774. Springer, Berlin.
90. Melchior, P. J. (1958). Earth tides. *Advances in Geophys.* **4,** 391–443.
91. Pekeris, C. L. (1951). Effect of the quadratic terms in the differential equations of atmospheric oscillations. *Natl. Adv. Com. Aeronaut. Tech. Notes* **2314.**
92. Sawada, R. (1954). The atmospheric lunar tides. *Meteorol. Pap.* **2,** (3); (1956). The atmospheric lunar tides and the temperature profile in the upper atmosphere. *Geophys. Mag.* **27,** 213–236.
93. Details of this analysis will be published elsewhere.
94. Mügge, R., and Möller, F. (1932). Zur Berechnung von Strahlungsströmen und Temperaturänderungen in Atmosphären von beliebigem Aufbau. *Z. Geophys.* **8,** 53–64.
95. Möller, F. (1957). Strahlung in der unteren Atmosphäre. *In* "Handbuch der Physik" (S. Flügge, ed.), Vol. 48, pp. 155–253. Springer, Berlin.
96. Sváva-Kováts, J. (1938). Verteilung der Luftfeuchte auf der Erde. *Ann. Hydrog.* **66,** 373–378; Penndorf, R. (1952). Wasserdampf. *In* Landolt-Börnstein "Zahlenwerte

und Funktionen aus Physik, Chemie, Astronomie, Geophysik und Technik" (J. Bartels, P. ten Bruggencate, ed.), Vol. 3 (Astronomie und Geophysik), pp. 568–571, Springer, Berlin.

97. Yamamoto, G. (1949). Average vertical distribution of water vapor in the atmosphere. *Sci. Repts. Tôhoku Univ. Fifth Ser.* **1**, 76–79; Flohn, H. (1951). Zur vertikalen Verteilung des Wasserdampfes in der Atmosphäre. *Z. Meteorol.* **5**, 148–152; Möller, F. (1953). Mittlere Dampfdruck- und Dampfdichteabnahme mit der Höhe oder Temperatur. In "Linke's Meteorol. Taschenbuch" (F. Baur, ed.), Leipzig, Akad. Verlagsges. Geest u. Portig, Vol. 2, p. 484.
98. Sen, H. K., and White, M. L. (1955). Thermal and gravitational excitation of atmospheric oscillations. *J. Geophys. Res.* **60**, 483–495.
99. Holmberg, E. R. R. (1952). A suggested explanation of the present value of the velocity of rotation of the earth. *Monthly Not. Roy. Astr. Soc. Geophys. Suppl.* **6**, 325–330.
100. The author is much obliged to Dr. W. Kertz, Geophys. Inst. Göttingen, for verbal communication of this value.
101. Johnson, F. S. (1953). High-altitude diurnal temperature changes due to ozone absorption. *Bull. Am. Meteorol. Soc.* **34**, 106–110.
102. Pressman, J. (1955). Diurnal temperature variations in the middle atmosphere. *Bull. Am. Meteorol. Soc.* **36**, 220–223.
103. Greenhow, J. S., and Neufeld, E. L. (1955). Diurnal and seasonal wind variations in the upper atmosphere. *Phil. Mag.* (7) **46**, 549–562; (1956). The height variation of upper atmospheric winds. *Phil. Mag.* (8) **1**, 1157–1171.
104. Haurwitz B., London, J., Sepúlveda, G. M., and Siebert, M. (1957). Solar activity and atmospheric tides. *J. Geophys. Res.* **62**, 489–491.
105. Thrane, P. (1958). The diurnal and the semidiurnal atmospheric solar tide. *Tellus* **10**, 415–429.
106. Siebert, M. (1959). Die solaren Gezeiten im Barogramm des Juli 1959. *Z. Geophys.* **25**, 109–112.
107. Pekeris, C. L., and Alterman, Z. (1959). A method of solving the nonlinear differential equations of atmospheric tides with applications to an atmosphere of constant temperature. *In* "The atmosphere and the sea in motion." (B. Bolin, ed.) pp. 268–276. The Rockefeller Inst. Press, New York.
108. White, M. L. (1960). Atmospheric tides and ionospheric electrodynamics. *J. Geophys. Res.* **65**, 153–171; (1960). Thermal and gravitational atmospheric oscillations—ionospheric dynamo effects included. *J. Atmos. Terr. Phys.* **17**, 220–245.

Further references can be found in [14] and [29].

GENERALIZED HARMONIC ANALYSIS

J. Van Isacker

Royal Meteorological Institute, Uccle, Belgium

1. Introduction

Time series, and more precisely sequences of quantitative experimental data, are very frequent in meteorology and geophysics. Long series of accurate data are becoming available at an ever growing rate and require refined methods of statistical treatment. This statistical treatment serves essentially to show the internal organization of the time series and the mutual interaction between different elements but may also be used for prediction purposes. These methods have been largely developed during the last years by communication engineers as well as by pure mathematicians. This paper is an attempt to extract from these developments a method best suited for use in geophysical research. As a matter of fact, communication engineers are mainly interested in continuous functions with very fast fluctuations such as potentials or currents which must be automatically controlled by electronic devices which are also continuous [1, 2].

The time scale of geophysical phenomena is so much larger that the sequences of observations are discrete more often than not. A numerical treatment is also more suitable with the punched card equipment now so widely available.

Modern stochastic theories cover not only continuous and discrete cases but much more general ones also [3, 4, 5, 6]. Unfortunately, their mathematical difficulty does not make them very attractive for practical use.

The restriction to discrete sequences permits important simplifications and eliminates some problems of convergence. It seems thus useful to make this assumption from the very beginning in order to obtain an elementary but nevertheless correct account of the subject.

2. Stochastic Sequence

Assume an infinite sequence or time series y_m; $m = -\infty$ to $+\infty$, which may be obtained as the result of an experience or a set of observations. This experience may be reproduced over again giving new results y_m', y_m'' . . . which are generally very different from each other due to some disturbing factors which are not important for the studied phenomenon.

We are thus led to consider not only one time series but an "ensemble" of sequences, called a stochastic time series, associated with a law of probability. This stochastic time series may be characterized, following N. Wiener [1], by a parameter α in such a way that for each value of α, $y_m(\alpha)$ is one possible experimental sequence. It is also necessary to define some probability law for α, for instance, a distribution function $F(\alpha)$. From this, we can deduce the moments of first order, second order, and so on:

$$\bar{y}_m = \int y_m(\alpha) dF(\alpha) \tag{2.1}$$

$$\overline{y_m y_n} = \int y_m(\alpha) y_n(\alpha) dF(\alpha) \tag{2.2}$$

The upper bar designates always a mathematical expectancy. The integral in (2.1) and (2.2) is a Lebègue-Stieltjes integral. Existence of moments of the two first orders is always assumed, but we shall need also moments of fourth order and generally assume the existence of all used moments.

There are many other definitions of stochastic time series; for instance, we may use a set of distribution laws

$$F_k(y_1, \ldots, y_k ; m_1, \ldots m_k) \tag{2.3}$$

The meaning of this function is the probability of the simultaneous events

$$y_{m_1} \leqslant y_1, \quad y_{m_2} \leqslant y_2, \ldots, \quad y_{m_k} \leqslant y_k$$

The laws F_k must verify the following conditions:

(i) F is symmetrical for the couples of variables y_i, m_i.

(ii)

$$F_k(y_1, \ldots, y_k; m_1, \ldots, m_k) = F_{k+1}(y_1, \ldots, y_k, \infty; m_1, \ldots, m_{k+1}) \tag{2.4}$$

Moments may then be obtained by the integrals:

$$\bar{y}_m = \int \alpha dF_1(\alpha; m) \tag{2.5}$$

$$\overline{y_m y_n} = \iint \alpha\beta d^2 F_2(\alpha, \beta; m, n) \tag{2.6}$$

The two definitions are not completely identical, in fact it has been proved [3] that there exists a stochastic process like that of Wiener's having the distribution laws F_k, but this stochastic process is not entirely defined by these laws.

The time series considered hereafter will always be stationary; that is, all probability laws will be invariant for any time shift. For instance, the distribution laws F_k depend only on the differences $m_i - m_j$ and not on the absolute values of m_i. As a consequence, the first-order moment $\overline{y_m}$ will not depend on m and the second-order moment $\overline{y_m y_n}$, called auto-covariant, depends only on the difference $m - n$.

3. Determination of the Auto-Covariant and the Power Spectrum

Taking a stationary stochastic time series of zero mean value, we obtain

$$\overline{y_m} = 0 \tag{3.1}$$

$$C_m \equiv \overline{y_n y_{n+m}} = \frac{1}{2\pi}\int_{-\pi}^{+\pi} f(k) e^{ikm} dk = C_{-m} \tag{3.2}$$

$$f(k) = \sum_{-\infty}^{+\infty} C_m e^{-ikm} = f(-k) \geqslant 0 \tag{3.3}$$

Practical determination of the auto-covariant C_m and the spectrum $f(k)$ presents two problems:

(1) The covariant and the spectrum are determined by stochastic means, which, in principle, may be estimated only by repetition of a great amount of independent experiments. Practically, a small amount and even one experiment only will be available. This last case is particularly frequent in the observational sciences such as geophysics.

We must suppose that the function y_m verifies certain ergodism conditions and especially that

$$\lim{}^* \frac{1}{N} \sum_{n=1}^{N} y_n y_{n+m} = \overline{y_p y_{p+m}} \equiv C_m \tag{3.4}$$

This ergodism condition assumes thus the equivalence between a stochastic mean on a large set of independent experiments using one couple of y's only from each experiment and a Reynolds mean on all the couples $y_n y_{m+n}$ of one experiment only.

*The limit is here defined as a limit in quadratic mean (or limit in the mean as called by N. Wiener); that is,

$$\lim X_N = X \quad \text{means} \quad \lim_{N\to\infty} \overline{(X_N - X)^2} = 0,$$

where X_N is some set of random numbers while X may be random or not.

(2) Any observation may only produce a finite number of values for y. It is thus necessary to be able to estimate the error committed in the limits of a definite value of N in the preceding relation.

If we want to estimate the standard error, we must compute the expression

$$\sigma^2 = \overline{\left(\frac{1}{N}\sum_{n=1}^{N} y_n y_{n+m} - C_m\right)^2} = \frac{1}{N^2}\sum_n \sum_p \overline{y_n y_{n+m} y_p y_{p+m}} - C_m{}^2 \tag{3.5}$$

The knowledge of the error committed in the estimation of C_m requires the knowledge of the mean of an expression of the fourth degree in y.

Besides the length of the necessary calculations for this last estimate, we do not know its precision unless we calculate the mean of the products of the eighth degree, and so on. It is thus essentially necessary to postulate a hypothesis on the nature of the stochastic function y. We will assume here that the stochastic function is Laplacian, i.e.

$$\overline{e^{\Sigma\alpha_m y_m}} = e^{\frac{1}{2}\Sigma\Sigma\alpha_m\alpha_n C_{m-n}} \tag{3.6}$$

as soon as $\sum |\alpha_m|$ exists.

We infer easily that the mean of any monomial of the third degree in y is zero and that

$$\overline{y_m y_n y_p y_q} = C_{m-n}C_{p-q} + C_{m-p}\,C_{n-q} + C_{m-q}C_{n-p} \tag{3.7}$$

The verification that the ergodism condition is fulfilled is:

$$\sigma^2 = \lim \overline{\left(\frac{1}{N}\sum_{1}^{N} y_n y_{n+m} - C_m\right)^2} = \lim \frac{1}{N^2}\sum_n \sum_p (C_{n-p}^2 + C_{n-p-m}C_{n-p+m})$$

$$\leqslant \lim \frac{1}{N}\sum_{-\infty}^{+\infty} (C_p{}^2 + |C_{p-m}C_{p+m}|)$$

If we assume that $\sum C_p{}^2$ is finite, the condition is verified, but this last expression cannot be experimentally determined. This difficulty is very important in practice. Indeed, the asymptotic behavior of the auto-covariance function C_m is determined by the unknown long-range fluctuations of the stochastic function y. We will see that the determination of the power spectrum $f(k)$ does not meet the same difficulties.

4. Practical Determination of the Power Spectrum

Let us consider the expression

$$F(k) = \frac{1}{4\Lambda}\sum_{n=-\Lambda+\frac{1}{2}}^{\Lambda-\frac{1}{2}}\ \sum_{m=-\lambda+\frac{1}{2}}^{\lambda-\frac{1}{2}} e^{-imk}\left(1+\cos\frac{m\pi}{\lambda}\right)\left(1+\cos\frac{n\pi}{\Lambda}\right) y_n y_{n+m} \tag{4.1}$$

Its mean value is

$$\overline{F(k)} = \frac{1}{4\Lambda} \sum_m \sum_n e^{-imk} \left(1 + \cos \frac{m\pi}{\lambda}\right)\left(1 + \cos \frac{n\pi}{\Lambda}\right) C_m$$

The summation on n takes the form

$$\overline{F(k)} = \frac{1}{2} \sum_m e^{-imk} \left(1 + \cos \frac{m\pi}{\lambda}\right) C_m$$

Replacing the auto-covariant C_m by its expression (3.2) in function of the spectrum $f(k)$ yields:

$$\overline{F(k)} = \frac{1}{4\pi} \sum_m \int_{-\pi}^{+\pi} f(l) e^{im(l-k)} \left(1 + \cos \frac{m\pi}{\lambda}\right) dl \tag{4.2}$$

$$= \int_{-\pi}^{+\pi} V(k - l, \lambda) f(l) dl$$

in which V is a weighting function of the form

$$V(r, \lambda) = \frac{\sin \lambda r}{8\pi} \left(\frac{2}{\sin \frac{1}{2} r} - \frac{1}{\sin \frac{1}{2} (r + \pi/\lambda)} - \frac{1}{\sin \frac{1}{2} (r - \pi/\lambda)} \right) \tag{4.3}$$

$$\int_{-\pi}^{+\pi} V(r, \lambda) dr = 1 \tag{4.4}$$

Note that for large values of λ, we have

$$V(r, \lambda) \approx \frac{-\pi \sin r\lambda}{2r(r^2\lambda^2 - \pi^2)} \tag{4.5}$$

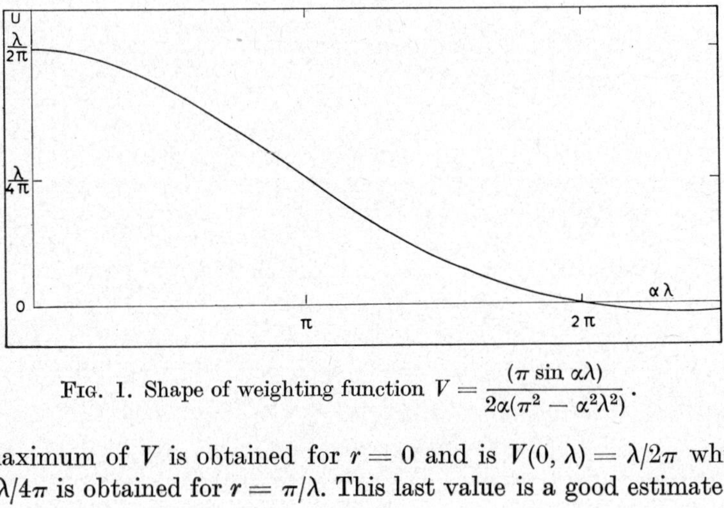

FIG. 1. Shape of weighting function $V = \dfrac{(\pi \sin \alpha\lambda)}{2\alpha(\pi^2 - \alpha^2\lambda^2)}$.

The maximum of V is obtained for $r = 0$ and is $V(0, \lambda) = \lambda/2\pi$ while the value $\lambda/4\pi$ is obtained for $r = \pi/\lambda$. This last value is a good estimate of the sharpness of the spectrum obtained by the formula (4.1), in that way that two points, the abscissas of which differ from $\varDelta r = 2\pi/\lambda$, may be considered as distinct.

The precision of the calculation may be estimated by computing the mean square error

$$\sigma^2 = \overline{F(k)^2} - \overline{F(k)}^2$$

If the stochastic function y_m is Laplacian, we have

$$\sigma^2 = \frac{1}{16\varLambda^2} \sum_m \sum_p \sum_n \sum_q e^{-i(m+p)k} \left(1 + \cos\frac{m\pi}{\lambda}\right)\left(1 + \cos\frac{p\pi}{\lambda}\right) \left(1 + \cos\frac{n\pi}{\varLambda}\right)\left(1 + \cos\frac{q\pi}{\varLambda}\right) \times (C_{n-q}\, C_{n-q+m-p} + C_{n-q+m}\, C_{n-q-p}) \tag{4.6}$$

which easily gives

$$\sigma^2 = \frac{4\pi^2}{\varLambda^2} \iint J(r) f(s)\, V^2(r - s, \varLambda)\, [V(k + s, \lambda)\, V(k - s, \lambda) + V(r - k, \lambda)\, V(s - k, \lambda)] dr ds. \tag{4.7}$$

Notice that $\varLambda$ is always larger than λ. We can thus simplify the preceding expression, putting $r = s$ except in $V^2(r - s, \varLambda)$. We can then integrate in s, and using

$$\int V^2(r - s, \varLambda) ds = \frac{3\varLambda}{8\pi} \tag{4.8}$$

we obtain

$$\sigma^2 = \frac{3\pi}{2\varLambda} \int_{-\pi}^{+\pi} f^2(r)\, [V(k + r, \lambda)\, V(r - k, \lambda) + V^2(r - k, \lambda)] dr$$

but

$$V(k + r, \lambda)\, V(r - k, \lambda) \approx 0 \qquad \text{except for } r = k = 0$$

it follows that

$$\sigma^2 = \frac{3\pi}{2\varLambda} \int f^2(r)\, V^2(r - k, \lambda) dr \approx \frac{9}{16} \times \frac{\lambda}{\varLambda} f^2(k) \qquad \text{for } k \neq 0$$

From this last relation we obtain the result

$$\frac{\sigma}{f(k)} = \frac{3}{4}\sqrt{\frac{\lambda}{\varLambda}} \tag{4.9}$$

Formula (4.1) may be used as it is for computing the spectrum; one needs to know $2(\lambda + \varLambda) + 1$ values of y; the coefficients $(1 + \cos n\pi/\varLambda)$ serve essentially to diminish the end effects due to the artificial truncation of the y

function while the coefficients $(1 + \cos m\pi/\lambda)$ induce a smoothing of the spectrum which reduces the probable error in the calculation.

Given the fact that λ must always be kept very small with respect to Λ the function $(1 + \cos n\pi/\Lambda)$ is not very sensitive to a variation of n of the order of λ; it is then evident that we may write the formula (4.1) in the practically equivalent form:

$$F(k) = \frac{1}{2\Lambda} \sum_n \sum_m \mathrm{e}^{-imk} \left(1 + \cos \frac{mk}{\lambda}\right) z_n \, z_{n+m} \tag{4.10}$$

with

$$z_n = \cos \frac{n\pi}{2\Lambda} y_n$$

This transformation reduces considerably the length of the calculations.

Table I gives for the minimum amount N of observations the values λ, Λ as well as the number ν of distinct points between 0 and π of the power spectrum as function of precision.

TABLE I

$\sigma/f(k)$	λ	Λ	N	ν
1%	2.5	14,062.5	28,131	1
5%	2.5	562.5	1,126	1
	20.5	4,612.5	9,267	10
10%	2.5	140.5	284	1
	20.5	1,153.5	2,349	10
	200.5	11,278.5	22,959	100

We shall recall here that the calculation of the correlogram is always delicate when one has no assurance of absence of long-range fluctuation in y; otherwise, the computed autocovariances will systematically depend on the number N of observations. Particularly $\sum_1^N y_n^2/N$ is a monotonically increasing function of N.

On the contrary, the error on the power spectrum $f(k)$ may be estimated on a reasonable basis. This error has a systematic part which reduces to a smoothing of the curve and to an accidental error, easily estimated.

Figures 2 and 3 give an example of a power spectrum computed following this method, using two years of twelve hourly observations of atmospheric pressure at Uccle.

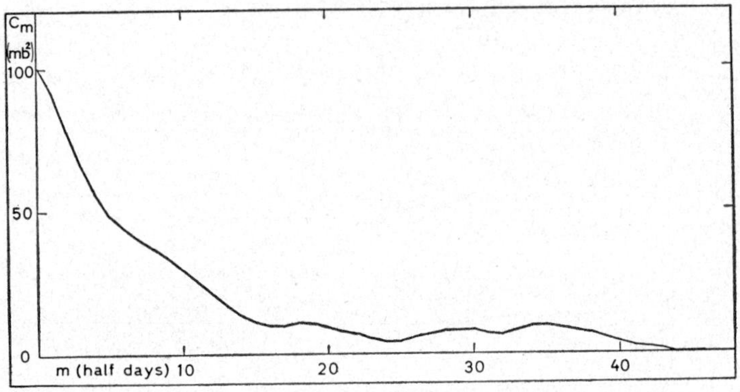

FIG. 2. Autocovariant C_m of atmospheric pressure fluctuations at Uccle during the years 1952–1953.

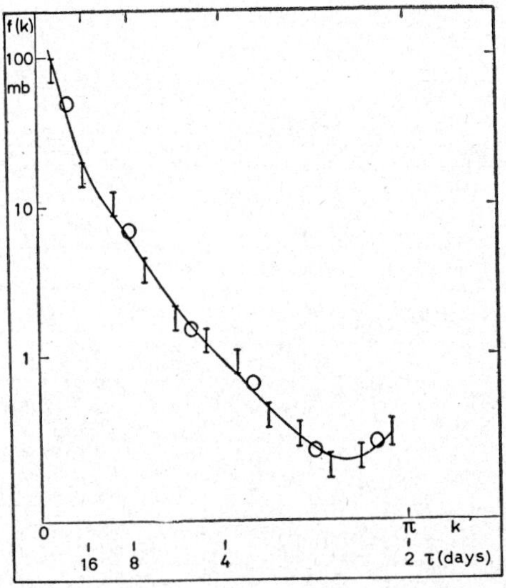

FIG. 3. Power spectrum (log . scale) of pressure fluctuations at Uccle (1952–1953). Computed with $\Lambda = 746.5$, $\lambda = 29.5$ (vertical strips of length equal to twice the standard error σ) and $\lambda = 14.5$ (circles of radius σ).

5. Covariance and Co-Spectrum of two Stochastic Functions

If we have two series of numbers y_n and z_n of zero mean value, we may compute a covariance

$$\overline{yz} = \lim_{N\to\infty} \frac{1}{N} \sum_1^N y_n z_n \tag{5.1}$$

or a correlation coefficient

$$\rho = \overline{yz}/\sqrt{\overline{y^2} \, . \, \overline{z^2}} \tag{5.2}$$

The tables of Pierson give the levels of significance of this coefficient when two assumptions are verified: (i) the distribution function of the couple of variables y, z is Laplacian. (ii) the y_n (resp. z_n) are statistically independent. It is evident that this last condition is not the case when y_n and z_n are auto-correlated time series.

In this circumstance the tables of Pierson lose all value and the correlation coefficients become unreliable. Especially, they will systematically depend on the number N of observations. This problem is of course parallel to the determination of the auto-covariant, and we propose to present the possibility of replacing the coefficient of covariance by a co-spectrum, the statistical behavior of which should be more satisfactory and should yield information at least equivalent to that of the covariance.

From the following definitions

$$\begin{aligned} C_m &\equiv \overline{y_n y_{n+m}} = \frac{1}{2\pi}\int_{-\pi}^{+\pi} f(k) e^{ikm}\, dk \\ D_m &\equiv \overline{z_n z_{n+m}} = \frac{1}{2\pi}\int_{-\pi}^{+\pi} g(k) e^{ikm}\, dk \\ \Gamma_m &\equiv \overline{y_n z_{n+m}} = \frac{1}{2\pi}\int_{-\pi}^{+\pi} \phi(k) e^{ikm}\, dk \end{aligned} \tag{5.3}$$

we obtain easily

$$\Phi(k) = \sum\sum \frac{1}{4\Lambda} e^{ikm}\left(1+\cos\frac{m\pi}{\lambda}\right)\left(1+\cos\frac{n\pi}{\Lambda}\right) y_n\, z_{n+m} \tag{5.4}$$

$$\overline{\Phi(k)} = \int_{-\pi}^{+\pi} V(k-l, \lambda)\phi(l)\, dl$$

with

$$\begin{aligned} \sigma^2 &= \overline{|\,\Phi(k)\,|^2} - |\,\overline{\Phi(k)}\,|^2 \\ &= \frac{1}{16\Lambda^2}\sum\sum\sum\sum e^{-ik(m-p)}\left(1+\cos\frac{m\pi}{\lambda}\right)\left(1+\cos\frac{p\pi}{\lambda}\right)\left(1+\cos\frac{n\pi}{\Lambda}\right) \\ &\qquad \left(1+\cos\frac{q\pi}{\Lambda}\right)\times(C_{n-q}\, D_{n-q+m-p} + \Gamma_{n-p-q}\, \Gamma_{q-n-m}) \end{aligned}$$

The introduction of spectrums and co-spectrums results in

$$\sigma^2 = \sigma_{\mathrm{I}}^2 + \sigma_{\mathrm{II}}^2$$

$$\sigma_{\mathrm{I}}^2 = \frac{1}{64\Lambda^2\pi^2}\iint \sum\sum\sum\sum \mathrm{e}^{-ik(m-p)+ir(n-q)+is(n-q+m-p)}$$

$$\times\left(1+\cos\frac{m\pi}{\lambda}\right)\left(1+\cos\frac{p\pi}{\lambda}\right)\left(1+\cos\frac{n\pi}{\Lambda}\right)\left(1+\cos\frac{q\pi}{\Lambda}\right)f(r)\,g(s)\,drds$$

$$\approx \frac{9\lambda}{16\Lambda} f(k)\,g(k).$$

We can also obtain

$$\sigma_{\mathrm{II}}^2 = \frac{9\lambda}{16\Lambda}\,|\,\phi(k)\,|^2$$

Finally,

$$\sigma \approx \frac{3}{4}\sqrt{\frac{\lambda}{\Lambda}}\sqrt{f(k)\,g(k) + |\,\phi(k)\,|^2} \tag{5.5}$$

results.

$\Phi(k)$ defined by (5.4) is thus a convenient estimation of the cospectrum $\phi(k)$ with a standard error determined by (5.5).

6. Generalized Harmonic Analysis

As a further simplification we assume that $\sum_{-\infty}^{+\infty} |\,C_m\,|$ is finite; we can easily deduce that the power spectrum $f(k)$ is uniformly continuous in the domain $-\pi$ to $+\pi$. This process will allow us, in the future, to reverse summations and integrations.

Under such conditions the Fourier transformation of the auto-covariant C_m presents no difficulty. However, it is quite different with the transformation of the time series y_m. It is nonetheless possible to express y_m in the form of a Stieltjès integral

$$y_m = \frac{1}{2\pi}\int_{-\pi}^{+\pi} \mathrm{e}^{ikm} d\rho(k) \tag{6.1}$$

Here, $\rho(k)$ is a complex non-stationary stochastic function, and the integral has to be considered as the limit in mean square

$$y_m = \lim_{\Delta k\to 0} \frac{1}{2\pi}\sum_{\Delta k} \mathrm{e}^{ikm}\Delta\rho(k) \tag{6.2}$$

$$= \lim_{N\to\infty}\frac{1}{2\pi}\sum_{n=-N}^{N-1} \mathrm{e}^{i(kn\pi/N)}\left[\rho\left(\frac{[n+1]\pi}{N}\right) - \rho\left(\frac{n\pi}{N}\right)\right]$$

We must first determine the reality of the function $\rho(k)$ and then the validity of the formula (6.1).
Let us put down

$$\rho_N(k) = \int_0^k d\xi \sum_{-N}^{+N} y_m e^{-i\xi m} \tag{6.3}$$

and

$$\rho(k) = \lim_{N\to\infty} \rho_N(k) = -\rho^*(-k) \tag{6.4}$$

As proof of the existence of this limit we refer to the stochastic equivalent of the criterion of Cauchy, i.e., we demonstrate that

$$\lim_{\substack{N\to\infty \\ M\to\infty}} \overline{|\rho_N(k) - \rho_M(k)|^2} = 0 \tag{6.5}$$

It is then successively found

$$\lim_{M,N\to\infty} \overline{\rho_M(k)\rho_N^*(l)} = \lim \int_0^k d\xi \int_0^l d\eta \sum_{m=-M}^{M} \sum_{n=-N}^{N} C_{m-n} e^{-i\xi m + i\eta n}$$

$$= \lim \frac{1}{2\pi} \int_0^k d\xi \int_0^l d\eta \int_{-\pi}^{+\pi} dr\, f(r) \sum \sum e^{-i(\xi-r)m + i(\eta-r)n}$$

$$= \lim \frac{1}{2\pi} \int_{-\pi}^{+\pi} dr\, f(r) \int_0^k d\xi \int_0^l d\eta \frac{\sin(\xi - r)(M+\frac{1}{2})}{\sin\frac{1}{2}(\xi - r)} \cdot \frac{\sin(\eta - r)(N+\frac{1}{2})}{\sin\frac{1}{2}(\eta - r)}$$

We can readily determine the Dirichlet integrals. The integration with respect to ζ and η gives, after approaching the limit:

$$\lim_{M,N\to\infty} \overline{\rho_M(k)\,\rho_N^*(l)} = 0 \qquad \text{if } k\,.\,l \leqslant 0 \tag{6.6}$$

$$= 2\pi \int_0^Q f(r)dr \qquad \text{if } k\,.\,l \geqslant 0$$

where Q has the value of k or l, whichever is smaller in modulus.

Introducing this result in (6.5), we obtain

$$\lim_{N,M\to\infty} \overline{|\rho_N(k) - \rho_M(k)|^2} = \lim_{N,M\to 0} \{\overline{|\rho_N(k)|^2} + \overline{|\rho_M(k)|^2} - \overline{\rho_N(k)\rho_M^*(k)} - \overline{\rho_N^*(k)\rho_M(k)}\} = 0$$

Thus, $\rho(k)$ is real and (6.6) results in

$$\overline{\rho(k)\rho^*(l)} = 2\pi \int_0^Q f(r)dr \tag{6.7}$$

further,

$$\overline{\rho(k)y_m} = \lim_{N\to\infty} \frac{1}{2\pi} \int_0^k d\xi \int_{-\pi}^{+\pi} dr\, f(r) \sum_{-N}^{+N} e^{in(r-\xi)-imr}$$

$$(6.8) \qquad = \lim_{N\to\infty} \frac{1}{2\pi} \int_0^k d\xi \int_{-\pi}^{+\pi} dr\, f(r) \frac{\sin (r-\xi)(N+\frac{1}{2})}{\sin \frac{1}{2}(r-\xi)} e^{-imr}$$

$$= \int_0^k f(r) e^{-imr}\, dr$$

From this we deduce

$$(6.9) \qquad \Delta\rho(k)\Delta\rho^*(l) = 2\pi \int_{\Delta s} f(r)\, dr$$

where Δs is the intersection of Δk and Δl, and also

$$(6.10) \qquad \overline{\Delta\rho(k) y_m} = \int_{\Delta k} f(r) e^{-imr} dr$$

These last two results allow us to demonstrate the transformation formula (6.1):

$$\overline{\left| y_m - \frac{1}{2\pi} \int_{-\pi}^{+\pi} e^{ikm} d\rho(k) \right|^2} = \lim_{\Delta k\to 0} \overline{\left| y_m - \frac{1}{2\pi} \sum_{\Delta k} e^{ikm} \Delta\rho(k) \right|^2}$$

$$= \lim \left\{ C_0 - \frac{1}{2\pi} \sum e^{ikm}\, \overline{y_m \Delta\rho(k)} - \frac{1}{2\pi} \sum e^{-ikm}\, \overline{y_m \Delta\rho^*(k)} \right.$$

$$\left. + \frac{1}{4\pi^2} \sum\sum e^{i(k-l)m}\, \overline{\Delta\rho(k)\Delta\rho^*(l)} \right\}$$

$$= C_0 - \frac{1}{2\pi} \int_{-\pi}^{+\pi} f(r) dr = 0$$

It needs to be pointed out that if $\rho_N(k)$ has a limit this is not the case with its derivative

$$(6.11) \qquad \frac{\partial \rho_N(k)}{\partial k} = \sum_{-N}^{+N} y_m e^{-ikm}$$

The generalized harmonic analysis permits the decomposition of the time series into components of different frequencies:

$$y_m(k\Delta) = \frac{1}{2\pi}\int_{k-\Delta}^{k+\Delta} e^{i\xi m} d\rho(\xi) + \frac{1}{2\pi}\int_{-k-\Delta}^{-k+\Delta} e^{i\xi m} d\rho(\xi) \tag{6.12}$$

$$= \frac{1}{2\pi}\int_{k-\Delta}^{k+\Delta} [e^{i\xi m} d\rho(\xi) + e^{-i\xi m} d\rho^*(\xi)]$$

where k and Δ are positive.

Therefrom we obtain using (6.9)

$$\overline{y_m{}^2(k, \Delta)} = \frac{1}{\pi}\int_{k-\Delta}^{k+\Delta} f(\xi) d\xi \tag{6.13}$$

$$\overline{y_m(k, \Delta) y_n(k', \Delta')} = 0 \quad \text{if } k \neq k' \text{ and } \Delta + \Delta' < | k - k' | \tag{6.14}$$

This last relation proves that the stochastic series $y_m(k, \Delta)$ and $y_n(k', \Delta')$ are noncorrelated, but we may not deduce therefrom that they are independent, unless they are Laplacian.

Considering two time series y_m and z_m, we have

$$\overline{y_m(k, \Delta) z_m(k, \Delta)} = \frac{1}{2\pi}\int_{k-\Delta}^{k+\Delta} [\phi(\xi) + \phi^*(\xi)] d\xi \tag{6.15}$$

and more generally

$$\overline{y_m(k, \Delta) z_n(k, \Delta)} = \frac{1}{2\pi}\int_{k-\Delta}^{k+\Delta} [\phi(\xi) e^{i\xi(m-n)} + \phi^*(\xi) e^{-i\xi(m-n)}] d\xi \tag{6.16}$$

These two last formulas allow us to interpret

$$\{[\phi(k) + \phi^*(k)]/2\pi\} dk$$

as a partial covariance of the series y and z, in the spectral domain dk.

If we set

$$\phi(k) = | \phi(k) | e^{ik\alpha}$$

we note that a shifting of the series z from α with respect to the series y will result for the partial covariance in the maximum value $| \phi(k) |$. This operation is, however, only possible when α is an integer. Then, $| \phi(k) |$ may be considered as a maximal covariance and $k\alpha$ as a difference in phase.

7. Filters

An "R filter" [3] is a linear operation transforming a stationary time series into another stationary time series

$$z_m = R(y_m) = \sum_{-\infty}^{+\infty} R_p y_{m+p} \tag{7.1}$$

with

$$\sum | R_p | < C < \infty$$

The series of the R_p, being absolutely convergent, has a uniformly continuous transform:

$$R_p = \frac{1}{2\pi} \int_{-\pi}^{+\pi} G(k) e^{ikp} dk \tag{7.2}$$

The function $G(k)$ is called "filter gain."

$$G(k) = \sum_{-\infty}^{+\infty} R_p e^{-ikp} \tag{7.3}$$

The R filters have several interesting properties given as follows:

(i) The sum of two R filters is an R filter. Indeed

$$R_1(y_m) + R_2(y_m) = \sum (R_{1,p} + R_{2,p}) y_{m+p} = R(y_m) \tag{7.4}$$

therefrom

$$\sum | R_p | \leqslant \sum | R_{1,p} | + \sum | R_{2,p} | < C_1 + C_2 < \infty$$

and

$$G(k) = G_1(k) + G_2(k) \tag{7.5}$$

(ii) The product of two R filters is an R filter. Indeed

$$R_1(R_2(y_m)) = \sum R_{1,p} \sum R_{2,q} y_{m+p+q} = \sum_r (\sum_p R_{1,p} R_{2,r-p}) y_{m+r} = R(y_m) \tag{7.6}$$

therefrom

$$\sum | R_p | = \sum_p | \sum_q R_{1,q} R_{2,p-q} | \leqslant \sum_p \sum_q | R_{1,q} || R_{2,p-q} | < C_1 C_2 < \infty$$

and

$$G(k) = \sum \sum e^{-ikm} R_{1,n} R_{2,m-n} = \sum \sum e^{-ikn} R_{1,n} \,.\, e^{-ikp} R_{2,p} = G_1(k) G_2(k) \tag{7.7}$$

(iii)

$$R(y_m) = \frac{1}{2\pi} \int_{-\pi}^{+\pi} e^{ikm} G(k) d\rho(k)$$

the integral being considered as a limit in mean square.

We have indeed:

$$\overline{| \sum R_p y_{m+p} - \frac{1}{2\pi} \int e^{ikm} G(k) d\rho(k) |^2} \tag{7.8}$$

$$= \frac{1}{4\pi^2} \overline{|\sum R_p \int e^{ik(m+p)} d\rho(k) - \int e^{ikm} G(k) d\rho(k) |^2}$$

$$= \frac{1}{2\pi} \sum\sum R_p R_q \int e^{ik(p-q)} f(k) dk + \frac{1}{2\pi} \int |G(k)|^2 f(k) dk$$

$$- \frac{1}{2\pi} \sum R_p \int e^{ikp} G^*(k) f(k) dk - \frac{1}{2\pi} \sum R_p \int e^{-ikp} G(k) f(k) dk$$

$$= 0$$

because the four terms are equal in modulus according to the definition of $G(k)$.

The most frequently used filters are the bandpass filters. They are defined by their gain:

$$G(k) = 1 \text{ for } k_1 > |k| > k_2 \geqslant 0 \qquad (7.9)$$
$$= 0 \text{ otherwise}$$

We see immediately that they are not R filters because $G(k)$ is not continuous. On the other hand, we have

$$R_p = \frac{1}{\pi} \int_{k_2}^{k_1} \cos kp dk = \frac{\sin k_1 p - \sin k_2 p}{\pi p} \qquad (7.10)$$

and

$$\sum |R_p| = \infty$$

Although, this bandpass filter may be obtained as limit of R filters, particularly as limit of the series of filters

$$R_N(y_m) = \sum_{-N}^{+N} \frac{\sin k_1 p - \sin k_2 p}{\pi p} y_{m+p} \qquad (7.11)$$

One can easily demonstrate that the sequence of stochastic series $R_N |y_m|$ converges in mean square toward $y_m(k, \Delta)$, where $k = (k_1 + k_2)/2$ and $\Delta = (k_1 - k_2)/2$.

8. Practical Construction of an Optimal Filter

It has been shown that a bandpass filter may be obtained as limit of a series of finite extension filters R_N [7]. These filters are the only ones of practical usefulness and we will now construct finite filters approximating a perfect filter for a given extension. For estimating the quality of a filter we choose the following criterion:

Suppose R_N a filter of finite extension:

$$R_N(y_m) = \sum_{-N}^{+N} R_{N,p} y_{m+p} \qquad (8.1)$$

with

$$R_{N,p} = R_{N,-p} = R^*_{N,p}$$

and $y_m(k, \Delta)$, the component of the time series y_m corresponding to the band $k - \Delta$, $k + \Delta$. We will define the best filter R_N for the time series y_m as the one which minimizes the mean square deviation:

$$\sigma^2 = \overline{| R_N(y_m) - y_m(k, \Delta) |^2} \tag{8.2}$$

The introduction of the power spectrum $f(k)$ and the gain $G_N(k)$ permits putting (8.2) in the form

$$\sigma^2 = \frac{1}{4\pi^2} \overline{\left| \int_{-\pi}^{+\pi} G_N(\zeta) e^{i\zeta m} d\rho(\zeta) - \int_{k-\Delta}^{k+\Delta} e^{i\zeta m} d\rho(\zeta) - \int_{-k-\Delta}^{-k+\Delta} e^{i\zeta m} d\rho(\zeta) \right|^2}$$

$$= \frac{1}{2\pi} \left\{ 2 \int_0^{\pi} G_N{}^2(\zeta)(f\zeta) d\zeta - 4 \int_{k-\Delta}^{k+\Delta} G_N(\zeta) f(\zeta) d\zeta + 2 \int_{k-\Delta}^{k+\Delta} f(\zeta) d\zeta \right\}$$

$$= \frac{1}{\pi} \int_0^{\pi} \left[G_N(\zeta) - D(\zeta) \right]^2 f(\zeta) d\zeta \tag{8.3}$$

where

$$D(\xi) = 1 \text{ for } | \xi - k | < \Delta$$
$$= 0 \text{ for } | \xi - k | > \Delta$$

Notice that

$$G_N(k) = \sum_{-N}^{+N} R_{N,p}\, e^{ikp}$$
$$= \sum_{-N}^{+N} R_{N,p} \cos kp$$
$$= \sum_{-N}^{+N} R_{N,p} T_p (\cos k)$$
$$= \sum_{0}^{N} S_p \cos^p k \tag{8.4}$$

Here, T_p is the Tchebycheff polynomial of degree p. The problem thus reduces to minimizing the expression

$$\sigma^2 = \frac{1}{\pi} \int_0^{\pi} [G_N(\zeta) - D(\zeta)]^2 f(\zeta) d\zeta$$

by a polynomial of the Nth degree in $\cos \zeta$. The solution is readily obtained by changing the variables

$$\eta = \cos \zeta \qquad E(\eta) = D(\zeta) \qquad f(\zeta) = \sqrt{1 - \eta^2}\, g(\eta)$$

$$\sigma^2 = \frac{1}{\pi} \int_{-1}^{+1} [S_p \eta^p - E(\eta)]^2 g(\eta) d\eta \tag{8.5}$$

The minimum deviation will be obtained for the values of S_p, verifying the system of equations:

$$\sum_{p=0}^{N} S_p \int_{-1}^{+1} \eta^{p+q} g(\eta) d\eta = \int_{-1}^{+1} \eta^q E(\eta) g(\eta) d\eta \tag{8.6}$$

for p and $q = 0.1 \ldots N$.

The solution of this equation system will be made easier by the previous construction of the $N + 1$ polynomials, orthogonalized by the relation

$$\int_{-1}^{+1} \mathscr{P}_m(\eta) \mathscr{P}_n(\eta) g(\eta) d\eta = \delta_{m,\,n} \tag{8.7}$$

We have then successively

$$E(\eta) = \sum_{0}^{\infty} e_m \mathscr{P}_m(\eta) \qquad e_m = \int_{-1}^{+1} E(\eta) \mathscr{P}_m(\eta) g(\eta) d\eta \tag{8.8}$$

$$G_N(\xi) = \sum_{0}^{N} e_m \mathscr{P}_m(\eta) \tag{8.9}$$

$$\sigma^2 = \frac{1}{\pi} \sum_{N+1}^{\infty} e_m{}^2 \tag{8.10}$$

Especially for $f(\zeta) = 1$, the polynomials $\mathscr{P}_m$ are proportional to the Tcheby-cheff polynomials.

These formulas are easily demonstrated when introducing in (8.5) the following expansion

$$G_N(\zeta) = \sum_{0}^{N} S_p \eta^p = \sum_{0}^{N} \gamma_p \mathscr{P}_p(\eta)$$

we have

$$\sigma^2 = \frac{1}{\pi} \int_{-1}^{+1} \left(\sum_{0}^{N} \gamma_m \mathscr{P}_m - \sum_{0}^{\infty} e_m \mathscr{P}_m \right)^2 g(\eta) d\eta$$

$$= \frac{1}{\pi} \left[\sum_{0}^{N} (\gamma_m - e_m)^2 + \sum_{N+1}^{\infty} e_m{}^2 \right]$$

the minimum of which being, of course, obtained for $\gamma_m = e_m$.

As an example, Fig. 4 permits the comparison of the bandpass filter gain ($k = \Delta = \pi/4$) to the optimum filter gain for $f(k) = 1$.

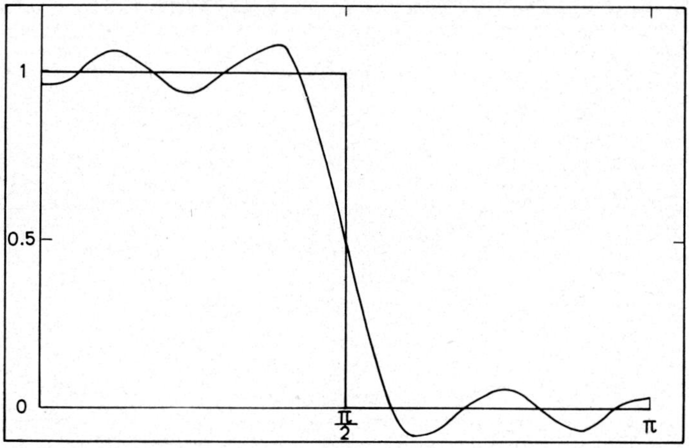

FIG. 4. Optimum filter gain ($N = 8, f(k) = 1$) compared with the perfect low-pass filter gain ($k_1 = \pi/2$, $k_2 = 0$).

Notice that the factorization of the filter gain $G(\zeta)$ as a product of first and second degree polynomials in $\eta = \cos\zeta$ transforms the filter itself in a product of very simple ones.

9. Statistical Prevision by the Method of N. Wiener

The spectral analysis of stochastic functions is not only useful for the understanding of physical phenomena, but also has a direct practical use. It can particularly provide a purely statistical forecast. However, one should not consider this process as the most desirable; on the contrary it should be used only when classical methods are not available, either for lack of initial data or because the phenomenon is not easily reducible to a strictly causal process.

When a stochastic function is known in a semi-infinite domain, continuous ($t = -\infty$ to t_0) or discrete ($m = -\infty$ to m_0) one can look for a prevision formula, tying functionally $y(\tau)$, ($\tau > t_0$) to $y(t)$ or $y_\mu(\mu > m_0)$ to y_m. The error is often estimated by the root mean square deviation. Thus the determination of a function (in a given class) which minimizes that standard deviation becomes important.

Let us consider in particular the class of linear functions:

$$\tilde{y}_\mu = \sum_{n=0}^{\infty} K_n y_{m_0-n} \tag{9.1}$$

The root mean square deviation which is to be minimized will be

$$(9.2) \qquad \sigma^2 = (\tilde{y}_\mu - y_\mu)^2 = C_0 - 2 \sum_0^\infty K_n C_{\mu - m_0 + n} + \sum_0^\infty\sum K_m K_p C_{n-p}$$

It will be minimized for

$$(9.3) \qquad C_{\mu - m_0 + n} = \sum_0^\infty K_p C_{n-p} \qquad (\text{for } n \geqslant 0)$$

The minimum error has then the value of

$$(9.4) \qquad \sigma^2 = C_0 - \sum_0^\infty K_n C_{\mu - m_0 + n}$$

The solution of the equation (9.3) can be obtained in different ways; the most elegant seems to be that one of Wiener. This is detailed below. The Wiener method relies on a factorization of the auto-correlation function, which permits the solution of the equation by a Fourier transformation. Notice first that equation (9.3) cannot be solved directly by a Fourier transformation. Indeed, when putting $\mu - m_0 = a$ and multiplying the two terms of (9.3) by e^{ink}, we obtain

$$\sum_{n=0}^\infty C_{a+n}\, e^{ink} = \sum_{p=0}^\infty \sum_{n=0}^\infty K_p\, e^{ipk} C_{n-p} e^{i(n-p)k}$$

$$= \sum_{p=0}^\infty K_p e^{ipk} \sum_{m=-p}^\infty C_m e^{imk}$$

This means that the summations on m and p may not be separated. The factorization of the auto-covariant C_m consists in searching two series $C_{I,m}$ and $C_{E,m}$ verifying simultaneously

$$C_{I,m} = 0 \text{ for } m < 0$$

$$(9.5) \qquad C_{E,m} = 0 \text{ for } m > 0$$

$$C_m = \sum_{l=0}^\infty C_{E,-l}\, C_{I,l+m}$$

We assume now that the series C_I and C_E are known so that equation (9.3) may be deduced from the relation

$$(9.6) \qquad C_{I,a+m} = \sum_0^m K_p\, C_{I,m-p} \text{ for } m \geqslant 0$$

Multiplying the preceding relations by $C_{E,n-m}$ and making the summations for $m \geqslant n \geqslant 0$

$$\sum_{m=n}^\infty C_{E,n-m}\, C_{I,a+m} = \sum_{p=0}^m K_p \sum_{m=n}^\infty C_{E,n-m}\, C_{I,m-p}$$

gives

$$C_{a+n} = \sum_{p=0}^\infty K_p \sum_{q=0}^\infty C_{E,-q}\, C_{I,q+n-p}$$

$$= \sum_{p=0}^{\infty} K_p \, C_{n-p}$$

Finally, equation (9.6) may be solved by a Fourier transformation. Let us consider the complex variable $\zeta = e^{ik}$. The Fourier transformation (3.3) may be written in the equivalent form:

$$f(k) = j(\zeta) \tag{9.7}$$

$$j(\zeta) = \sum_{-\infty}^{+\infty} C_m \, \zeta^{-m} \tag{9.8}$$

$$C_m = \frac{1}{2\pi i} \oint j(\zeta) \, \zeta^{-m} \, d\zeta \tag{9.9}$$

the integral being taken in the positive sense along the unit circle. We notice immediately that the Fourier transform $k(\zeta)$ of a series, such as K_m, of which the elements with m negative are zero, is analytical and has its singularities inside the unit circle and reciprocally. It will be the same for $j_I(\zeta)$ transform of C_I. In contrast the transform of C_E will have its singularities outside the unit circle.

Let us consider now equation (9.6). Multiplying by ζ^{-m} and summing on m, we obtain

$$\begin{aligned} \sum_{m=0}^{\infty} C_{I,\,a+m} \, \zeta^{-m} &= \sum_{p=0}^{m} K_p \, \zeta^{-p} \sum_{m=0}^{\infty} \zeta^{p-m} \, C_{I,\,m-p} \\ &= \sum_{p=0}^{\infty} K_p \, \zeta^{-p} \sum_{n=0}^{\infty} \zeta^{-n} \, C_{I,\,n} \\ &= k(\zeta) \, j_I(\zeta) \end{aligned} \tag{9.10}$$

and therefrom

$$k(\zeta) = \frac{\sum_{m=0}^{\infty} C_{I,\,a+m} \, \zeta^{-m}}{j_I(\zeta)} \tag{9.11}$$

Notice though that $k(\zeta)$ must have its singularities inside the unit circle; it is thus necessary that $j_I(\zeta)$ should not have a zero either outside or on the circle. We will see that it happens that way. Equation (9.11) gives a solution of the problem; we still can define the precision of this solution by computing the mean square error. We have

$$\begin{aligned} \sigma^2 &= C_0 - \sum_0^{\infty} K_m \, C_{a+m} \\ &= \sum_0^{\infty} C_{E,-l} \, C_{I,\,l} - \sum_{m=0}^{\infty} K_m \sum_{l=0}^{\infty} C_{E,-l} \, C_{I,\,l-a-m} \\ &= \sum_0^{\infty} C_{E,-l} \, C_{I,\,l} - \sum_{l=a}^{\infty} \sum_{m=0}^{l-a} K_m \, C_{E,-l} \, C_{I,\,l-a-m} \end{aligned}$$

because $(l - a - m)$ must be $\geqslant 0$; using the relation (9.6) it gives

$$\sigma^2 = \sum_0^\infty C_{E,-l}\, C_{I,l} - \sum_{l=0}^{\infty} C_{E,-l}\, C_{I,l}$$

and finally

$$\sigma^2 = \sum_0^{a-1} C_{E,-l}\, C_{I,l} \tag{9.12}$$

There is still a need to demonstrate the possibility of factorizing C_m. We have

$$C_m = \sum_0^\infty C_{E,-l}\, C_{I,l+m}$$

Multiplying by ζ^{-m} and summing for $m = -\infty$ to $+\infty$, we find

$$j(\zeta) = \sum_{l=0}^{\infty} \sum_{m=-\infty}^{+\infty} C_{E,-l}\, \zeta^l\, C_{I,l+m}\, \zeta^{-l-m} = j_E(\zeta)\, j_I(\zeta) \tag{9.13}$$

We must express an analytical function, regular, real, and positive on the unit circle as the product of two regular analytical functions, respectively, in and outside the circle. The problem is easily reduced to the one of Dirichlet.

Let us consider the logarithms

$$\log j(\zeta) = \log j_E(\zeta) + \log j_I(\zeta) \tag{9.14}$$

$j(\zeta)$ being real and positive on the circle, the same is true for $\sqrt{j(\zeta)}$ and we formulate

$$j_I(\zeta) = j_E(\zeta) = \sqrt{j(\zeta)} \text{ for } |\zeta| = 1 \tag{9.15}$$

The real part of log $j_E(\zeta)$ will be an harmonic function in the unit circle, taking on the latter the value $\frac{1}{2}\log j(\zeta)$. Also, $\log j_I(\zeta)$ will be the solution of the corresponding outside problem. The solution is rather easily expressed in the analytical form:

$$\log j_E(\zeta) = \frac{1}{2\pi i} \oint \tfrac{1}{2} \log j(\eta)\, \frac{\eta + \zeta}{\eta - \zeta}\, \frac{d\eta}{\eta} \text{ for } |\zeta| \leqslant 1 \tag{9.16}$$

$$\log j_I(\zeta) = \frac{1}{2\pi i} \oint \tfrac{1}{2} \log j(\eta)\, \frac{\zeta + \eta}{\zeta - \eta}\, \frac{d\eta}{\eta} \text{ for } |\zeta| \geqslant 1 \tag{9.17}$$

If we develop $\log j_E$ in series of Laurent

$$\log j(\zeta) = \sum_{-\infty}^{+\infty} s_m\, \zeta^{-m}$$

we obtain

$$\log j_E(\zeta) = \tfrac{1}{2}s_0 + \sum_1^\infty s_{-m}\, \zeta^m; \quad \log j_I(\zeta) = \tfrac{1}{2}s_0 + \sum_1^\infty s_m\, \zeta^{-m} \tag{9.18}$$

and therefore j_I has no root outside or on the unit circle.

10. Practical Determination of the Forecast Formula

The formulas (9.16) and (9.17) are not readily applied numerically. We can indicate, however, computational procedure which is accurate enough and reasonably easy to apply. Having computed the spectrum $f(k)$ by the indicated method (note that it is necessary to interpolate it for $k = 0$), we obtain easily the logarithm

$$\log f(k) = \sum_{-\infty}^{+\infty} s_m \, e^{-ikm} \tag{10.1}$$

The s_m are computed by a new Fourier transformation

$$s_m = \frac{1}{2\pi} \int_{-\pi}^{+\pi} \log f(k) \, e^{ikm} \, dk \tag{10.2}$$

We have then

$$j_I = \exp\left(\tfrac{1}{2}s_0 + \sum_{1}^{\infty} s_m \, \zeta^{-m}\right) = \sum_{0}^{\infty} C_{I,m} \, \zeta^{-m} \tag{10.3}$$

and also

$$j_I^{-1} = \exp\left(-\tfrac{1}{2}s_0 - \sum_{1}^{\infty} s_m \, \zeta^{-m}\right) = \sum_{0}^{\infty} \rho_m \, \zeta^{-m} \tag{10.4}$$

the $C_{I,m}$ and s_m being computed by development of the exponential in series.

We have then

$$k(\zeta) = \sum_{m=0}^{\infty} K_m \, \zeta^{-m} = \sum_{m=0}^{\infty} C_{I,\,a+m} \, \zeta^m \, j_I^{-1}(\zeta) \tag{10.5}$$

the k_m being easily obtained by development of the last member following the decreasing powers of ζ. This method has the advantage of involving only the calculation of real numbers, particularly $C_{I,\,m} = C^*_{I,\,m} = C_{E,-m}$. We deduce from it the value of the minimum error:

$$\sigma^2 = \sum_{0}^{a-1} C^2_{I,\,l} \tag{10.6}$$

This method has been applied to the already mentioned example for obtaining a statistical forecast formula of the atmospheric pressure in Uccle.

Table II gives, in function of a forecasting range a (days), the values of the coefficients K_m as well as of the error σ; the latter can be compared to the error σ' given by the simplest formula of persistence:

$$\tilde{y}_{m+a} = \frac{C_a y_m}{C_0}$$

$$\sigma' = \sqrt{\frac{C_0{}^2 - C_a{}^2}{C_0}}$$

TABLE II

m	$a = 1$	$a = 2$	$a = 3$	$a = 4$	$a = 5$
0	1.269	1.165	0.917	0.812	0.605
1	− 0.445	− 0.561	− 0.353	− 0.423	− 0.281
2	0.004	0.165	− 0.014	0.089	0.046
3	0.160	− 0.019	0.085	0.049	0.091
4	− 0.222	− 0.101	− 0.098	− 0.039	− 0.077
5	0.181	0.161	0.165	0.103	0.089
6	− 0.069	− 0.046	− 0.063	− 0.058	− 0.037
7	0.042	0.017	0.005	0.019	
8	− 0.036	− 0.044	− 0.020		
9	0.002	0.022			
10	0.019				
σ(mb)	2.75	4.45	5.5	6.1	6.5
σ'(mb)	4.14	6.36	7.7	8.3	8.8

11. Time Series with Periodic Component

The absolute convergence condition of the series $\sum C_m$ involves some physically important restrictions. Particularly, we see immediately that the time series may not have any precisely periodic component. Let us, for instance, consider the series

$$z_m = y_m + a \cos m\alpha \qquad |\alpha| < \pi \tag{11.1}$$

we will have

$$\lim \frac{1}{M} \sum_0^M z_m z_{m+n} = C_n + \frac{a^2}{2} \cos n\alpha \tag{11.2}$$

and we see that the last term is not tending to zero for $n \to \infty$. The situation would not be better if the additional term comported a stochastic phase ζ_m:

$$z_m = y_m + a \cos [m\alpha + \zeta_m] \tag{11.3}$$

Supposing y and ζ independent and ζ Laplacian and of zero mean, we would have

$$\begin{aligned} \lim \frac{1}{M} \sum_0^M z_m z_{m+n} &= C_n + \frac{a^2}{2} \overline{\cos (n\alpha + \zeta_{m+n} - \zeta_m)} \\ &= C_n + \frac{a^2}{2} \cos n\alpha e^{-\gamma_0 + \gamma_n} \end{aligned} \tag{11.4}$$

where $\quad \gamma_n = \overline{\zeta_m \zeta_{m+n}}$.

The application of the computing method presented in Section 4 to time series type (11.1) gives, however, interesting results. First we notice that formula (4.2) becomes:

$$\overline{F_z(k)} = \overline{F_y(k)} + \frac{1}{8\Lambda} \sum_{n=\Lambda+\frac{1}{2}}^{\Lambda-\frac{1}{2}} \sum_{m=-\lambda+\frac{1}{2}}^{\lambda-\frac{1}{2}} e^{-imk}\left(1+\cos\frac{m\pi}{\lambda}\right)\left(1+\cos\frac{n\pi}{\Lambda}\right) \times a^2[\cos(2n+m)\alpha + \cos m\alpha]$$

$$= \overline{F_y(k)} + \frac{a^2}{8\Lambda} \sum_m e^{imk} \cos m\alpha \left(1+\cos\frac{m\pi}{\lambda}\right)[2\Lambda + 4\pi V(2\alpha, \Lambda)]$$

$$= \overline{F_y(k)} + \frac{a^2\pi}{4\Lambda}[2\Lambda + 4\pi V(2\alpha, \Lambda)]\,[V(k-\alpha,\lambda) + V(k+\alpha.\ \lambda)]$$

$$\approx \overline{F_y(k)} + \frac{a^2\pi}{2}[V(k-\alpha,\lambda) + V(k+\alpha,\lambda)]$$

The ray $k = \pm\,\alpha$ is replaced by the continuous function $a^2\pi V(k+\alpha, \lambda)/2$ the maximum amplitude of which has the value $a^2\lambda/4$ for $|\,k\,| = |\,\alpha\,|$. We conclude that the presence in the $F(k)$ computed spectrum of an isolated peak the amplitude of which linearily varies with λ, permits the discovery of the presence of a periodic component.

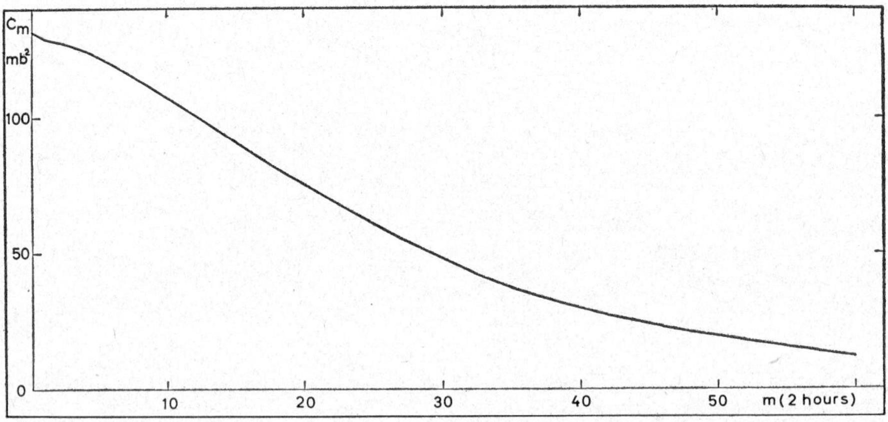

FIG. 5. Auto-covariant C_m of atmospheric pressure fluctuations at Upsala during the year 1952.

The random error $\sigma(k)$ is not affected by the presence of the periodic term because this is not stochastic. Finally, the presence of a pseudo-periodic term, type (11.3), will also be discovered by this method, the maximum amplitude lying between $a^2\lambda/4$ and $a^2\lambda/4 \cdot e^{-\gamma_0}$. This method has been applied for detection of the atmospheric tide. Four thousand bi-hourly observations of the atmospheric pressure at Upsala (1952) permitted the computation of the covariant of Fig. 5 and then the spectrum of Fig. 6. In the latter we distinguish easily the semi-diurnal component, but the tide amplitude is only 0.2 mb. This semi-diurnal tide may also be detected in the mean hourly pressure pattern as showed in Fig. 7.

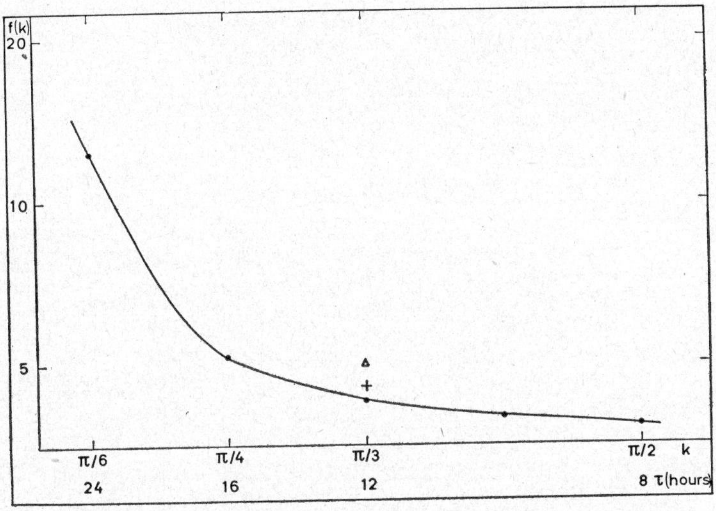

FIG. 6. Part of the power spectrum (log. scale) of pressure fluctuation at Upsala (1952). Computed with $\Lambda = 2000.5$, $\lambda = 14.5$. For $k = \pi/3$, which corresponds to a period of 12 hours, the computations have been done also with $\lambda = 29.5$ (cross) and $\lambda = 59.5$ (triangle) showing the existence of a periodic component.

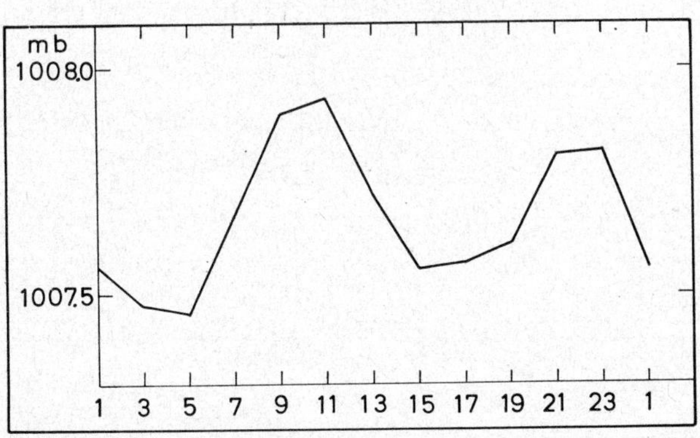

FIG. 7. Mean hourly pressure fluctuation at Upsala for the whole year 1952 showing a well-marked semi-diurnal tide oscillation.

List of Symbols

y_m	Discrete time series
$R(y_m)$	Filtered time series
$\bar{y}$	Mathematical expectation of the random variable y
$C_m = \overline{y_n y_{n+m}}$	Covariant of the time series y_m
$f(k)$	Power spectrum of the time series y_m
$\rho(k)$	Generalized integral Fourier transform of the time series y_m
σ	Normal dispersion parameter
$\vert y \vert$	Absolute value (or modulus) of y
y^*	Complex conjugate of y
Δk	Finite interval of k
$\Delta f(k)$	Corresponding variation of $f(k)$
$\delta_{m,n}$	Kronecker symbol
T_p	Tchebycheff polynomial: $T_p(\cos \xi) = \cos p\xi$

References

1. Wiener, N. (1949). "Extrapolation, Interpolation and Smoothing of Stationary Time-Series with Engineering Application." Wiley, New York.
2. Aigrain, P. (1951). "La Cybernétique." Revue d'optique, édit. Paris.
3. Blanc-Lapierre, A. and Fortet, R. (1953). "Théorie des fonctions aléatoires." Masson, Paris.
4. Cramer, H. (1940). On the theory of stationary random processes. *Am. Math.* **41**, 215.
5. Moyal, J., (1949). Stochastic processes and statistical physics. Symp. on Stochastic Processes. *J. Roy. Statist. Soc.* **11**, 150.
6. Wold, H. (1948). On prediction in stationary time-series. *Ann. Math. Statist.* **19**, 558.
7. Leith Holloway, J. Jr. (1958). Smoothing and Filtering of time-series and space-fields. *Advances in Geophys.* **4**, 351.

TEMPERATURE AND WIND IN THE LOWER STRATOSPHERE

H. A. Panofsky

The Pennsylvania State University, University Park, Pennsylvania

1. Introduction

1.1. Data Availability

With the introduction of more and more man-made vehicles and particles into the lower stratosphere, it is becoming increasingly important to obtain as much information as possible about the meteorological characteristics of this region, especially of the wind. Fortunately, the development of plastic radiosonde balloons has added greatly to our knowledge of this region since about 1951. However, good records are still available only in North America and Europe. Even there the records are short, and many of our current ideas will need modification as additional data become available.

1.2. Definitions

The lower boundary of the stratosphere is the tropopause. This is usually defined as the first level above which the lapse rate is less than 2°C/km for a distance of 2 km or more. This definition, of course, includes the possibility that the lapse rate above the tropopause is zero (isothermal) or negative (inversion).

There is no generally accepted definition of the top of the stratosphere. For the purpose of this discussion, we will define the "lower stratosphere" as that region above the tropopause in which observations are frequently obtained by radiosondes. Thus, the top of the lower stratosphere is about 30 km or 10 mb. This level also is usually near the base of a strong inversion.

1.3. Significance of the Stratosphere

In contrast to conditions in the troposphere, the temperature in the lower stratosphere varies slowly with height unless it is invaded by air from higher levels, in which case strong inversions develop. The temperature generally increases upward in the tropics, is nearly constant in moderate latitudes and polar regions in summer, and decreases in the surroundings of the winter pole.

Along with a change in vertical temperature gradient, the horizontal variation in temperature changes at the tropopause. In large regions of the lower stratosphere the temperature decreases equatorward, in contrast to the normal temperature gradient in the troposphere.

Fundamentally, the difference between the troposphere and stratosphere is that the distribution of temperature in the troposphere is controlled by the heating and cooling of the earth's surface and subsequent convection and vertical transport of latent heat, while the distribution of temperature in the stratosphere is determined mostly by the direct reaction of its gases to radiation and by the circulation in the stratosphere itself. Of course, since the atmosphere is a continuous fluid, the stratosphere and troposphere cannot act independently of each other, for changes in part of the fluid will affect its other parts.

1.4. Composition of the Stratosphere

The percentages of the permanent gases in the stratosphere are the same as in the troposphere. The relative humidity generally is low and decreases with height in the stratosphere. A typical mixing ratio is 0.04 part per thousand by weight. The ozone concentration reaches its maximum in the upper portion of the lower stratosphere, but even there its "mixing ratio" is usually less than 0.01 part per thousand. The special importance of ozone lies in the fact that, of all the gases in the stratosphere, only ozone can absorb significant amounts of solar radiation. Further, an extremely small amount of ozone can absorb a large amount of radiation.

Carbon dioxide is also important in calculations of stratospheric temperatures, since it emits radiation efficiently (as can ozone and water vapor). The mixing ratio for carbon dioxide is about 30 times as large as that for ozone. For further details regarding the composition and structure of the stratosphere one may consult Goody [1].

1.5. Some Practical Rules of Thumb

For most purposes, it can be assumed that the geostrophic wind equation is applicable in the stratosphere. In general, systematic deviations from geostrophic wind in the stratosphere are much smaller than errors of observation. Also, of course, the hydrostatic equation is valid. As a result, the "thermal wind" equations can be used in the stratosphere outside of the equatorial regions. A number of useful rules of thumb follow from the thermal wind equation:

(1) When contour height and temperature vary in the same sense along isobaric surfaces, the wind speed increases upward. For example, in most of the troposphere, and in the region near the winter pole in the stratosphere, both temperature and contour height decrease poleward, and the westerly wind speeds increase with height. Or, in the upper regions of the synoptic stratosphere in summer, both temperature and height increase poleward, so that the winds (easterlies in this case) again increase upward.

(2) When contour height and temperature vary in the opposite sense along isobaric surfaces, the wind decreases upward. For example, in middle latitudes in the lower stratosphere, height decreases poleward, but temperature increases poleward, resulting in westerlies, decreasing upward in intensity.

(3) Warm ridges and cold troughs increase in intensity upward.

(4) Warm troughs and cold ridges decrease in intensity upward.

1.6. Correspondence between Isobaric Surfaces and Heights

In the succeeding sections we shall be concerned with both isobaric surfaces and heights. Table I gives the general heights at which certain isobaric surfaces are likely to be found.

TABLE I. Approximate correspondence of heights and certain isobaric surfaces.

Isobaric surface, mb	200	100	50	25	10
Mean height, km	11.8	16.2	20.6	23.9	30.9
Typical range, km	10.5	15.0	19.0	23.0	28.0
	12.5	17.0	21.0	25.5	32.0

2. General Characteristics of Stratospheric Properties

2.1. The Temperature Distribution in Summer

Figure 1a, adapted from Kochanski [2], shows a typical temperature cross section through the troposphere and stratosphere. The distribution of temperature along horizontal or along isobaric surfaces is relatively simple in the lower stratosphere, at least above 16 km, the average height of the tropopause in the tropics. Generally, the temperature is coldest at the equator, warmest near the pole. The horizontal temperature gradient is greatest at about 16 km and decreases upward. The figure indicates the lowest temperature at 16 km to be at the equator, about — 73°C.

Figure 1 also shows that at about 30 km the temperature gradient is quite small due to the strong stratospheric inversion in the tropics. Polar regions are nearly isothermal with only a minor inversion indicated. Between 10 and 17 km the temperature structure is more complex because at any given time, part of this layer is in the stratosphere and part in the troposphere. In the stratospheric portion, the temperature gradients are quite small. Also, in the tropospheric region of the tropics there is little horizontal temperature gradient; however, below 11 km the temperature in the tropical troposphere is normally greater than that in the stratosphere at the same height. Above 11 km the tropical troposphere is colder than the stratosphere.

In summary, the temperature distribution in the summer stratosphere is as follows: At approximately 11 km the temperature at the pole is slightly warmer than the temperature at the equator. With increasing height, the temperature gradient increases to a maximum at approximately 16 km, and here the temperatures at the equator are 30°C lower than at the pole. Above 17 km the temperature gradient decreases, becoming negligible at 30 km. In the tropics, a strong stratospheric inversion overlies the tropopause, whereas in the middle and polar latitudes, only a slight inversion is found above the tropopause.

2.2. The Temperature Distribution in Winter

Figure 1b shows vertical and latitudinal temperature distribution in winter which is typical for the months of October through March. As might be expected, the distribution of temperature in the tropics is essentially the same in summer and winter. Outside the tropics the major difference from the summer pattern is the cold air pocket near the pole, and the strong poleward temperature gradient in the arctic latitudes. This region, the most interesting in the stratosphere, will be called the region of the "polar regime" (PR). Below a height of about 20 km, the polar regime is on the average confined to the polar region north of 55° latitude. Above that it broadens so that at an altitude of 30 km its influence covers the hemisphere from the pole to

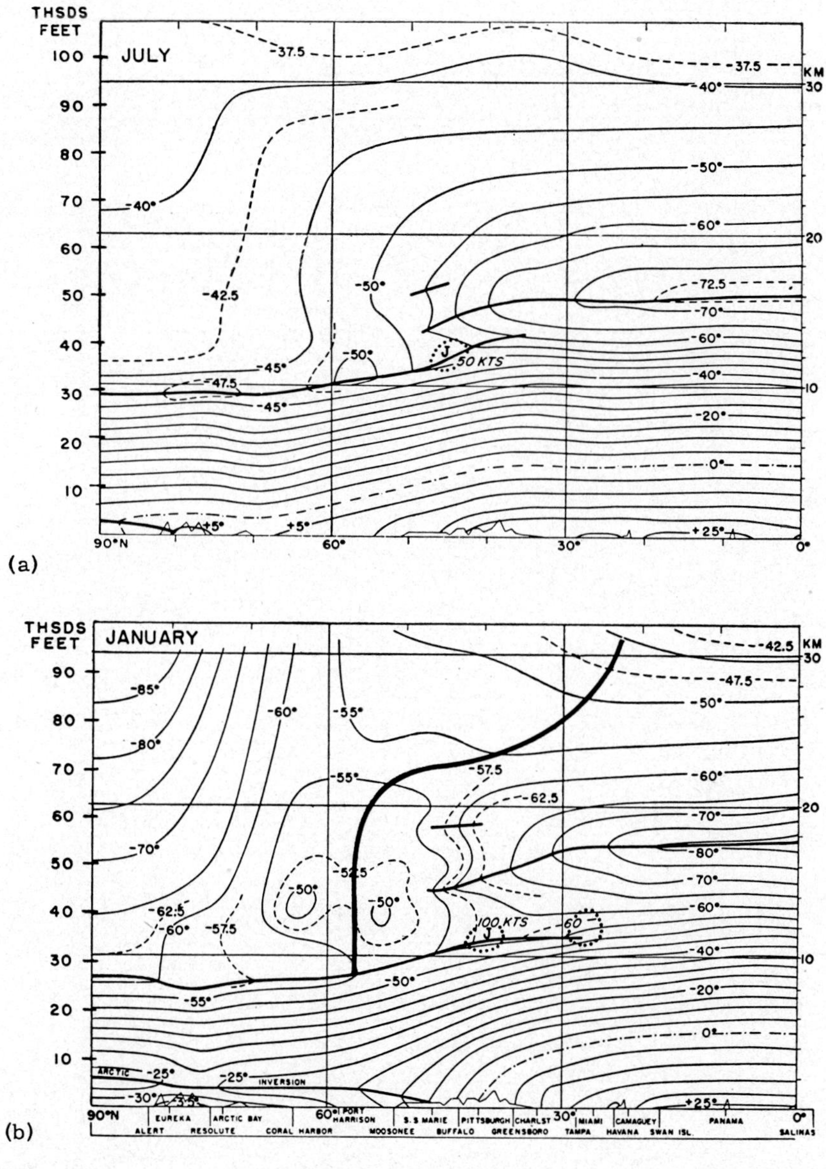

FIG. 1. Mean temperature cross section in (a) July and (b) January. Isotherms in °C. The mean positions of the tropopauses, jets and the PRB (polar regime boundary) are also shown (adapted from Kochanski [2]).

approximately 30° latitude. The cold PR exists in the polar regions during the winter because of the absence of solar radiation. In the portion of the stratosphere not covered by the PR, the temperature generally increases poleward from near the equator, as in summer. Again, this increase is greatest at about 17 km. Since the temperature increases from the equator toward the poles up to the PR and then decreases again in the PR, there must be a belt of relatively warm air which marks the boundary of the PR. This will be called the "polar regime boundary" (PRB). This boundary, essentially a vertical wall near latitude 55° from the tropopause up to about 20 km, spreads out above that level and reaches 30° latitude near 24 km. At latitude 45°, for example, the PRB is found at about 23 km. This means that, going up at latitude 45°, one finds poleward temperature gradients in the troposphere, equatorward gradients in the lower stratosphere, and poleward gradients again above 23 km. It will be seen in later sections that the PRB is an important boundary surface, not only for the temperature regime but particularly for the wind regime.

In summary, we have the following picture of the average winter temperature distribution: In the layer from about 12 km up to about 20 km, the temperature increases from a minimum near the equator to a maximum at 55° and decreases again to the pole. The temperature along the PRB is nearly constant at − 50°C. In the cold PR temperatures decrease with height up to at least 30 km, but in the equatorial region the temperature is lowest at about 17 km. Above 20 km, the temperature gradient is weak from the equator to the PRB, but from the PRB to the pole the temperature gradient is strong, the temperature being lowest near the pole.

2.3. The Distribution of Zonal Wind in Summer

In this section, the distribution of wind with latitude and height is considered. It should be emphasized that this section deals with average conditions only. The flow patterns in individual periods may differ considerably from the average pattern; however, the variation from summer to summer appears to be considerably less than that from winter to winter. We will be concerned only with the zonal wind components. The meridional components show essential variations with longitude, but their average around a latitude circle is close to zero.

Except in the tropics, the distribution of zonal wind can be inferred from the temperature distribution already discussed in this chapter, and the wind distribution at the tropopause can be estimated by using the rules of thumb given in Section 1.5.

Figure 2 shows schematically the vertical and latitudinal zonal wind distribution in summer. The zero isopleth separates the pattern into two regimes, the easterly regime (ER) and the regime of westerlies. Since westerlies, decreasing upward, are typical for a large portion of the lower stratosphere in all seasons the regime of westerlies is called the "lower stratosphere regime" (LSR). In the summer, the LSR is quite small and extends from the tropopause up to about 50 mb (20 km).

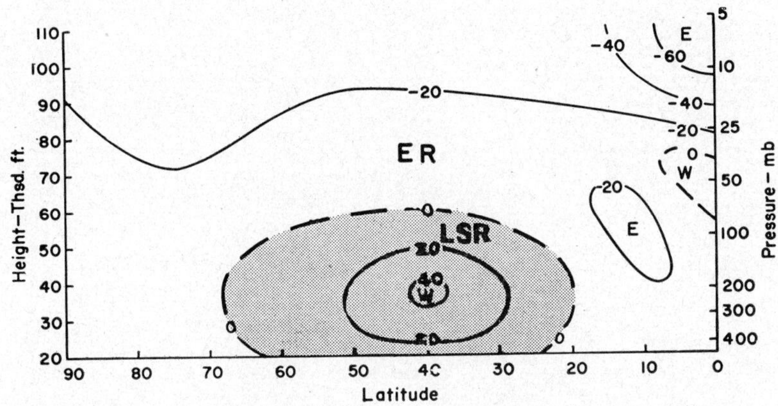

FIG. 2. Idealized distribution of zonal wind in summer. Negative numbers are easterly wind speed in knots (1 knot = 0.51 meters/sec); positive numbers are westerly speeds; 10,000 ft ≅ 3 km.

The ER extends vertically throughout the stratosphere in the tropics up to about 20 km and covers all latitudes above that level. The ER contains two speed maxima, the largest with speeds of 30 meters/sec, near the equator at about 10 mb. The strong easterly jet near 10 mb is called the region of "Krakatoa Easterlies" since these currents carried the debris from the 1883 explosion of the volcano Krakatoa around the world.

Imbedded in the ER near the equator is a peculiar current of weak westerlies. These winds are called "Berson Westerlies." Recent studies have shown that this belt does not remain at a fixed height, but moves downward at the rate of about 1 km/month, followed by another westerly jet a few km higher.

The separation between LSR and ER in middle latitudes is important for a reason besides the obvious change in direction: The current patterns in the LSR are closely related to the tropospheric current patterns underneath, whereas the easterly currents appear to bear no simple relationship to tropospheric patterns.

The use of the rules of thumb given in Section 1.5 can be demonstrated by comparing Fig. 2 with Fig. 1a. For example, in the LSR, we have westerlies, indicating that contour heights on isobaric surfaces decrease poleward. Since temperatures increase poleward in this region, it follows that the wind speed decreases upward. In the region of easterlies above 20 km, both heights and temperatures increase poleward, so that the easterly wind speeds increase upward. From another point of view, the upward intensification of these easterlies can be ascribed to the "warm" high near the pole.

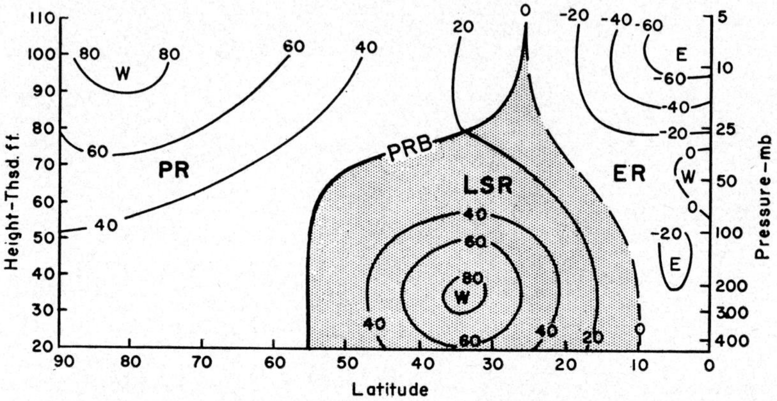

FIG. 3. Idealized distribution of zonal wind in winter. Negative numbers are easterly wind speed in knots (1 knot = 0.51 meters/sec); positive numbers are westerly speeds; 10,000 ft. ≅ 3 km.

2.4. *The Distribution of Zonal Wind in Winter*

Figure 3 illustrates the distribution of zonal winds in winter. We now have to distinguish three regions; the ER region (easterlies), the LSR (westerlies decreasing with height), and the PR (cyclonic flow about a center near the pole, increasing with increasing height). The boundary between the ER and the LSR is the isopleth of zero zonal wind which is, in winter, more or less vertical. The boundary between PR and LSR is the PRB, already mentioned in Section 2.2. Above about 25 mb, the zero isopleth and the PRB merge, so that the PR is directly adjacent to the ER.

As compared with summer, the LSR is stronger and larger in winter, starting from very fast winds near 250 mb, and terminating in much weaker west winds at about 25 mb on the average. The ER in tropical regions is similar in summer and winter, even including the Berson Westerlies near the equator. The remainder of the region occupied by the summer ER is mostly taken over by the PR with its strong cyclonic flow at great heights. This cyclone will be called the "polar vortex."

The faster circulation in the winter LSR reflects the stronger tropospheric circulation underneath. The PR, on the other hand, is stronger than the underlying flow pattern and is essentially a stratospheric phenomenon, although vestiges of it are found at 500 mb.

As mentioned already, the PRB separates the region of westerlies increasing upward from the region of decreasing currents. At about 100 mb the westerlies are equally strong on both sides of the PRB. At greater heights, the polar circulation is stronger than the LSR circulation, and at lower heights the LSR circulation is stronger.

Altogether then, we have four centers of maximum wind in the winter: the strong westerly jet in middle latitudes in the tropopause region, the cyclonic jet near 10 mb in the Arctic, a weak westerly jet near 50 mb in the vicinity of the equator, a strong easterly jet at 10 mb near the equator.

To summarize the zonal circulation in winter: The tropical easterlies are similar to their summer equivalents, with a weak westerly belt at 50 mb; in middle latitudes, the westerlies decrease from strong westerlies near 250 mb to weak westerlies at 25 mb and increase again at higher levels; in the Arctic, the cyclonic currents increase gradually from the low speeds near the tropopause to very high speeds (50 meters/sec average) at 10 mb. It should be pointed out, however, that there are tremendous variations from winter to winter, so that the "normal" conditions are not well known and may not be representative of conditions in any one winter.

2.5. Annual Variation of Temperature

The annual variation of temperature south of latitude 50°N is generally less than 15°C at levels below 50 mb. It may have either sign; at 100 mb in Europe winters are colder than summers, whereas in the United States, July temperatures at 100 mb tend to be lower than January temperatures. North of 50°N the annual variation of temperature increases rapidly with both latitude and height, and the coldest temperatures occur in the winter.

2.6. Annual Variation of Wind

As pointed out previously, the PR, which is a winter feature, differs significantly from the rest of the stratosphere below 25 mb in that the temperature decreases generally poleward and the wind speed generally increases upward.

Since the temperature decreases equatorward outside of the cold core, the PRB is marked by a belt of relatively warm temperature. This boundary (nearly vertical above the tropopause to 25 mb) is farthest south about the time of the winter solstice, when it is located on the average at latitude 55°N; but its latitudinal position is not the same at all longitudes, and it also varies with time.

During the late winter and early spring, the relatively warm boundary withdraws toward the pole, and by the end of March—sometimes the end of April—becomes a warm center near the pole. In the summer no polar vortex exists, and temperatures generally increase poleward. The PR probably starts to make its appearance again in September, but by the end of October the average position of the PRB at 100 mb perhaps has reached its most equatorward position, 55°N.

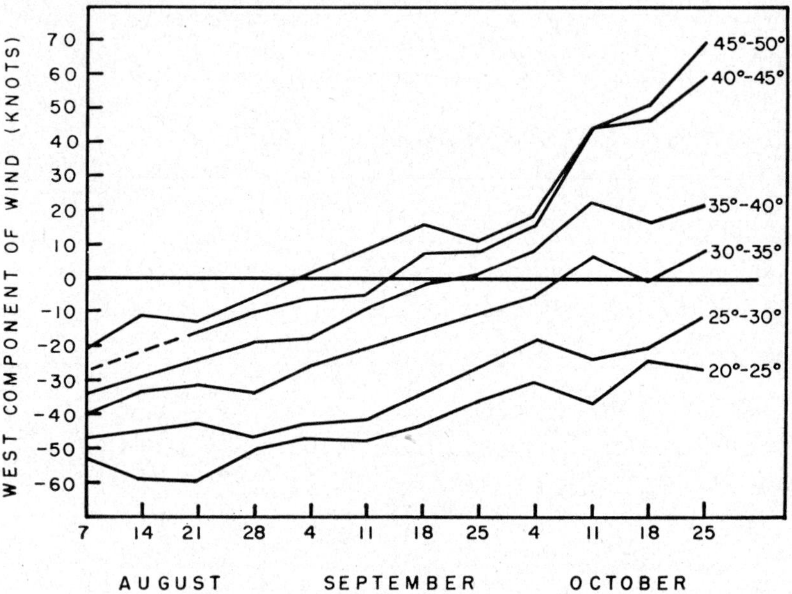

FIG. 4. The progression of weekly mean 30-km west-wind components for the period August 4 through October 28, 1957 (from Borden [3]).

Much of the lower stratosphere is characterized by winds with easterly components in summer and winds with westerly components in winter. In January the easterlies are concentrated in a small subtropical volume above 50 mb, but by May they have spread over most of the stratosphere above 40 mb. In July, easterlies predominate over virtually the whole synoptic stratosphere, except in the middle latitude between 250 and 50 mb. The withdrawal of the easterlies from July to October is even more rapid than their arrival; by the end of October they occupy a volume about equal to that in January. The change from easterlies to westerlies in middle latitudes in late summer and early fall is shown in Fig. 4 taken from Borden [3].

Of course, the region in which westerlies prevail all year also has important annual variations. The mean speed in the belt of maximum westerlies varies from about 50 meters/sec in winter to 25 meters/sec in summer. April and October maxima on mean maps are also close to 25–30 meters/sec.

According to Kochanski [4], the wind variability generally has the following characteristics:

(1) The variability is somewhat greater in winter than in summer, and it is greater near the tropopause (maximum-wind level) than at other levels in the troposphere or synoptic stratosphere.

(2) The variability is greatest in the middle-latitude regions influenced by jet streams, particularly where the jet stream itself undergoes large variations in position or strength.

(3) In jet stream regions, average 24- and 48-hr vector changes are greater than the average departure of daily winds from the climatic mean. Hence, although climatology is a poor forecast tool near jets, persistence is usually even worse for periods of 24 hr or more.

(4) Wind variability generally decreases with height from the tropopause up to 25 mb and then increases again. Thus normals generally give a good estimate of the flow between 100 and 10 mb outside of the warm belt in winter. Variability is small at all latitudes outside the tropics at and above 100 mb in the lower stratosphere during summer.

(5) Prominent changes take place in the region influenced by the warm belt and the PR in winter, so that normals can be quite unreliable. However these major developments are slow and appear to be prominent only upon comparison of charts one or more weeks apart. Hence, persistence of flow patterns can be expected for short periods.

In summary, the greatest variability of wind vectors is generally found in regions of fastest winds.

3. Synoptic Properties between the Tropopause and 20 km

3.1. The Lower Stratosphere in Summer

In the region between the tropopause and 20 km, from about 30° latitude north to the poles, we have the behavior which was designated by LSR (lower stratosphere regime); from about 30°S, on the average, easterlies prevail at all levels.

Figures 5–8 [4] show time cross sections through various stations in summer. At the southernmost station, Patrick Air Force Base, which is close to the southern boundary of the LSR, easterlies prevail at all levels; occasionally, however, the boundary between easterlies and westerlies passes this station near the base of the stratosphere. At Belmar, the typical LSR is found at all times from the tropopause upward to a level averaging perhaps 60 mb but varying from 110 to 25 mb. The same distribution of zonal winds persists at the stations further north, although, at the station selected at latitude 61°N, meridional wind components are strong. A significant difference between the three northern stations is the difference of wind speed at the base of the stratosphere. These winds are strong in the south (Belmar, New Jersey) and decrease with increasing latitude. The wind shear between this surface and the top of the lower stratosphere also decreases northward. Generally the winds are weak between 100 and 50 mb.

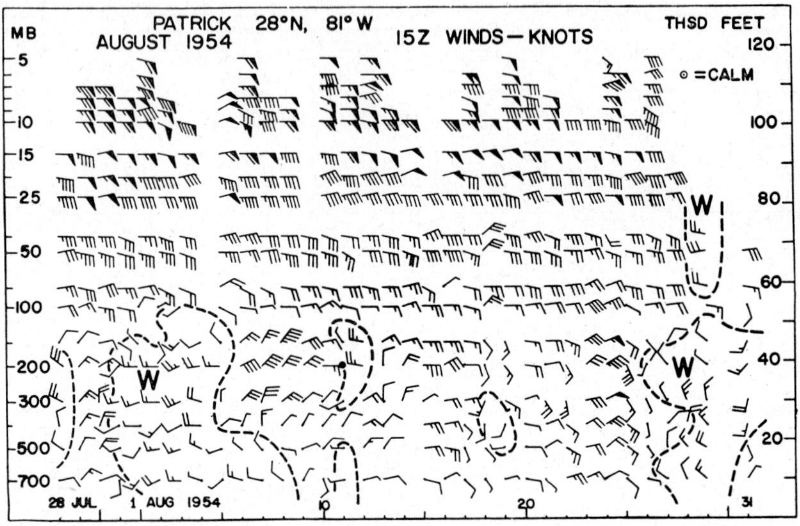

FIG. 5. Observed daily winds from 700 to 5 mb at Patrick Air Force Base, August, 1954 (after Kochanski [4]).

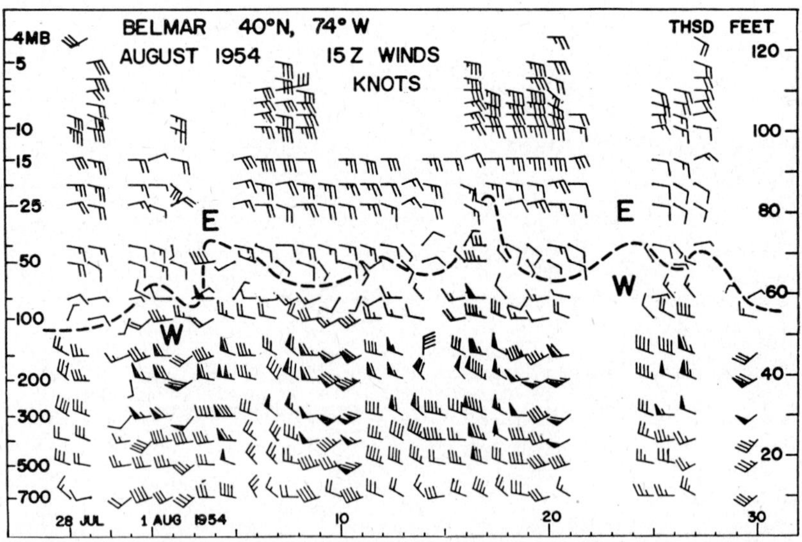

FIG. 6. Observed daily wind from 700 to 4 mb at Belmar, New Jersey, August, 1954 (after Kochanski [4]).

The synoptic features on isobaric charts in the lower stratosphere closely correspond to the simultaneous features in the upper troposphere. Since, however, contours and isotherms tend to be parallel, and the temperature gradient in the lower stratosphere is opposed to that in the troposphere, we

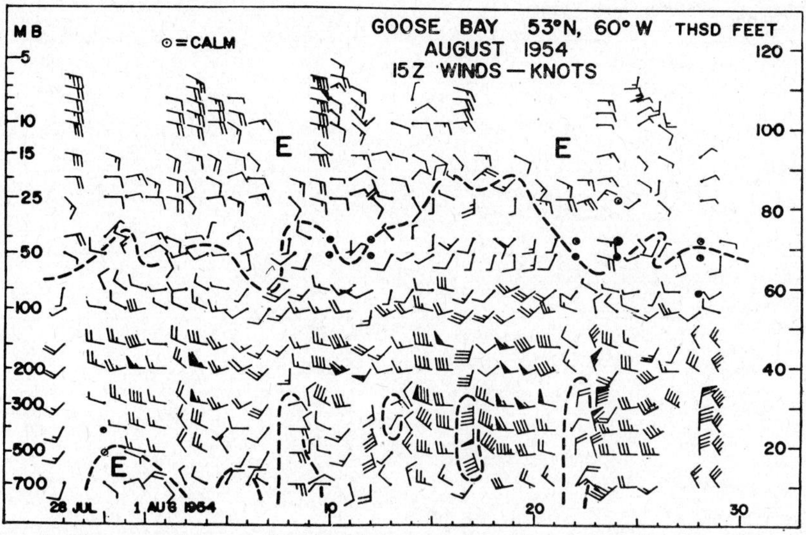

FIG. 7. Observed daily winds from 700 to 5 mb at Goose Bay, August, 1954 (after Kochanski [4]).

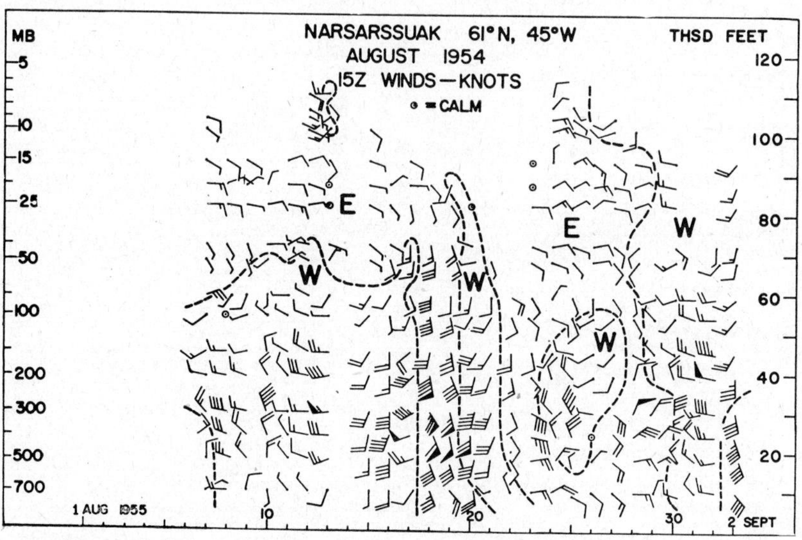

FIG. 8. Observed daily winds from 700 to 7 mb at Narsarssuak, August, 1954 (after Kochanski [4]).

find that cold troughs and warm ridges in the troposphere are situated under warm troughs or cold ridges in the lower stratosphere. In other words, troughs and ridges intensify upward to the tropopause and become weaker above. Thus the contour patterns and gradients found at an isobaric level well below the

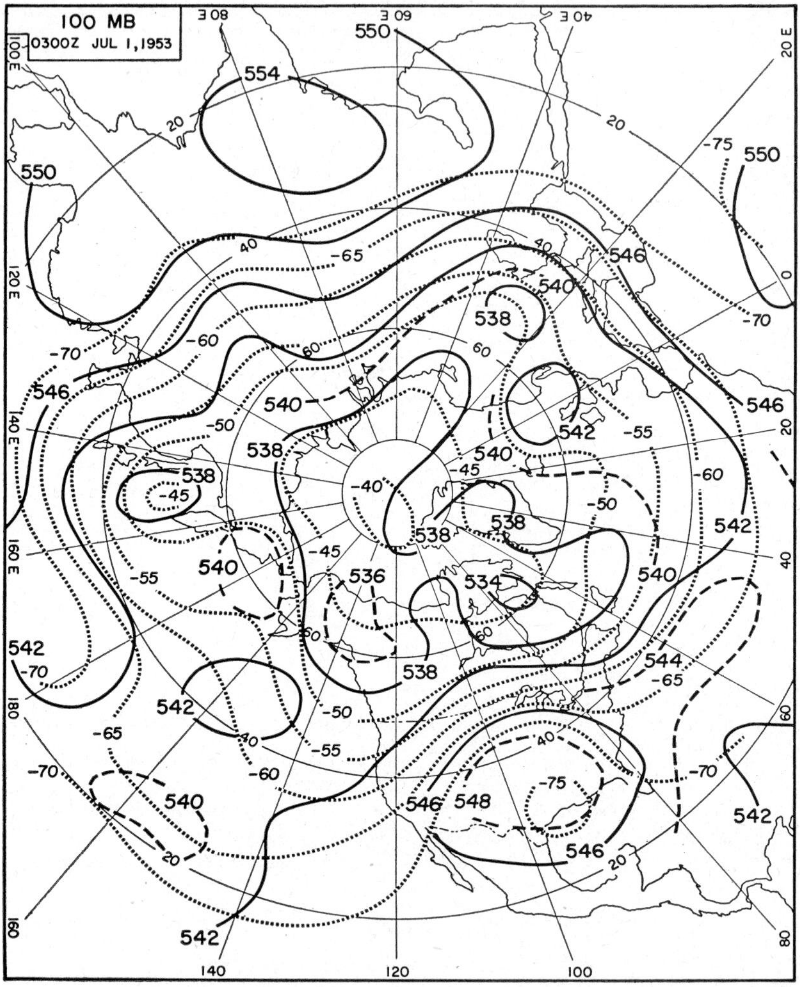

FIG. 9. A 100-mb northern hemisphere chart for July 1, 1953. Temperatures in °C, contour height in hundreds of geopotential feet (100 ft = 30 meters) (after Moreland [6]).

tropopause may bear a strong resemblance to patterns and gradients at an isobaric surface well above the tropopause. In particular, 100-mb contour patterns closely resemble 500-mb patterns [5]. For example, there exists a band of closed highs at low latitude at both 500 and 100 mb; however, the band of highs at 100 mb is located on the average perhaps 5° latitude further north than at 500 mb. Also at both 100 and 500 mb the principal zone of westerlies is situated, on the average, in the region of the surface polar front.

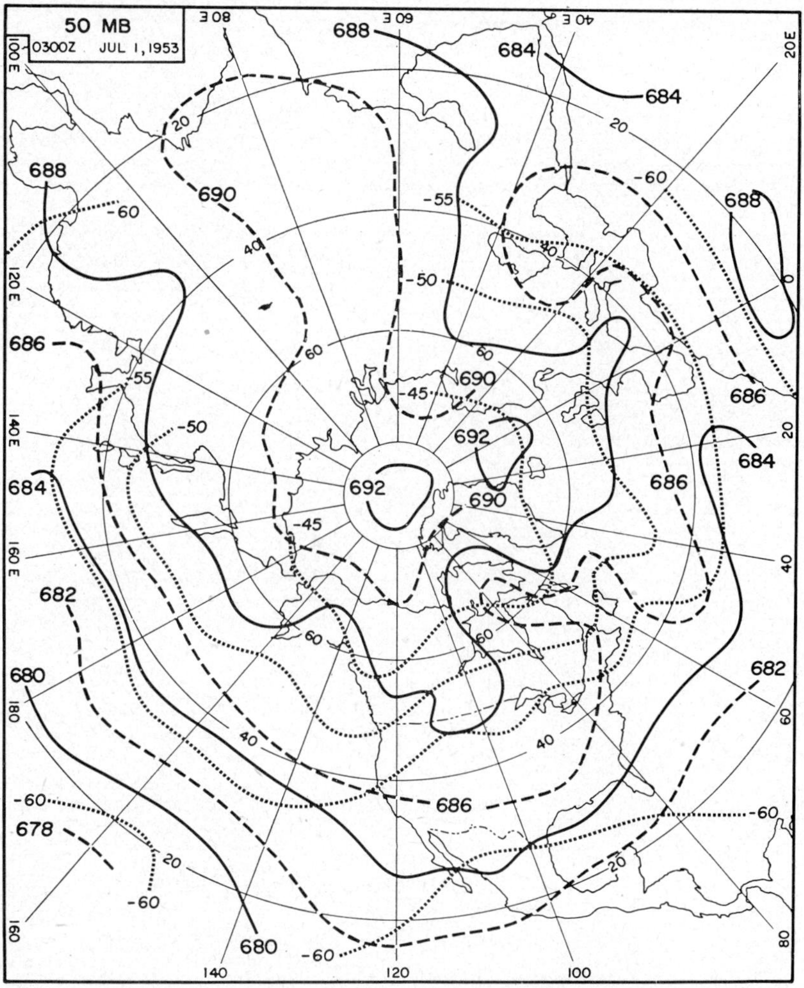

FIG. 10. A 50-mb northern hemisphere chart for July 1, 1953. Temperatures in °C, contour height in hundreds of geopotential feet (100 ft = 30 meters) (after Moreland [6]).

Figures 9 and 10 show some typical northern hemisphere charts for summer at 100 mb (16 km) and 50 mb (20 km). At 100 mb, the flow is weak westerly everywhere outside the tropics. The ridges and troughs in the patterns in middle latitudes generally occur at the same longitudes at 100 mb as they do at 500 mb in the troposphere (about 6 km). At 50 mb, the flow is predominantly easterly, but the winds are not strong. Clearly, however, the winds must increase in strength upward in view of the fact that both contour heights and temperatures increase poleward.

The maps shown were taken from a series of daily charts [6]. Comparison of successive charts indicates remarkably little change in the patterns from one day to the next. One has to wait about a week before any clear-cut changes become apparent, which cannot be ascribed to error of analysis due to lack of observations. Thus, for example, during the week following July 1, a trough, appearing from the north, gradually replaced the high in the United States.

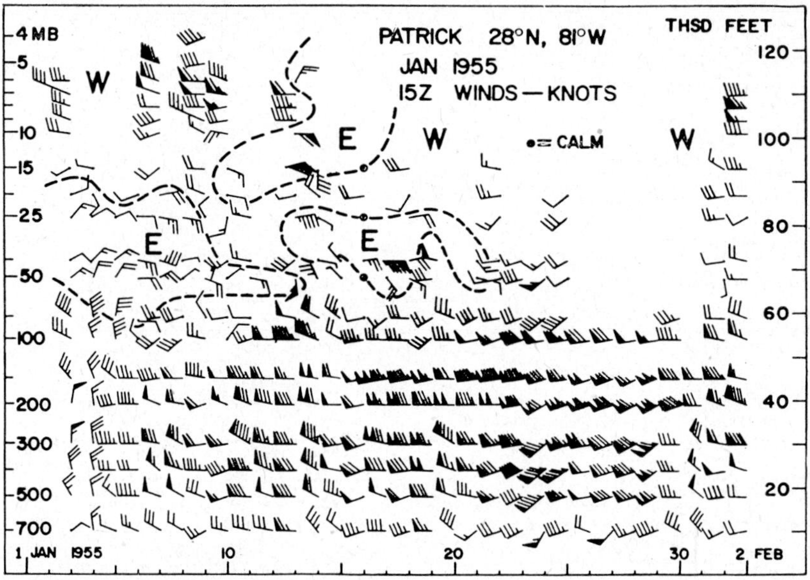

FIG. 11. Observed daily winds from 700 to 4 mb at Patrick Air Force Base, January, 1955 (after Kochanski [4]).

3.2. The Lower Stratosphere in Middle Latitudes in Winter

As in summer, the middle-latitude belt in winter is dominated by the LSR; i.e., westerlies decreasing upward in intensity. However, this regime extends northward only to the PRB (polar regime boundary), the average latitude of which is 55°N, but which may fluctuate considerably. The LSR belt extends somewhat further south in winter than in summer, and the subtropical anticyclones are located about 10° further south in winter than in summer. The result is that the winds at Patrick Air Force Base at latitude 28°N, are easterly in summer but westerly in winter, in the lower stratosphere, as is seen by comparing Figs. 5 and 11.

The LSR, in addition to extending somewhat further toward the equator in winter than in summer, extends also to a slightly higher level, the top averaging perhaps 25 mb. Otherwise the main distinction of the winter LSR is that the westerlies in it are on the average about 20 meters/sec faster than the corresponding summer winds. This can be seen by comparing Figs. 12–14 with the corresponding Figs. 6–8. Occasionally, the winds are still strong westerly at the top of the winter LSR, and, of course, at its bottom we find the well-known jet streams which often exceed 50 meters/sec.

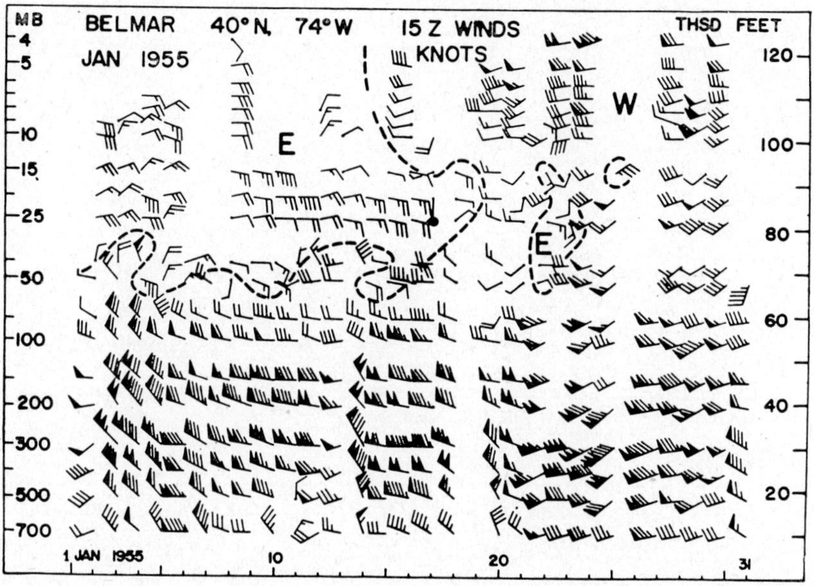

FIG. 12. Observed daily winds from 700 to 4 mb at Belmar, New Jersey, January, 1955 (after Kochanski [4]).

Superimposed on the westerly winds in the LSR are troughs and ridges which are associated with the troughs and ridges in the troposphere such as those shown on 500-mb charts. The tropopause is cold over a warm ridge in the troposphere and warm over a cold tropospheric trough. Hence, in the lower stratosphere the ridges are cold and the troughs warm. This is true both in winter and in summer, and for this reason both the contour and isotherm amplitudes decrease with elevation in the stratosphere. By 100 mb, the ridges and troughs are flatter, and the isotherms are more in phase with the contours than at the base of the stratosphere. Since the patterns at 500 mb are also flatter than those at the tropopause, we find in winter as in summer, a distinct similarity between 500-and 100-mb flow in the LSR.

There are, however, several systematic differences between 500-and 100-mb charts even in the LSR region. In the first place, of course, the temperature gradient in the stratosphere is opposite to that in the troposphere. Secondly, small tropospheric systems no longer make their appearance at 100 mb. As a

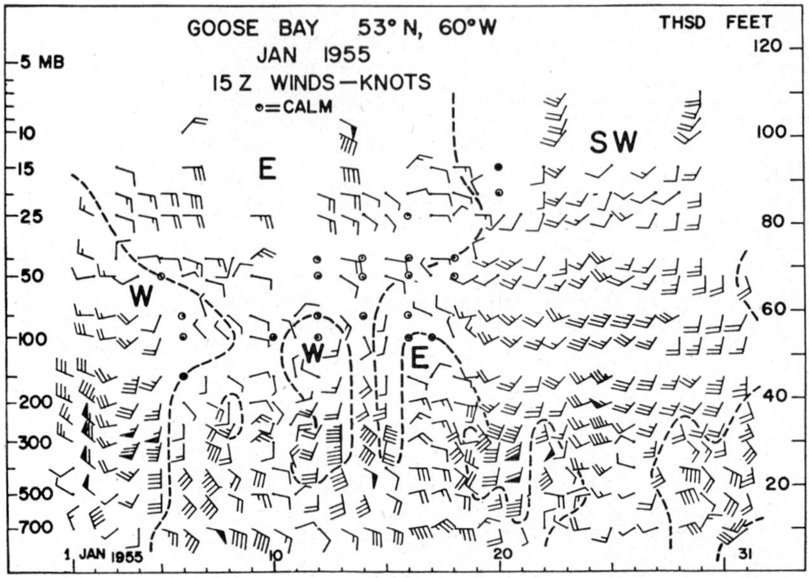

FIG. 13. Observed daily winds from 700 to 7 mb at Goose Bay, January, 1955 (after Kochanski [4]).

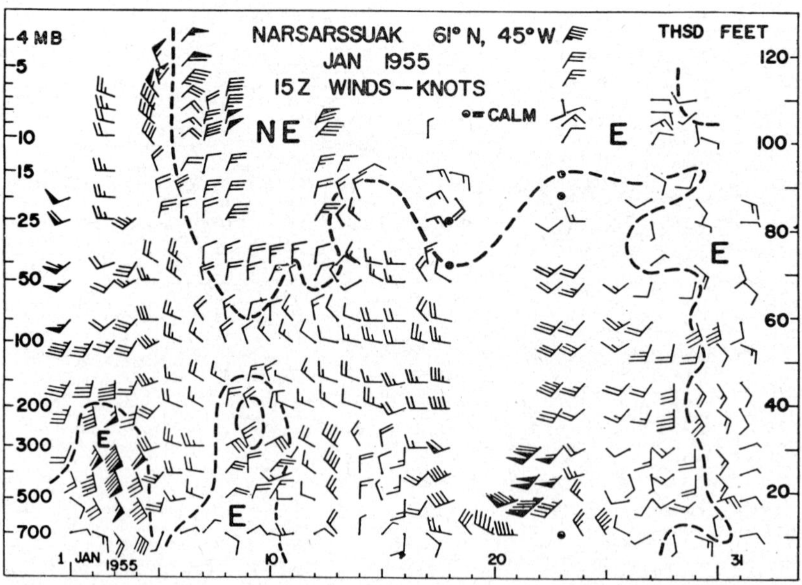

FIG. 14. Observed daily winds from 700 to 4 mb at Narsarssuak, January, 1955 (after Kochanski [4]).

result, *smoothed* 500-mb charts (for the theory of smoothing, see Holloway [7]) resemble 100-mb charts most closely [5]. Troughs and ridges occur at about the same longitudes at the two levels, and the speeds are comparable. At the southern edge of the belt, however, a discrepancy in speed appears. Here the tropopause is so high that compensation between the stratospheric and tropospheric temperature fields is not complete between 500 and 100 mb; as a result, winds average perhaps 7 meters/sec faster at 100 mb than at 500 mb near latitude 30°N.

The relation between 100-mb speeds and 500-mb smoothed speeds is not the same in every season. In seasons in which the polar vortex is relatively far south, it raises the speeds in the LSR at 100 mb relative to those at 500 mb at a given latitude. In such cases, for example, the flow at 100 mb between latitudes 40° and 50°N may average as much as 10 meters/sec faster than the space-smoothed flow at 500 mb.

The use of 500-mb smoothed charts also has some prognostic value for 100-mb flow; some tentative studies have shown that a given 100-mb chart in the LSR resembles the 500 mb chart 12 hr previously more than it resembles the concurrent 500-mb chart. Of course, it is difficult to verify this statement because changes at 100 mb are so slow. Smoothed 500-mb charts resemble 100-mb flow not only in winter, but also perhaps even more closely in the spring and fall transitional periods.

At 100 mb, the winds are still predominantly westerly in middle latitudes with a basically zonal pattern; by 50 mb the flow is still westerly in the LSR, but the winds are weaker. These features are demonstrated by Figs. 15 and 16. Since the flow patterns increase in intensity downward, it is reasonable to suppose that they are controlled by tropospheric energy sources. As a result, the waves in the LSR patterns tend to travel eastward slowly, together with the long waves in the tropospheric flow. However, the movement is so slow that it is difficult to detect any systematic changes in 24 hr; it is therefore impossible to improve on a forecast of no change. Nevertheless, the changes are somewhat more rapid than in summer.

3.3. The Polar Vortex and the PRB at and below 20 km

As mentioned before, the exact position of the polar vortex and the PRB varies radically from season to season, and slow changes occur within each season. Here also a smoothed 500-mb chart can be used to locate the centers of stratospheric circulation above tropospheric centers, and usually to locate the principal ridges and troughs. Such a smoothed chart is then reasonably satisfactory for indicating wind direction in the PRB area; however, the wind speeds in the stratospheric polar regions tend to be faster than those in the troposphere, and they increase with height. For, whereas a cold stratosphere overlies a warm troposphere in the LSR, and vice versa, inside the PRB a cold stratospheric cyclone is likely to be situated above a cold tropospheric cyclone. Hence, the characteristic compensation of high-troposphere and low-stratosphere features of the LSR is likely to be absent in the polar regime.

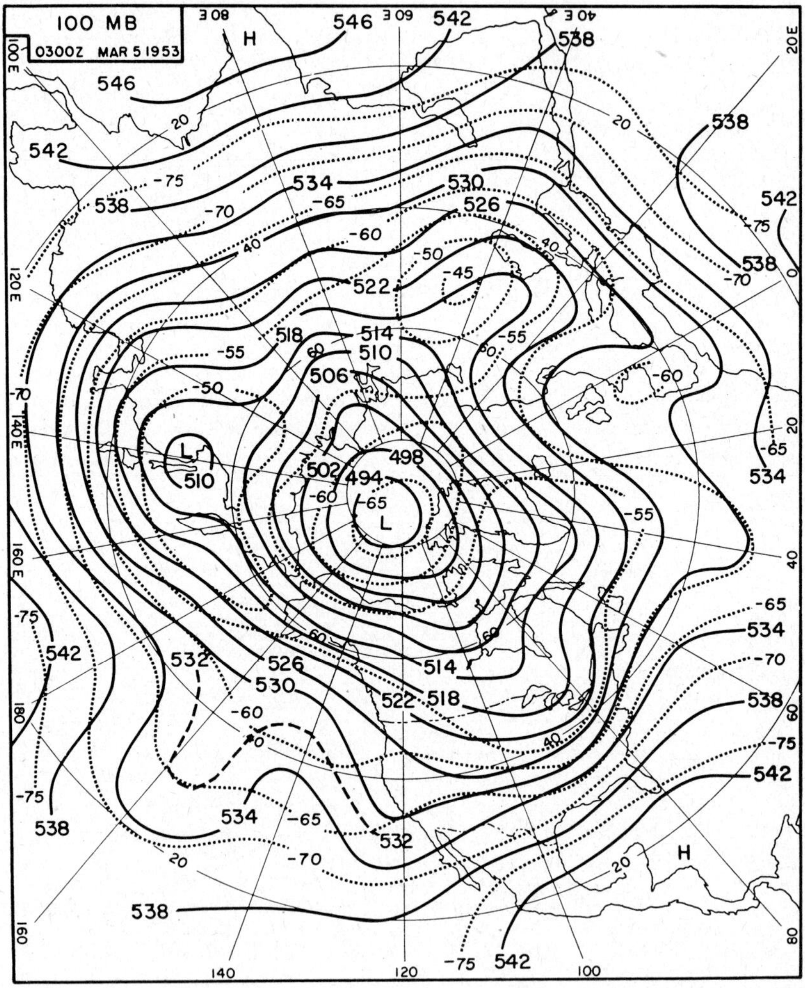

FIG. 15. A 100-mb northern hemisphere chart for March 5, 1953. Temperatures °C, contour height in hundreds of geopotential feet (100 ft = 30 meters) (after Moreland [6]).

Figures 15 and 16 show a pronounced cold polar vortex, which intensifies from 100 to 50 mb. It should be pointed out that the figures describe the flow in March, rather late in the season to find such strong polar vortices. Actually, the winter of 1953 (the only period for which a series of daily hemispheric high level charts has so far been published) was unusual in that the vortex was weak in January and February and reached full intensity only in March.

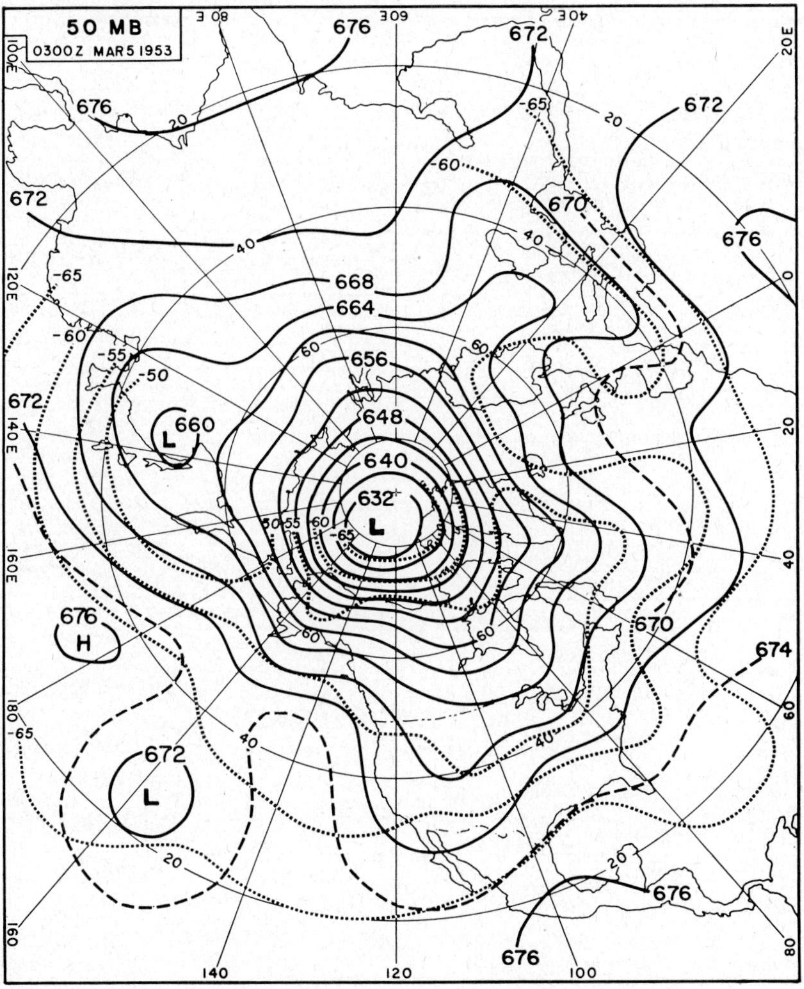

FIG. 16. A 50-mb northern hemisphere chart for March 5, 1953. Temperatures in °C. contour height in hundreds of geopotential feet (100 ft = 30 meters) (after Moreland [6].)

This is the reason why March 5 was selected for the figures. More commonly, the vortex is completely developed in December, and disappears more or less suddenly at any time in the next three months. Except for the time of the year, and perhaps for the too exact location of the vortex center over the pole, the situation depicted in Figs. 15 and 16 is quite characteristic: a single major

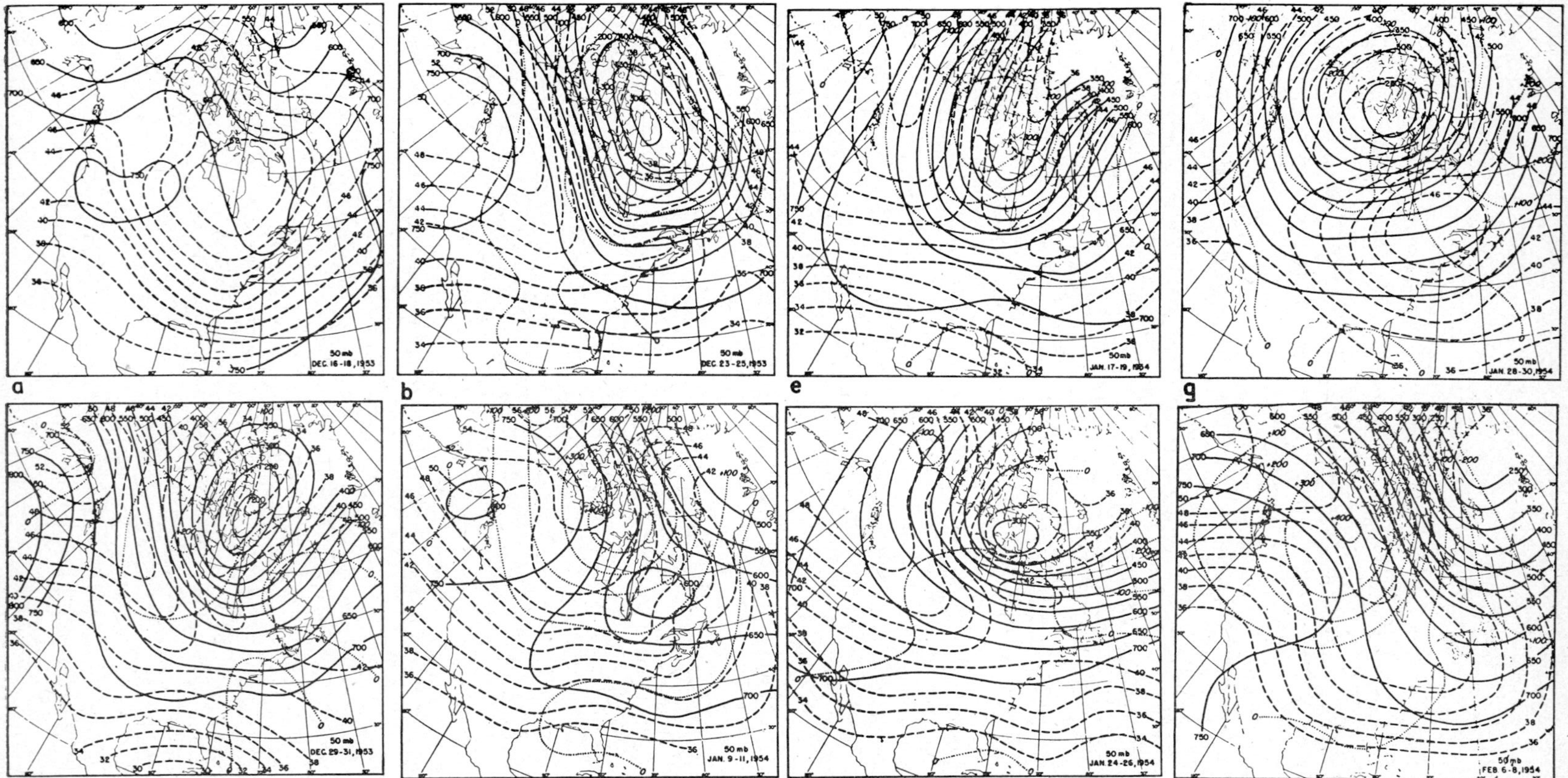

a b e g c d f h

FIG. 17. Three-day mean 50-mb charts for periods indicated. Solid lines are contour lines in tens of geopotential feet, with first digit (six) omitted (100 ft = 30 meters). Dashed lines are height differences between 100 and 50 mb in tens of geopotential feet, with the first digit omitted. These differences are proportional to the mean temperature between 100 and 50 mb; a difference of 100 ft corresponds to a difference of 1.5°C. Dotted lines are height changes from previous chart in tens of geopotential feet (after Austin and Krawitz [8]).

cyclonic cell, with minimum temperature in its center, surrounded by smaller lows. Also, there is a typical wedge over Alaska, a feature which intensifies upward and therefore belongs to the polar regime.

In other seasons, completely different configurations may occur. In general, day-to-day changes are small and are obscured by observational error. From Figs, 17a to 17h [8], showing a series of eight three-day mean patterns at weekly intervals, it is clear that considerable changes are occurring in the area of the PRB, which in this season is normally located close to the Canada-United States border. For example, the intensification of the polar vortex in the first week, and the subsequent filling by 4000 ft (1200 meters) are quite remarkable. Along with these variations go radical changes of wind direction as well as speed.

3.4. *Explosive Warming*

Perhaps the most interesting phenomenon of the polar stratosphere is the "explosive" warming (for a detailed discussion, see Craig and Hering [9], Teweles [10], Godson and Lee [11], and Teweles and Finger [12]) which occurs in some winters. This is, however, primarily a phenomenon of the stratosphere above 20 km, and will be discussed in more detail later. The effects of such warming on the lower stratosphere (100 and 50 mb) are somewhat less conspicuous. Figure 18 shows three 100-mb charts at weekly intervals, during the much-discussed warming of January, 1957, in which sections of the upper stratosphere experienced warming by as much as 50°C. All that is seen at 100 mb is a slow westward drift of the polar vortex, and a slow warming which eventually led to temperatures above − 40°C. Apparently, persistence forecasts for one to two days would not have led to too large errors in the flow. At 50 mb, the changes were more pronounced; the same westward drift of the center was accompanied by a filling of the primary vortex by 3600 ft, and along with it a warming to − 25°C. But, whereas the westerlies weaken during the warming, this period does not yet imply the return to the summer easterlies, which occurs in April or May.

In another famous situation, that of February, 1952 [13], the heights and temperatures at 50 mb in Canada became so high that, in the region often occupied by a cold cyclonic vortex, a strong warm anticyclone appeared instead, and for a two-week period easterly currents were found at 50 mb over most of the United States leading to a wind distribution in February usually found only in summer. Further, the 100-and 500-mb westerly indices became unusually low.

The so-called explosive changes are still slow by tropospheric standards. Furthermore, they rarely lead to dramatic changes at 100 mb, but they may affect the general regime there as well as in the troposphere in subsequent

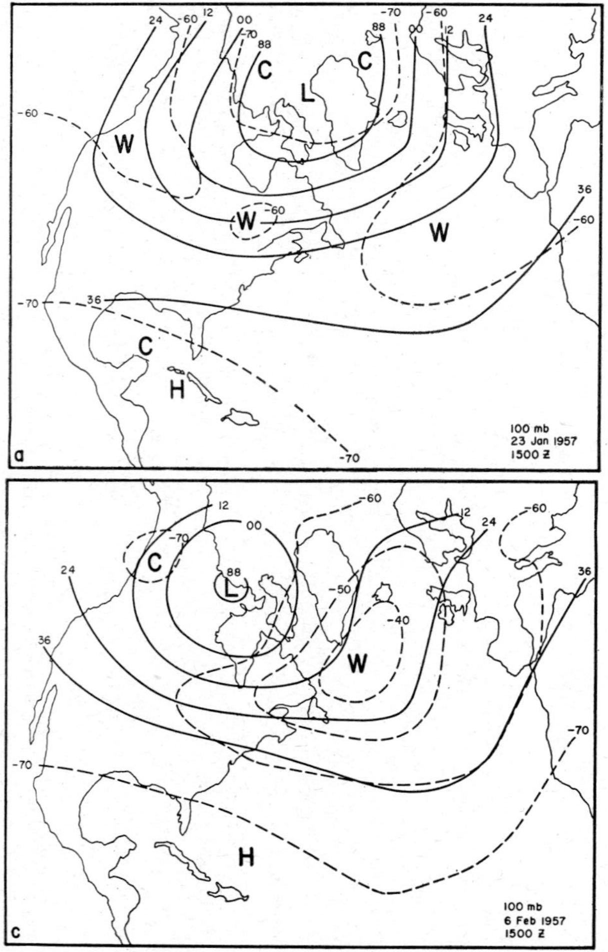

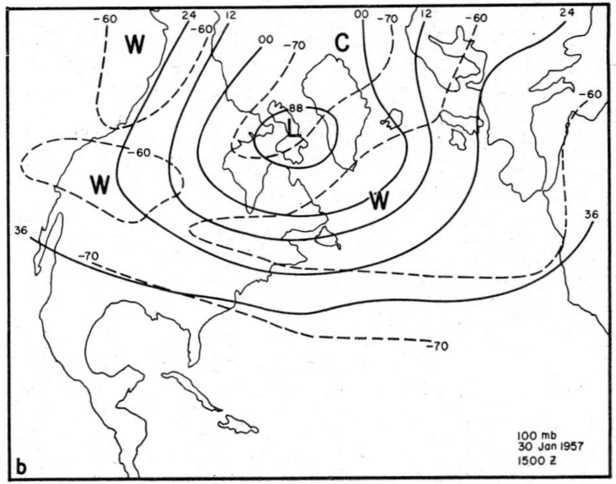

FIG. 18. Sequence of 100-mb charts showing "explosive" warming of January, 1957. Solid lines are contours in hundreds of feet (first digit omitted; 100 ft = 30 meters). Dashed lines are isotherms in °C (after Craig and Hering [9]).

weeks. The warming is usually pronounced at 50 mb and accompanied by increasing contour heights.

The cause of the relatively sudden warming is unknown, and was originally attributed to solar effects. The more generally accepted theory is something like this: in response to the cooling during the polar night, a strongly baroclinic region is formed just outside of the center of the polar vortex. As the temperature gradient becomes large, the circulation becomes unstable, and perturbations arise with vertical motions of the order of 1 to 5 cm/sec. With the large

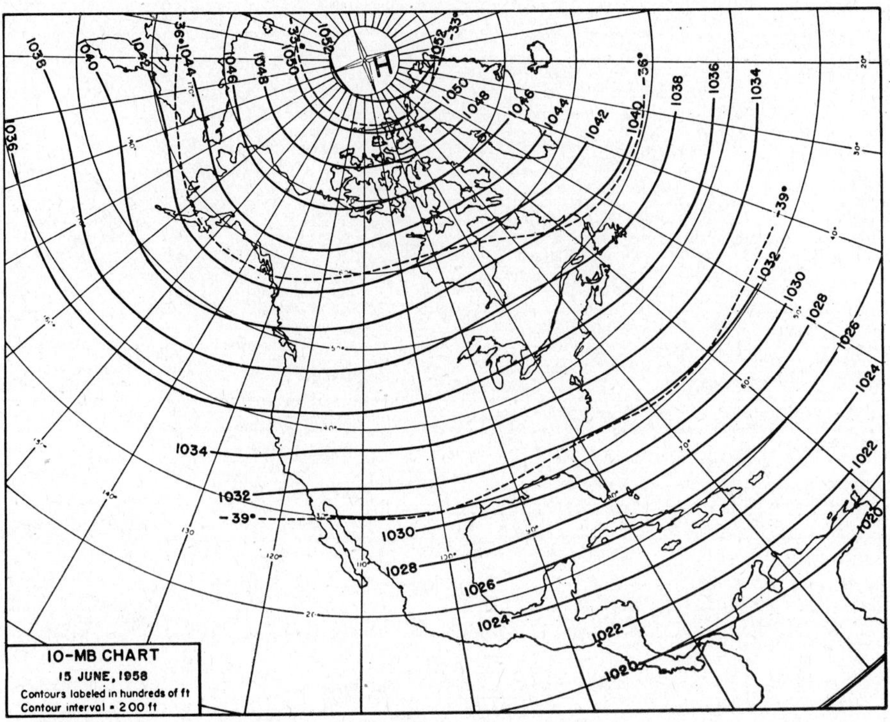

FIG. 19. A 10-mb chart, June 15, 1958. Contour lines (solid) in hundreds of geopotential feet (100 ft = 30 meters). Isotherms (dashed) in °C (after Teweles and Finger [14]).

stratospheric hydrostatic stability, these velocities are sufficient to account for most or all of the warming. Of course, once an unstable configuration has been reached, solar activity could still be the trigger which releases the instability. Statistical results on this point have so far not been particularly convincing.

These large downward air movements presumably provide the most efficient mechanism for bringing ozone and radioactive debris from the higher to the lower stratosphere, and are quite possibly responsible for the spring maxima of these atmospheric constituents at low levels.

Practically all developments in the polar stratosphere increase in intensity with increasing height. They are presumably not much influenced by tropospheric changes, although, as we have seen, the smoothed tropospheric charts

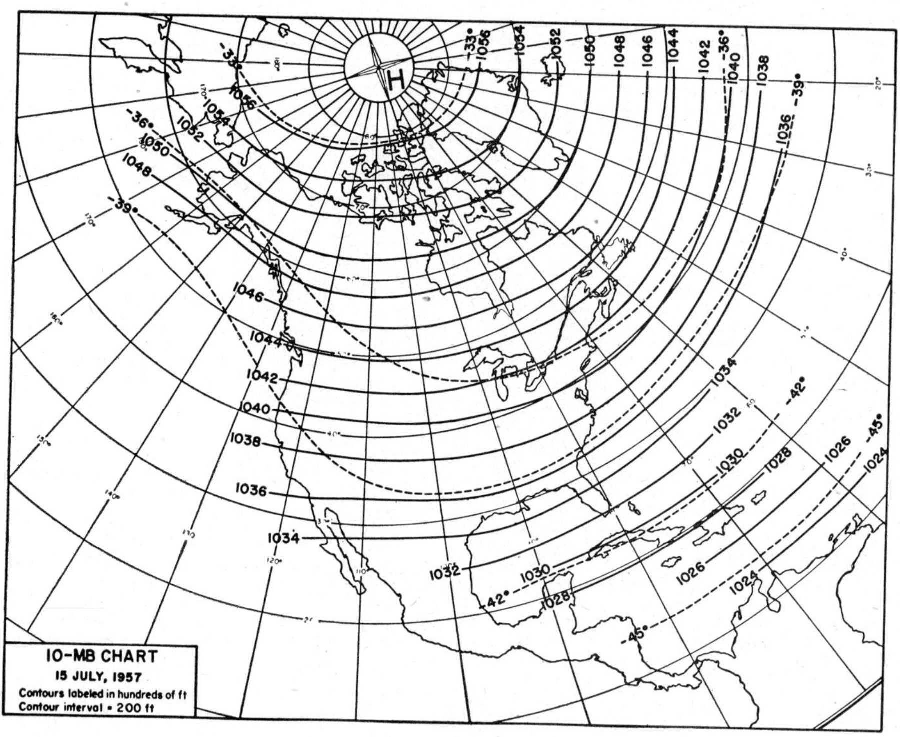

FIG. 20. A 10-mb chart, July 15, 1957. Contour lines (solid) in hundreds of Geopotential feet (100 ft = 30 meters). Isotherms (dashed) in °C (after Teweles and Finger [14]).

can be thought to show weakened versions of the stratospheric flow. It seems tempting to assume that the tropospheric flow in these latitudes is produced by the stronger stratospheric flow above, particularly since the strong northward temperature gradient in the polar stratosphere acts as a strong energy source, whereas no source of comparable intensity exists in the troposphere.

4. Synoptic Properties above 20 km

4.1. The Flow above 20 km in Summer

Above 20 km, the speed of all currents tends to increase with height. The lower boundary of this region is usually between 50 and 25 mb. It is characterized in all seasons by warm high or cold lows.

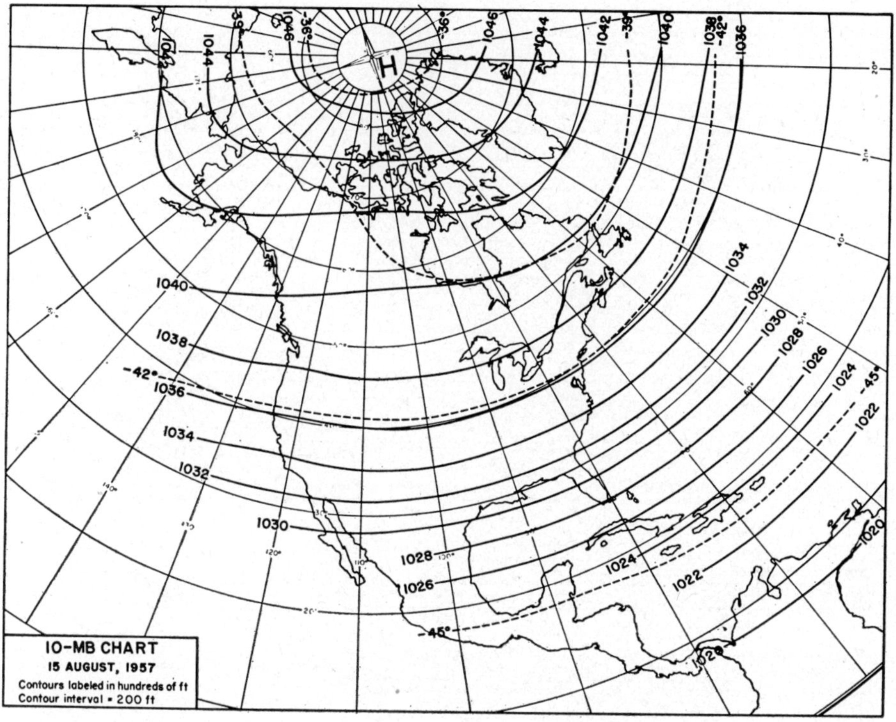

Fig. 21. A 10-mb chart, August 15, 1957. Contour lines (solid) in hundreds of Geopotential feet (100 ft = 30 meters). Isotherms (dashed) in °C (after Teweles and Finger [14]).

In the summer, the circulation is usually dominated by a single warm (— 33°C) high near the pole, so that the whole hemisphere is covered by easterly winds. Figures 19 to 21 from Teweles and Finger [14] show typical summer flow patterns at 10 mb.

The summer pattern is normally found only in June, July, and August. In April and September, high cells are found around latitudes 35° and 40°, and polar and mid-latitude regions are already dominated by westerlies and a cold polar vortex, a situation which we will see to be characteristic winter pattern. In the true winter pattern (October to March), of course, the high cells move even further south, to around 25°.

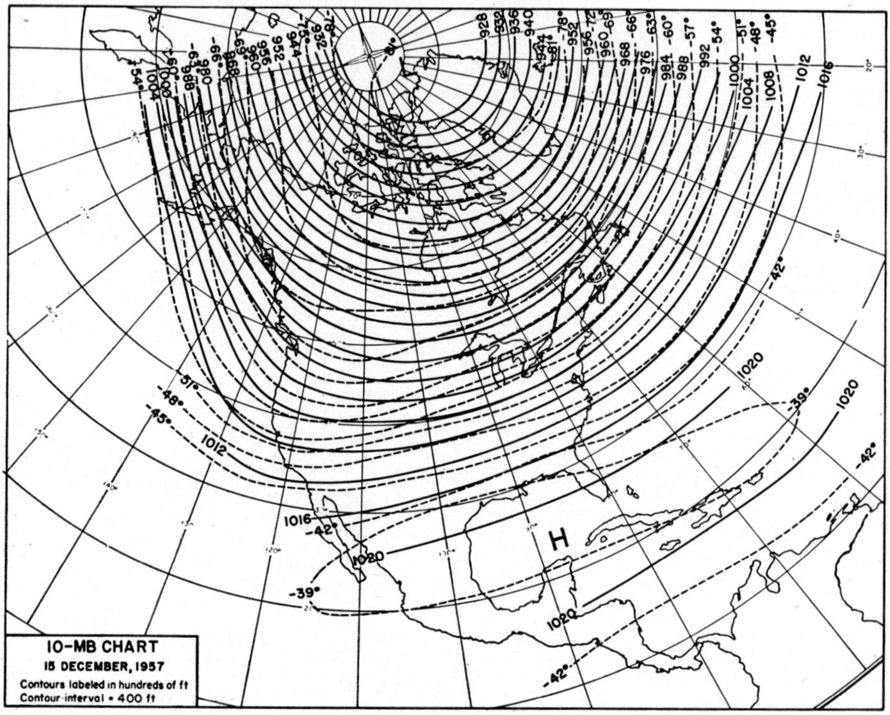

FIG. 22. A 10-mb chart, December 15, 1957. Contour lines (solid) in hundreds of geopotential feet (100 ft = 30 meters). Isotherms (dashed) in °C (after Teweles and Finger [14]).

The synoptic patterns in summer have been discussed by Hare [15], particularly in the Arctic. The assumption of easterlies of 15 meters/sec about latitude 45°, increasing to perhaps 30 meters/sec at 25°, should give the correct order of magnitude at 10 mb. Of course, if previous data are available, persistence is a particularly useful assumption in the upper stratosphere summer. There seems to be little variation from day to day, and from summer to summer.

4.2. The Flow above 20 km in Winter

The PR extends from the cold polar cyclonic center near the pole southward to the belt of warm high centers, usually around latitude 25° in winter.

This configuration yields westerlies which increase with height, generally between the poles and the tropics. The strongest westerlies are found about latitude 70° at 10 mb and are sometimes referred to as the "arctic jet."

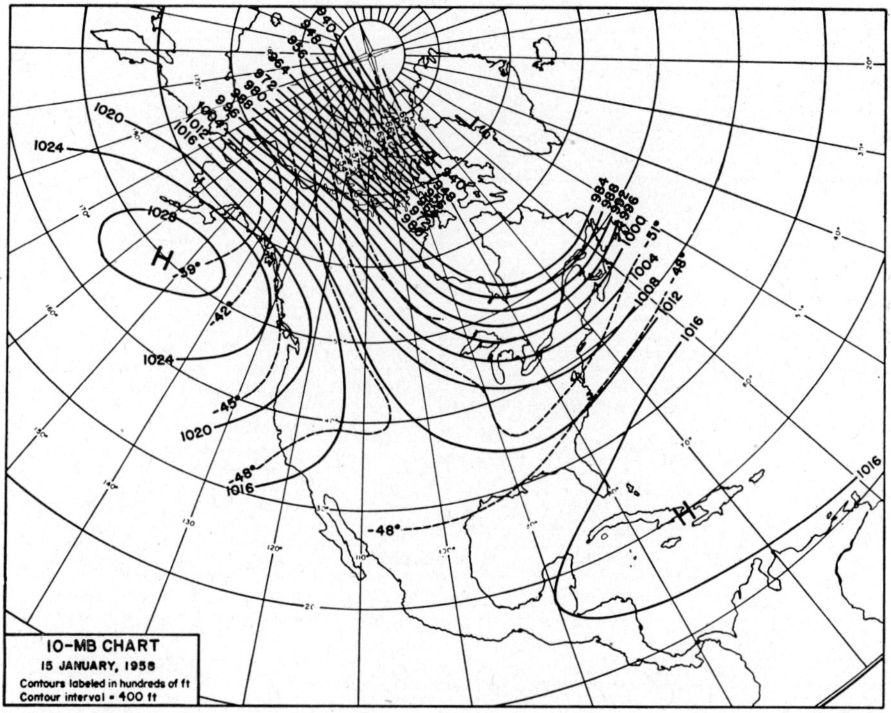

FIG. 23. A 10-mb chart, January 15, 1958. Contour lines (solid) in hundreds of geopotential feet (100 ft = 30 meters). Isotherms (dashed) in °C (after Teweles and Finger [14]).

The pattern is generally not a simple zonal one however. Particularly in the north Pacific and the Alaskan area, a warm high is frequently found as far north as latitude 50°, thus producing northerly wind in the western United States (see Fig. 23). In general, large changes occur over periods of the order of a month, as seen from Figs. 22–24; and, of course, larger variations occur from season to season. Still, for a day or two at a time, persistence is the most useful forecasting tool.

Extreme changes do occur, particularly after December, which, as already pointed out, can cause temperature increases of as much as 50°C in two weeks. These changes takes the form of temperature oscillations, usually culminating in a "final" or "explosive" warming (16). Such final warming is shown, for example, by Fig. 25 for Keflavik, Iceland. Here the temperature changes from — 70° to — 20°C between January 19 and February 1, 1957. The cross section also shows that the rapid warming occurs first and most strongly at the higher levels, but some effect is felt later as far down as 100 mb.

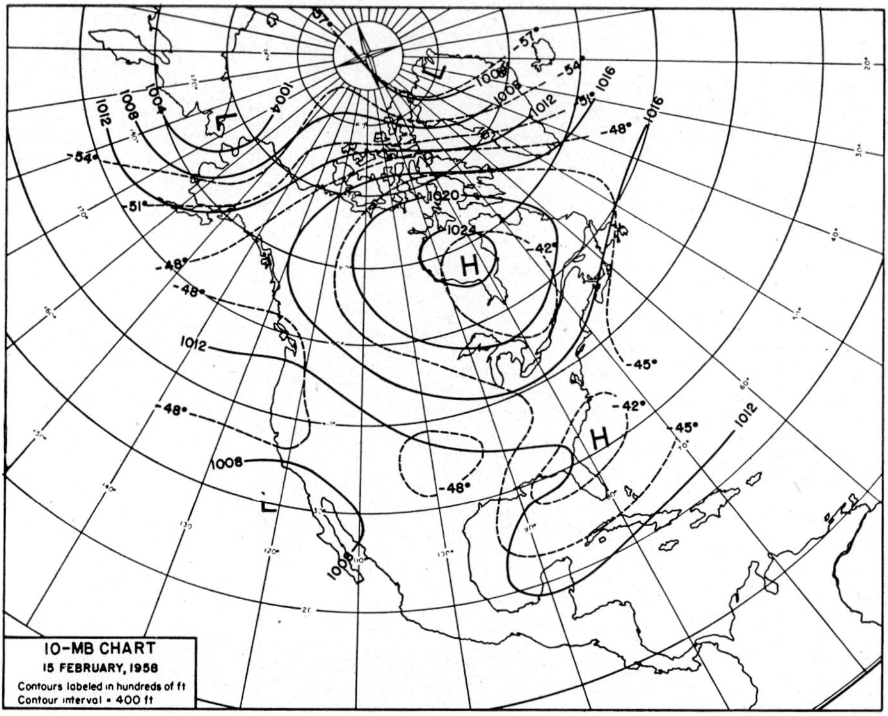

FIG. 24. A 10-mb chart, February 15, 1958. Contour lines (solid) in hundreds of geopotential feet (100 ft = 30 meters). Isotherms (dashed) in °C (after Teweles and Finger [14]).

The stratosphere up to 10 mb is normally nearly isothermal in the vertical. Above this level, the temperature increases rapidly upward to perhaps 1 mb (about 50 km), where the temperature is about the same as at sea level. This inversion layer has been called the mesoincline. Perhaps thermocline is a better designation, in analogy to the similar oceanographic phenomenon. The boundary between the thermocline and the isothermal layer is shown by a heavy line in Fig. 25. During periods of the "explosive" warming, this boundary apparently moves far down into the lower stratosphere. To put this another way: when there is subsidence in the thermocline, tremendous warming occurs underneath, which is due to the normal warming produced by subsidence but intensified by the fact that the temperature above 10 mb is quite warm initially. Thus, subsidence causes the inversion to move gradually to lower and lower levels, and with it the boundary surface.

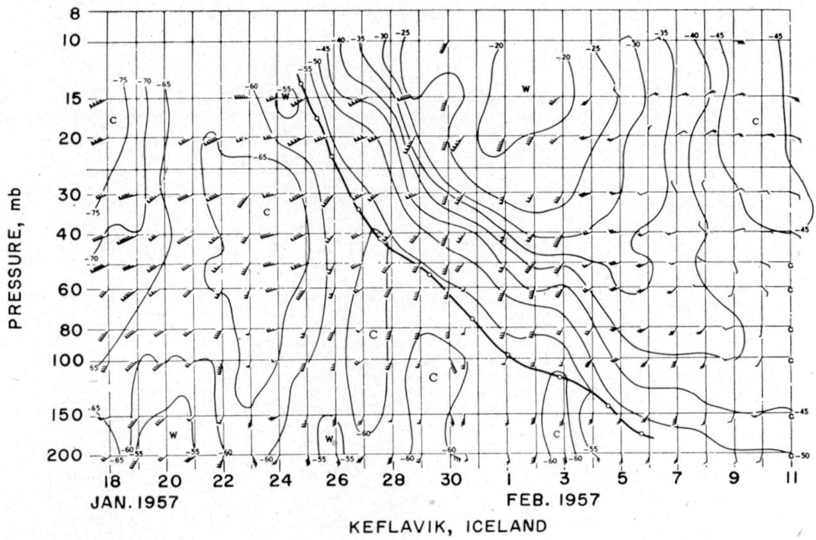

FIG. 25. Vertical time cross section, January 18 to February 11, 1957, for Keflavik, Iceland. Isotherms in °C, winds from bottom of diagram for south winds. Heavy line shows boundary between isothermal zone and thermocline (after Teweles [10]).

Along with the warming, anticyclogenesis occurs. Actual high centers are produced which increase in intensity upward from 50 mb, as shown for example in Fig. 26 at 25 mb. Generally, the cyclonic polar vortex is weakened. It has been claimed that the vortex does not re-establish itself after such fast warming; however, it is possible that strong remnant cold lows may still be found on the Asiatic side of the pole after warming on the American side since synoptic patterns over Siberia at high levels are still not well known.

4.3. Summary

In summary, the stratosphere from 20 to 30 km is a region of large-scale wind systems increasing in intensity upward, which may undergo profound changes over periods of the order of weeks and sometimes days. Westerlies normally prevail between the polar vortex and the high-pressure belt in the subtropics, but the actual flow patterns are often completely different from the normal flow patterns. It is doubtful that the beginning of changes which may occur relatively rapidly in the upper stratosphere can be anticipated from tropospheric changes. It is more likely that such changes spread downward and eventually affect the polar troposphere.

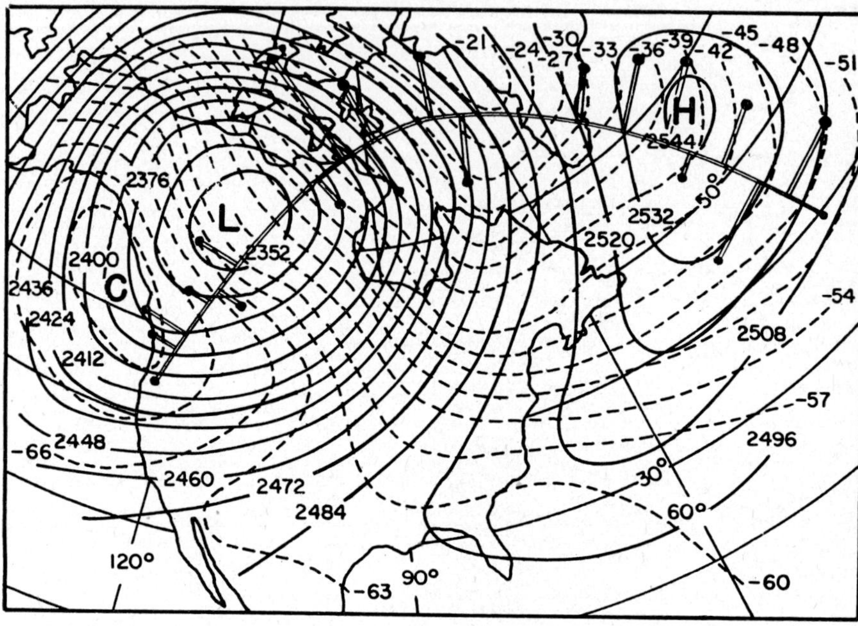

FIG. 26. A 25-mb chart, February 6, 1957. Solid lines are contours in tens of geopotential meters, dashed lines isotherms in °C (after Teweles [10]).

Summer patterns vary less from the normal than do winter patterns. The summer wind systems occur in about the same location each summer, but large differences may exist between different winters.

ACKNOWLEDGMENTS

This chapter is an only slightly modified version of a manual prepared for the Navy Weather Research Facility at Norfolk, Virginia, and was published with permission of its Officer-in-Charge, CDR Daniel F. Rex, U.S. Navy.

The author is extremely grateful to the staff of this organization for offering all possible help in preparing the manuscript; he would like to thank particularly Mr. Alvin Morris and LCDR W. L. Somervell for critically evaluating the manuscript and making many valuable suggestions.

REFERENCES

1. Goody, R. M. (1958). "The Physics of the Stratosphere," Cambridge Univ. Press, London and New York.
2. Kochanski, A. (1955). Cross sections of the mean zonal flow and temperature along 80°W, *J. Meteorol.* **12**, 95–106.

3. Borden, T. R., Jr. (1958). The wind field at 100,000 feet during the first six months of the IGY (July 1957–December 1957), *Contrib. Stratos. Meteorol. GRD Res. Notes No.* **1**, 51–70.
4. Kochanski, A. (1956). Wind, temperature and their variabilities to 120,000 feet, *Air Weather Service Tech. Rep.* pp. 105–142.
5. Julian, P. R., Krawitz, L., and Panofsky, H. A. (1959). The relation between height patterns at 500 mb and 100 mb, *Monthly Weather Rev.* **87**, 251–260.
6. Moreland, W. (1956). Daily series synoptic weather maps Northern Hemisphere, 100 mb and 50 mb charts, U.S. Weather Bureau.
7. Holloway, J. L., Jr. (1958). Smoothing and filtering, *Advances in Geophys.* **4**, 382–385.
8. Austin, J. M., and Krawitz, L. (1956). 50-millibar patterns and their relationship to tropospheric changes, *J. Meteorol.* **13**, 152–159.
9. Craig, R. A., and Hering, W. S. (1959). The stratospheric warming of January-February 1957, *J. Meteor.* **16**, 91–107.
10. Teweles, S. (1958). Anomalous warming of the stratosphere over North America in early 1957, *Monthly Weather Rev.* **86**, 377–396.
11. Godson, W., and Lee, R. (1958). High level fields of wind and temperature over the Canadian Arctic, *Beitr. Phys. Atmos.* **31**, 40–68.
12. Teweles, S., and Finger, F. G. (1958). An abrupt change in stratospheric circulation beginning in mid-January 1958, *Monthly Weather Rev.* **86**, 23–28.
13. Warnecke, G. (1952). Ein Beitrag zur Aerologie der arktischen Stratosphäre. *Meteorol. Abh.* **3**, issue 3, pp. 1–60.
14. Teweles, S., and Finger, F. G. (1958). Contours and isotherms at 10 millibars, July 1957 to June 1958, Unpublished report, U.S. Weather Bureau Office of Meteorological Research.
15. Hare, F. K. (1960). The summer circulation of the Arctic stratosphere below 30 km, *Quart. J. Roy. Meteorol. Soc.* **86**, 127–143.
16. Hare, F. K. (1960). The disturbed circulation of the Arctic stratosphere, *J. Meteor.* **17**, 36–51.

ARCTIC METEOROLOGY

*(A Ten-Year Review)**

A. D. Belmont

General Mills, Minneapolis, Minnesota

1. Introduction

In the eight years since Arctic meteorology was reviewed in the *Compendium of Meteorology* [1] there have been interesting advances in the subject. With the IGY having completed its original 18-month period it is to be expected that numerous additional features of the polar circulation will be discovered in the next five or ten years as the immense data files are slowly digested. Most of the IGY effort in polar meteorology was directed toward the Antarctic where far less is known than in the Arctic, but it is only by comparing the two polar regions that full benefit from the data of either can be realized. Complete Antarctic data for the years up to 1956 have recently been summarized

**Paper submitted February, 1959.*

most elegantly [2]. We intend here only to present some of the major new ideas concerning the Arctic circulation supplemented by a brief review of the literature to serve as background for the coming IGY contributions.

It has been our feeling for many years that the Arctic region is the scene of interesting features in the atmospheric circulation and therefore deserves special attention. It is true that the general circulation permits no region of the atmosphere to be isolated, but when we consider the unique properties of high latitudes it must be conceded that singularities may indeed exist here which cannot be studied from temperate latitude data. Consider, for example, that only in the polar regions do we have conditions of totally absent or continually present insolation. This fundamental property results in a complete seasonal reversal of the stratospheric thermal field and its circulation —a process tied to the polar regions but which affects the entire circulation. A second characteristic is the presence of the pole of rotation which serves well to point out the shifting asymmetry of the circulation at all levels. This also means that the Coriolis force is a maximum there. Total ozone in a vertical column also increases toward high latitudes. Where its maximum is and why are still unsolved. The presence of the geomagnetic poles means that any magnetic variations or associated effects whose nature is only speculation now, will have their maximum changes in the polar regions and that theories of possible solar-terrestrial relations might best be revealed at highest latitudes. For such reasons the Arctic is a region well worth special geophysical attention. Some of the extraordinary results reviewed here should stimulate further progress in this little known area, utilizing the increased amount of polar data from the IGY. Although our emphasis will be upon the Arctic area (north of 70° in the European sector, 60° elsewhere) it will be at times necessary to discuss lower latitude events in order to appreciate the general problem.

1.1 Station Network

Progress in meteorology has always been dependent upon direct observations. In the past twenty-five years since the advent of the radiosonde in the Arctic, a good station network has gradually grown.

The radiosonde was used along the Siberian Arctic coast in the 1930's and was introduced in 1932 to Alaska (Fairbanks) and to the Canadian Arctic in 1942 (Ft. Nelson). The first meteorological expedition to the Arctic Basin since that of the "Maud" was the Russian ice floe station "North Pole 1" of 1937–1938. It had no radiosondes, yet made important observations the results of which were not published until the post-war years. It was 1953 before many of the findings were translated and circulated in the West [3–5]. Following the war the importance of weather data from high latitudes for the requirements of aviation forecasts supported the establishment of the first North American Arctic stations in Greenland and Canada.

In addition, regular aerial weather reconnaissance flights at the 700-mb level between Fairbanks and the North Pole were begun in 1947. The altitude was later raised to 500 mb and the route varied in accordance with weather requirements. Beginning in 1950, dropsondes were released several times along the route from 500 mb giving the first routine upper-air data for this area.

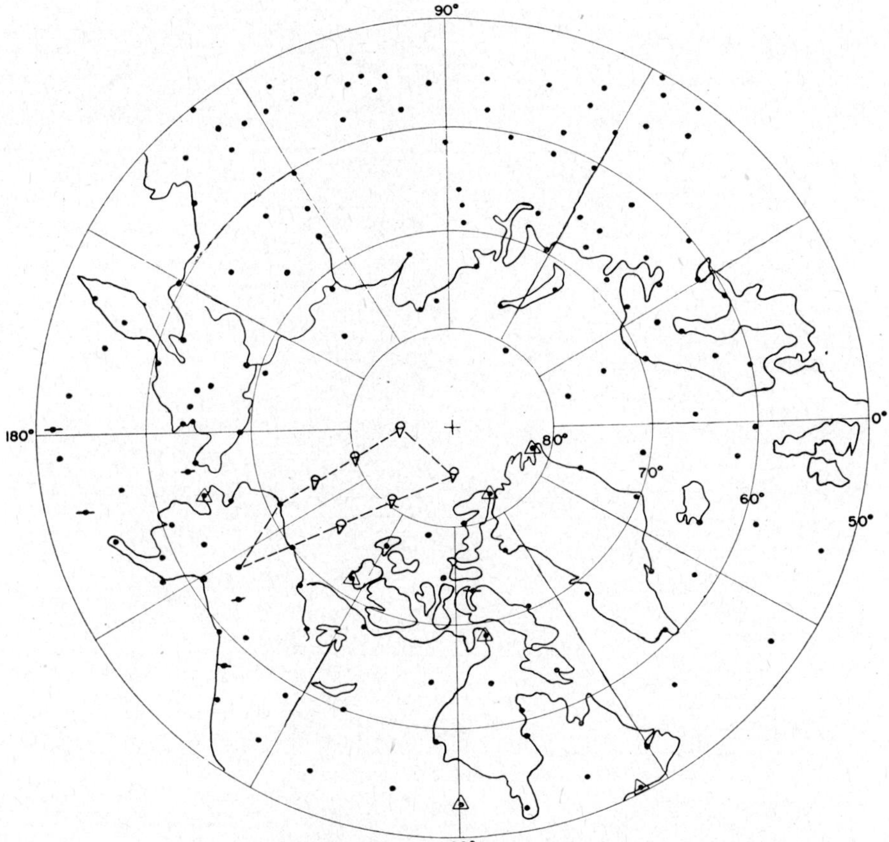

FIG. 1. Location of radiosonde stations, December 1958. Triangle marks stations added since 1950; bar shows those discontinued since 1950; parachutes mark approximate location of Ptarmigan flight dropsondes.

Figure 1 gives an impression of the density of radiosonde or rawinsonde stations in the Arctic as of December, 1958. Stations in North America which came into service since 1950 are indicated with a triangle and those discontinued since then with a bar. The routes of "Ptarmigan" reconnaissance flights are indicated and the approximate locations of their dropsondes.

The next step was the establishment of scientific stations on the ice, as had been proved practical by North Pole 1. The Russians had an ice floe station in the Arctic Basin again from 1950–1951 (North Pole 2) which took the first radiosonde observations in the polar basin during the summer of 1950 [6].

The next contribution to Arctic upper-air coverage came with the opening of "T–3," a "permanent" station on a large stable ice island from where rawinsondes were begun in the spring of 1952. The station originally was near the pole, but in the following two years drifted southward toward Alert and was finally abandoned in April, 1954 when it was no longer of sufficient

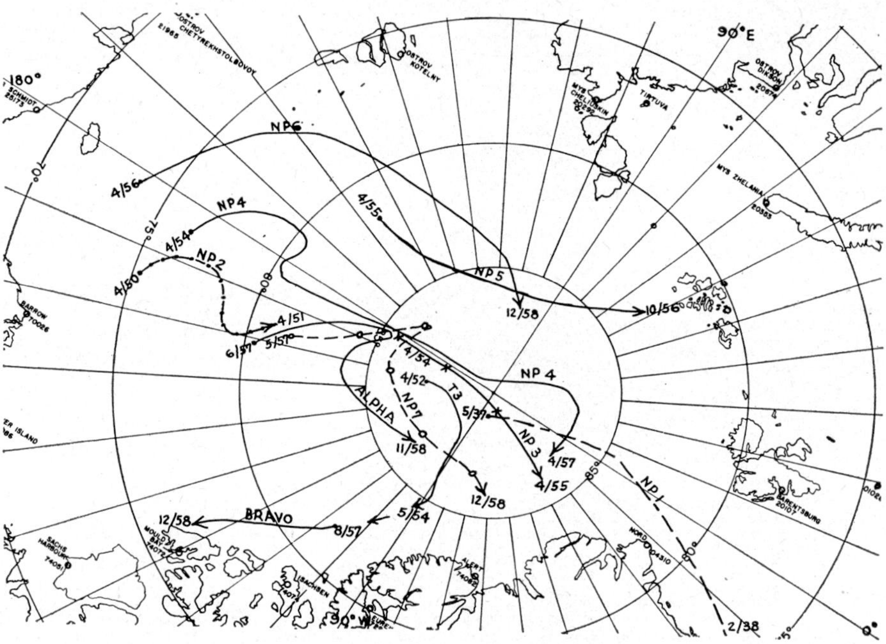

FIG. 2. Drifting Stations in the central Arctic.

meteorological value to warrant its upkeep. It was reoccupied during the summer of 1955, and again for the IGY in 1957 (called station "Bravo") but during neither of these periods were radiosondes resumed. Station "Alpha" was established on an ice floe in 1957 and is a radiosonde site. The Soviets increased their own ice floe program in April, 1952 (North Pole 3) and have had at least one geophysical station in operation since then. Complete information is lacking on the history of these stations, although from notices in *Polar Record*, *Polarforschung*, and reference [7] North Poles 4, 5, 6, and 7 were occupied for periods from 1954 to 1958. As of December, 1958, North Poles 6 and 7 began operation and transmitted radiosonde data. North Pole 6 differs from other Russian ice stations by being an ice island some 82-sq. km in extent and thus is comparable to T-3. The thickness of such islands is about 10 meters [8, 7].

Figure 2 gives the general drift area, in a stylized manner, of all ice islands. The map is based partly on locations contained in daily weather reports kindly provided by Mr. H. Wilson, of the Edmonton Meteorological Office; notices in *Polar Record*, especially in issues number 57 and 60; and in reference [7]. Table I lists their dates of operation and gives references to the results of their observations.

TABLE I. Arctic ice stations (as of December, 1958).

	Years of operation	Radiosonde (R/S) or Rawinsonde (R/W)	Reference to data and results
Russian Stations			
N.P. 1	5/1937–2/1938	—	[4]
N.P. 2	4/1950–4/1951	R/S	[6]
N.P. 3	4/1954–4/1955	R/S	
N.P. 4	4/1954–4/1957	R/W	[9, 10]
N.P. 5	4/1955–10/1956	—	[11, 12]
N.P. 6	4/1956–12/1958	R/W	[11, 12]
N.P. 7	4/1957–12/1958	R/W	
U.S. Stations			
T-3	6/1952–5/1954	R/W	[13, 14, 15]
T-3	Summer, 1955	—	
Alpha	6/1957–11/1958	R/W	
Bravo (T-3)	8/1957–12/1958	—	

Reference [7] is a convenient source which also summarizes some of the results of Arctic and Antarctic expeditions. Other general reviews of Soviet area geophysical results are given by references [6, 11, 12, 16] and notes in *Polar Record* and *Polarforschung*.

1.2. Data Sources

Raw data plotted on daily charts are often unavailable at a later time for research involving either short-term analyses or climatic summaries. Owing to the rarity of Arctic upper-air data, it was thought useful to summarize those sources which might be of interest for future researchers.

1.2.1. Mean Upper-Air Data. The following stations are each the subject of a standard monograph series presenting both upper-air and surface data, published by the U.S. Weather Bureau or the Canadian Meteorological Branch: Thule, Resolute, Alert, Eureka, Isachsen, and Mould Bay. A general compilation of mean data is contained in the U.S. Weather Bureau's "Monthly

Climatic Data for the World" which usually consists of the internationally transmitted "Temp" summaries for several standard levels only. Specialized data for the Arctic for various locations are presented by Flohn [17], for Alaska by Ratner [18], for Greenland (Central Station) by Bedel [19], for the Eurasian Arctic by Putnins and Stepanova [20], for Ice Island T-3 by Belmont [13], for Canada (to 1947 only) by Henry and Armstrong [21], and for Thule and Barrow by Tolefson [22].

1.2.2. Daily Values of Upper-Air Observations. These are available in the excellent but minuscule "Daily Upper Air Bulletin" of the U.S. Weather Bureau. Unfortunately, there is not yet a publication giving detailed monthly summaries of these data, although Alaska monthly values are given in the Bureau's "Climatological Data, National Summary." For a listing of individual observations from the station North Pole 2, see [6].

1.2.3. Mean Monthly Upper-Level Maps. The following national weather service serial publications are available: U.S.: "Monthly Weather Review" (contains contours and isotherms over North America for levels to 100 mb). Germany: "Grosswetterlagen Mitteleuropas" (contains contours for the Northern Hemisphere to 500 mb). A special series of individual mean monthly maps has been prepared [23] for the wind and temperature fields at 100, 50, and 25 mb for North America for the single months December, 1956 through February, 1957 and at 50 mb for January, 1953. The mean vector wind for the months July, 1957 through December, 1957 over North America has also been presented [24].

Longer-period means or "normal" surface charts have, for a long time, been prepared for the Arctic regions with considerable disagreement. This has been well reviewed elsewhere [1, 25]. However, there have been a few several-year means for the Arctic upper air, based on recently available observations, which will be discussed in some detail in Section 2.

1.2.4. Daily Upper-Level Maps. The U.S. "Daily Series, Synoptic Weather Maps" giving surface and 500-mb Northern Hemisphere maps, and the German "Täglicher Wetterbericht," containing hemispheric surface and 500-mb maps plus European area maps to 70° N for levels up to 100 mb, are well known. A special series of great value are the daily 100- and 50-mb hemispheric charts prepared for the seven months: January to May, July, and October of 1953 by the U.S. Weather Bureau. Daily Russian hemispheric maps for levels to 200 mb were prepared starting apparently in January 1958 and present good data coverage over the polar region.

1.3. Arctic Meteorological Research Groups

The history of Arctic research is naturally tied to that of availability of data. Prior to the last war this was mainly collected by expeditions, and with a few notable exceptions (Sverdrup, Hergesell) resulted in giving us only our first impressions of surface climate. The post-war period ushered in the first regular aerological coverage for the area upon which all recent progress is based. This newly available material inspired the creation of a research group in Arctic meteorology at the University of California in 1950. The main interest at UCLA was in upper-air circulation and especially the climatology of the thermal fields of the polar regions; a summary of the work done there is given by Belmont [14]. In 1954, the UCLA effort was transferred to McGill University, Montreal, where research still continues. With few exceptions the efforts there, until 1958, were directed toward statistical methods of describing the pressure field. In the past year, an interest has been added in the Arctic stratosphere, perhaps owing to the recent papers of Godson of the Canadian Meteorological Branch, Toronto, who serves as a consultant to the group, in describing some of the "sudden warmings" using Canadian data. Cooperative arrangements now permit high-level charts over North America (200, 100, and 25 mb) to be prepared at the Dorval Weather Central near Montreal and to be analyzed by the McGill group. The best available Arctic data are specially routed to this center so that interesting results can be anticipated in the near future. A summary of the work done to 1958 is given by Hare [26].

Shortly after the McGill group started, another program was begun at the University of Washington in Seattle. This group concentrates on the synoptic features of the area from the North Pacific to Greenland [27].

In addition to these North American Arctic meteorology centers, attention must be called to the Institut für Meteorologie und Geophysik, of the Freien Universität Berlin which, under the direction of Scherhag, has made some very notable contributions to Arctic meteorology in the course of its general research program. Since 1953, some six major contributions have so far appeared which have direct bearing on polar meteorology, although they may apply also to other latitudes. On an individual basis, Flohn, of the German Weather Service, has also contributed important papers to this area.

Soviet efforts in the Arctic are famed, but their meteorological research has not been generally available in the West even in the Russian language until very recently, and much less so in translation. From their journals it is evident that they actively continue research in polar meteorology, and it seems safe to estimate that until very recently there was a five-year time lag in receiving this information in English.

The results of the above centers will be summarized below. A review by Hare and Orvig [28] has just appeared which is a convenient summary to much work in Arctic meteorology up to about mid-1958. A major conference on the subject was held in Oslo in 1956 and is fully reported [A]. Further references may be found in special bibliographies on the subject [B-E].

Due to this very large amount of material, our attention in Sections 2 and 3 will be concerned with two major advances: the general circulation as shown by new mean stratospheric maps, and the phenomena of sudden temperature and wind changes. In Section 4 we will briefly mention papers which have appeared in some of the other phases of Arctic meteorology for the convenience of those wishing to locate recent work in particular aspects of the subject.

2. Mean Fields of Pressure and Temperature

Ideas of the mean pressure field over the polar regions have long been controversial from the days of the first sailing ships which observed the pressure and wind at high latitudes. Despite the great increase in our knowledge of temperate latitude meteorology during the past 100 years, it must be admitted that only in the past 10 years have regular observations from the central polar regions been furthered. Consequently, essential descriptions of the pressure, temperature, and motion fields in this area, especially in the stratosphere, are just recently becoming available.

The present radiosondes were not designed for use at levels at which we now require accurate data. Errors in the aneroid and thermal elements increase rapidly with height so that all present data can only give us tentative impressions of these levels [29,30, 31]. Despite the inconsistencies of instruments and of the application, if any, of radiation and lag errors, the patterns appear so generally consistent that the large-scale features should be reasonably good estimates. It is encouraging that these instrumental and observational deficiencies are at least recognized and that plans are underway in the U.S. for a completely new observing system. Fortunately, the days of the geographer-explorer for whom observations served mainly as an excuse for adventure, and yet whose data were the only basis for our knowledge, are decreasing, and the number of established scientific stations in polar regions, suitably manned, is increasing. When instrumental improvements are made our horizons will extend significantly into the stratosphere.

As upper-air observations for high latitudes were only begun in the past ten years, reasonably reliable mean charts for the Arctic upper air could not be prepared until recently. Mean maps prepared from data of 1948 or later are listed in Table II for convenient reference. There is no single series of charts showing all levels from the surface to the mid-stratosphere which is consistent both with respect to area and period of record. The two complementary series by Heastie [35] and Goldie [39] give mid-seasonal values for levels from 700 to 100 mb. However, a more complete set is the monumental series prepared by Scherhag's group in Berlin [25, 33, 37, 38, 40] which will be used here in the following discussion of the Arctic upper-air climatology.

TABLE II. "Normal" charts for the Arctic, based on data since 1948.

Source	Variables	Levels(mb)	Area	Frequency	Basic data
U.S. Weather Bureau, 1952 [32]	Height Temperature Thickness	Surface, 700, 500 Surface, 700, 500 700–1000	Northern Hemisphere	Monthly	Revised 1944 normals
Anonymous (Scherhag Group) 1953 [33]	Pressure	Surface	65–90°N	Monthly Annual	1949–1952
Kochanski, 1953 [34]	Height Temperature	200, 100, 50, 25 200, 100, 50, 25	Continental N. America	January, April, July, October	1948–1951
Heastie, 1955–1956 [35]	Height Thickness	700, 500, 300, 200, 150, 100 700–500 500–300 300–200 200–150 150–100	55–90°N	January, April, July, October	1949–1953
Namias, 1956 [36]	Height	Surface, 700	Northern Hemisphere	January, April, July, November	1948–1955
Jacobs (Scherhag Group), 1957 [25]	Height Thickness Temperature	1000, 850, 500, 300 500–1000 300–500 500, 300	65–90°N*	Monthly, Annual	1949–1953

*Plus most of the Atlantic hemisphere from 100°W to 50°E. Tabulated grid values also given.

TABLE II.—*continued.*

Source	Variable	Levels(mb)	Area	Frequency	Basic data
Wege (Scherhag Group), 1957 [37, 38]	Height Thickness Temperature Isotachs	200, 100, 50 200–300 100–200 50–100 200, 100, 50 200, 100, 50	Northern Hemisphere	Monthly	1949–1953
Goldie, 1958 [39]	Temperature and Standard deviation	700, 500, 300, 200, 150, 100	Northern Hemisphere	January, April, July, October	1941–1952
Wege *et al.* (Scherhag Group), 1958 [40]	Height Thickness Temperature Isotachs Density	30, 20, (15)* 30–50 20–30 (15–20) 30, 20, (15) 30, 20, (15) 200, 100, 50, 30, 20, (15)	Northern Hemisphere	Seasons	1955–1957
Wahl, 1958 [41]	Height Temperature	300, 200 500, 300, 200	Northern Hemisphere	Monthly	1950–1955

*15-mb data only for summer.

2.1. Mean Map Patterns

The Scherhag series which extends from the 1000 to 15 mb levels, and is mainly for 1949–1953 (see Table II), lacks surface charts for the hemisphere and excludes the Pacific hemisphere south of 65°N at 850, 500, and 300 mb. Their forty-year surface normals for the hemisphere and 500-mb were therefore used as supplements [33]. The 300 mb maps of Wahl also may be used [41]. Because of the only recent availability of stratospheric Arctic data the "mean" maps at highest levels for two years are not presented by these authors as anything but a first impression. Representative short-period stratospheric means are especially unlikely since the changes in intensity and pattern are frequently great from year to year.

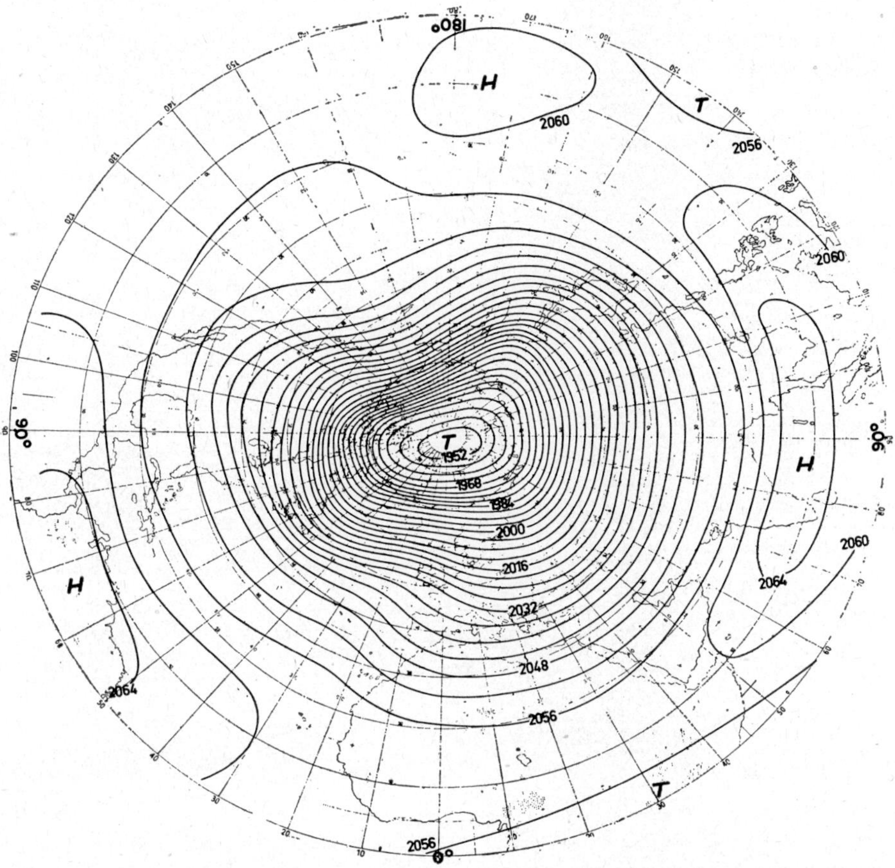

FIG. 3. 50-mb contours (geopot. dekameters), January (1949–1953), from Wege [37].

A full discussion of the charts is to be found in the respective publications, together with comprehensive bibliographies. Since this is the most consistent series of charts which include the stratosphere that can be expected for some time owing to the immense labor involved, we now include a brief summary of the first results of these new data.

Although we are concerned with the polar regions it is necessary to consider developments to the south as well, owing to the larger scale patterns at highest levels, which extend beyond the Arctic and are influenced by lower latitude processes.

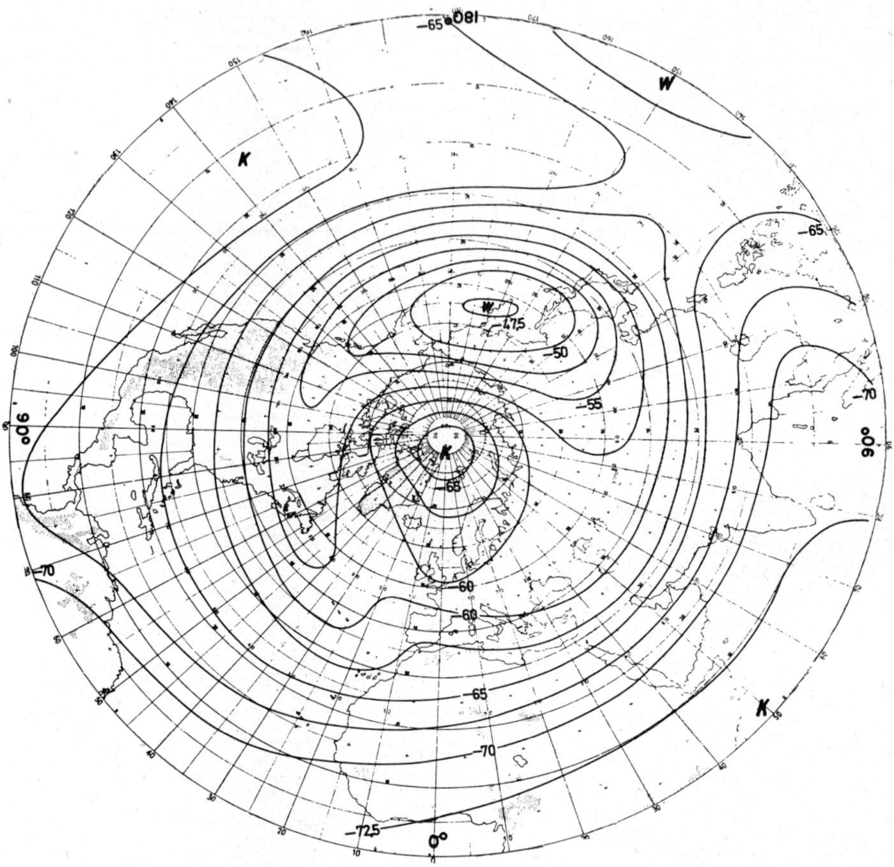

FIG. 4. 50-mb isotherms (°C), January (1949–1953), from Wege [37].

The major features of the mean pressure-temperature maps from the surface to 50 mb for the period 1949–1953 and to 20 mb (1955–1957) are worth examination. We limit our description to the mid-seasonal months, January, April, July, and October. It is naturally impossible to reproduce all the charts referred to, but, as an example, topography and temperature for 50 mb are shown in Figs. 3 to 10 through the courtesy of their authors; these have not been available prior to Wege's work.

2.1.1. January. The mean surface pattern shows a cold ridge across the Arctic connecting the intense high of Siberia with the far less intense one over

North America, and separating the deep, extensive Icelandic and Aleutian lows which cover the Northern Oceans. The Icelandic low extends from Baffin Bay to far along the European Arctic coast, to Novaya Zemlya and beyond,

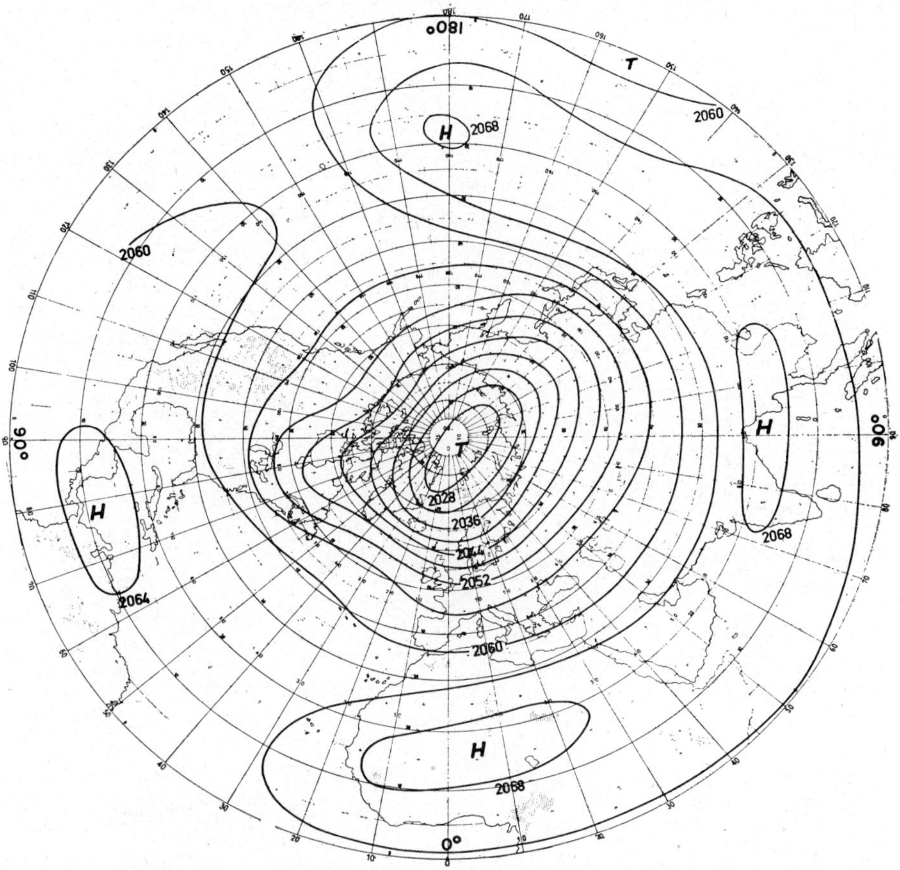

FIG. 5. 50-mb contours (geopot. dekameters), April (1949–1953), from Wege [37].

indicating that the Atlantic warming influence penetrates more deeply into the central Arctic at the surface than does the Pacific. At 850 mb the Icelandic low forms two centers, a large one over Baffin Bay and the other over Novaya Zemlya. The ridge across the Arctic is displaced toward Alaska and eastern Siberia. The central Arctic is almost dominated by the North Atlantic low.

At 500 mb the cold (— 52°C) Baffin low is reduced in area and a trough extends across the polar basin to an Okhotsk low, dividing the two ridges from the Bering Straits and the Norwegian Sea. The mean 500- to 1000-mb cold sources are thus located just north of Hudson Bay and at 130°E, 65°N in eastern Siberia.

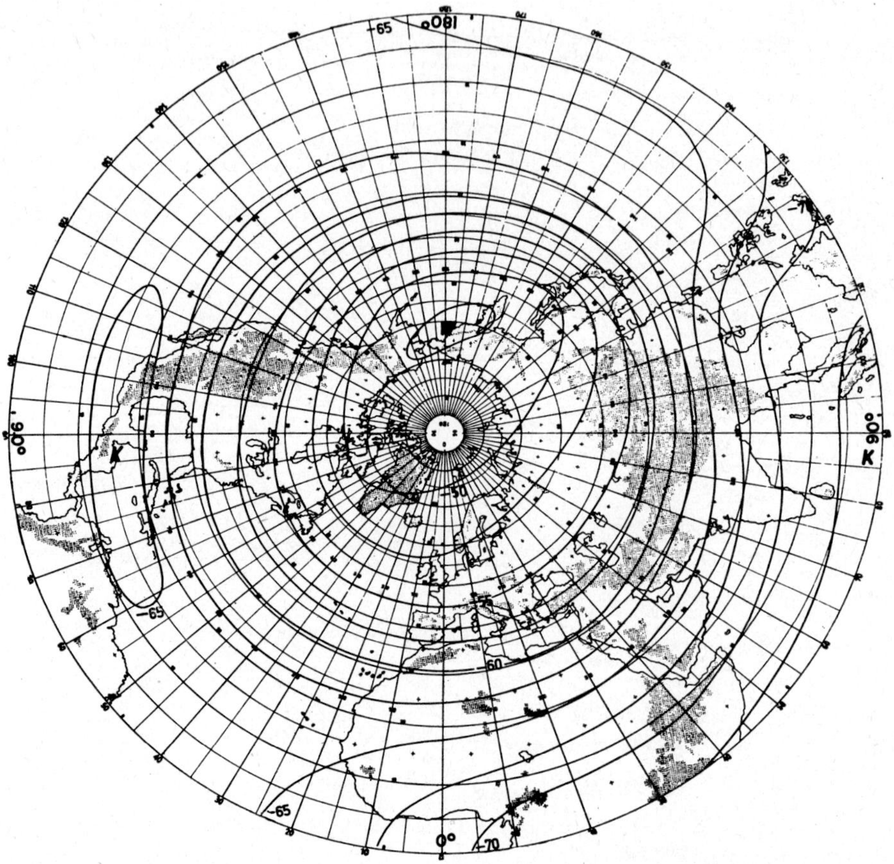

FIG. 6. 50-mb isotherms (°C), April (1949–1953), from Wege [37].

There is little change at 300 mb except for an increase in gradients. Central Arctic temperatures within the low center are about − 60°C, and it is noticeable that the thermal gradient is still strongest from the Atlantic where temperatures are 10°C warmer than on the Pacific side. However, the relative thickness from 300 to 200 mb is smallest over the Siberian side of the pole and largest along a wide belt from Baffin Island westward to Japan corresponding to a warm ring. This American–Pacific warm belt increases in intensity with height, and shifts its center to near Kamchatka.

At 200 mb the warm belt extends from south Greenland westward to Japan while the central Atlantic, central Arctic, and all of Europe are relatively cold.

This warm belt can be explained as due to the variation in height of the tropopause and to the difference in lapse rates with latitude. At high latitudes the temperature remains relatively isothermal above a low tropopause, while in mid-latitudes the temperature increases with height at 200 mb since this is

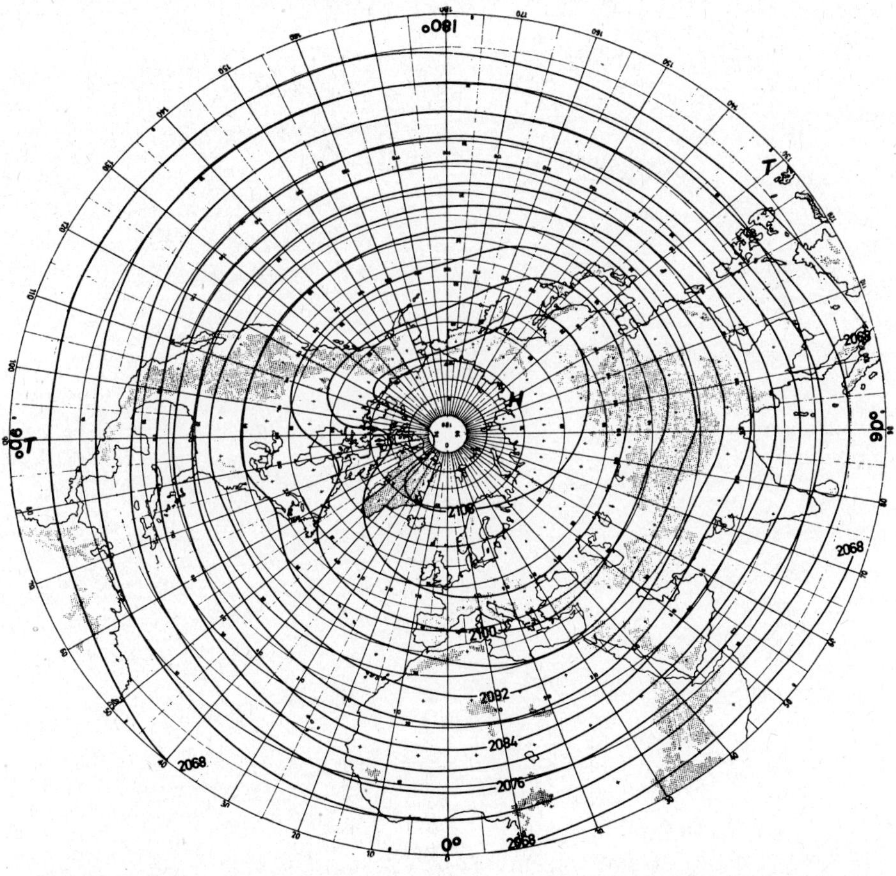

FIG. 7. 50-mb contours (geopot. dekameters), July (1949–1953), from Wege [37].

usually above the tropopause, and at low latitudes the temperature is still decreasing below the higher tropical tropopause. But it is difficult to see why Europe should not be similar in this regard to America. A second factor is the growth of the South Pacific subtropical anticyclone which becomes more intense and moves northward with height while no such phenomenon appears on the Atlantic side. At 200 mb it reaches about 50°N just east of Japan, at 100 mb it extends to 65°N across southern Alaska and eastern Siberia, and at 50 mb it is centered over Kamchatka and penetrates into the polar basin to about 75°N. The 200 mb cold center (— 60°C) over the Norwegian Sea remains fixed in position up to 50 mb (— 65°C) while the Pacific warm center (— 50°C) moves northward. The bipole low pattern at the 200-mb level is the same as at 300 mb, but at 100 mb the cold low becomes consolidated over the

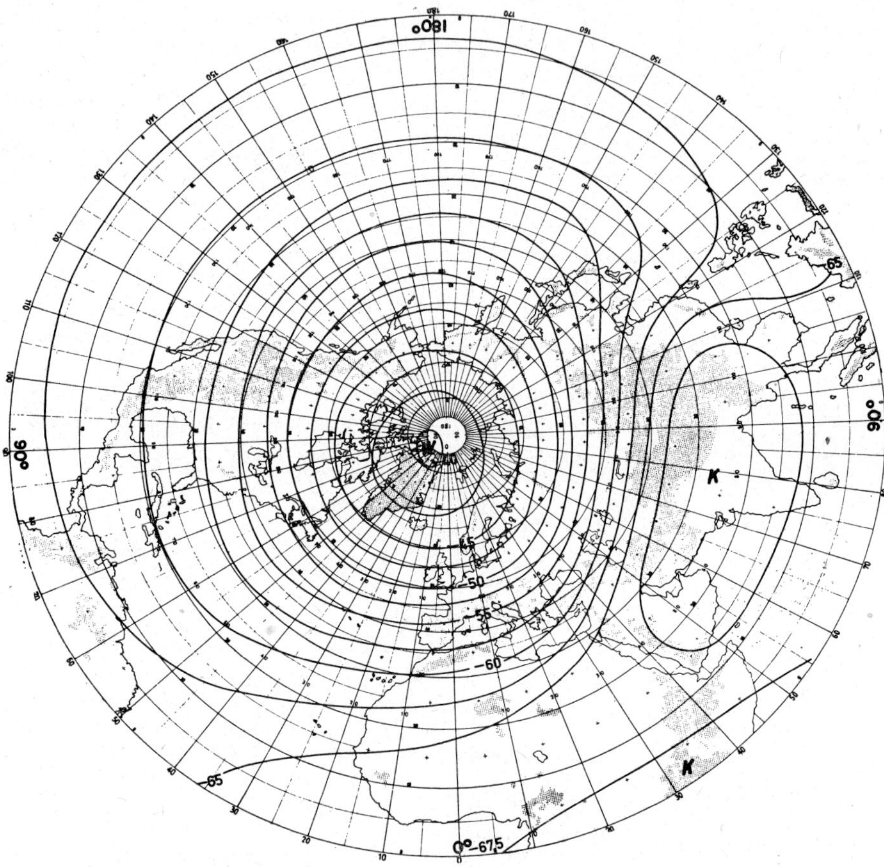

FIG. 8. 50-mb isotherms (°C), July (1949–1953), from Wege [37].

central Arctic with strong troughs toward its former centers. At 50 mb the effect of the warm Pacific is reflected in the shifting of the low toward the European side of the pole, with the Baffin Bay axis unchanged and the opposite end directed toward Novaya Zemlya. At 30 and 20 mb the low is found just north of Spitsbergen. From the two-year sample available at 30 and 20 mb the temperature maximum seems to recede southward again. This may be due to the more rapid warming with height at low latitudes than at higher latitudes.

Thus we have seen that the thermal patterns from the surface to 300 mb in which the Atlantic had greater influence on the Arctic than the Pacific, reverses in the stratosphere, so that at 200 mb upward the Pacific is the heat source rather than the Atlantic. The tropospheric asymmetry may be

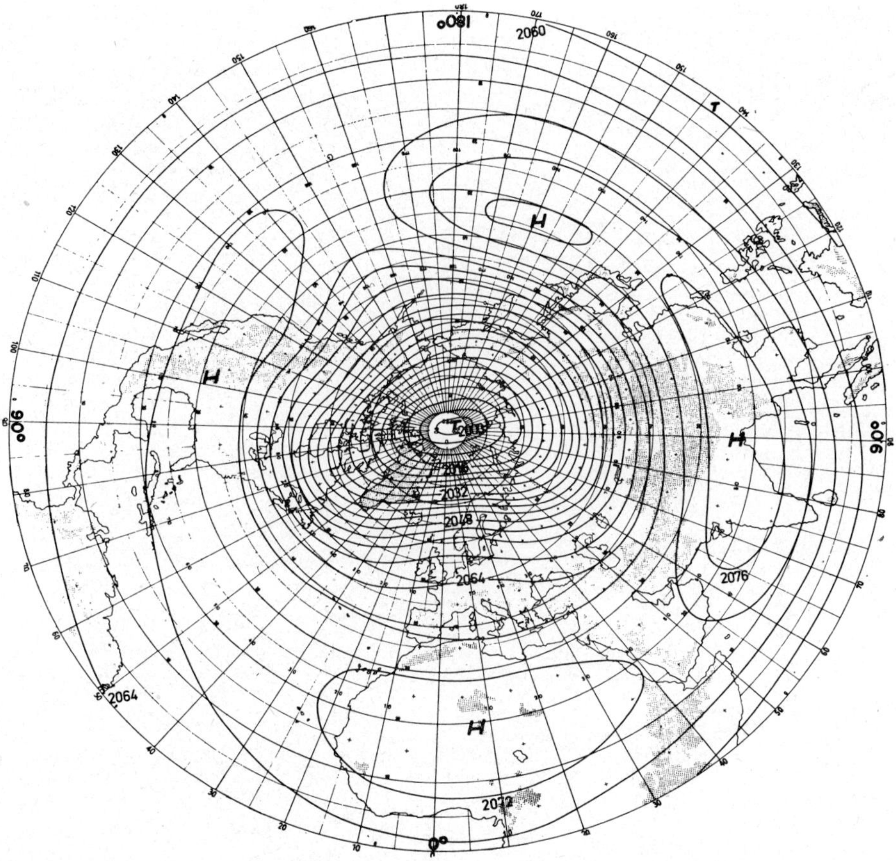

FIG. 9. 50-mb contours (geopot. dekameters), October (1949–1953), from Wege [37].

attributed to the greater penetration of the warm Gulf Stream into the European Arctic latitudes than is possible on the Pacific side due to the Siberian-Alaskan blockade of the central Arctic from the equivalent Kurushiro Current.

Wege suggests that the stratospheric warming in the Pacific may be due to subsidence in the strongly diffluent delta off the Asian east coast which in turn is ultimately due to the effects of Asian topography. The disappearance at 20 mb of the warm center east of Asia implies the lack of subsidence at that level.

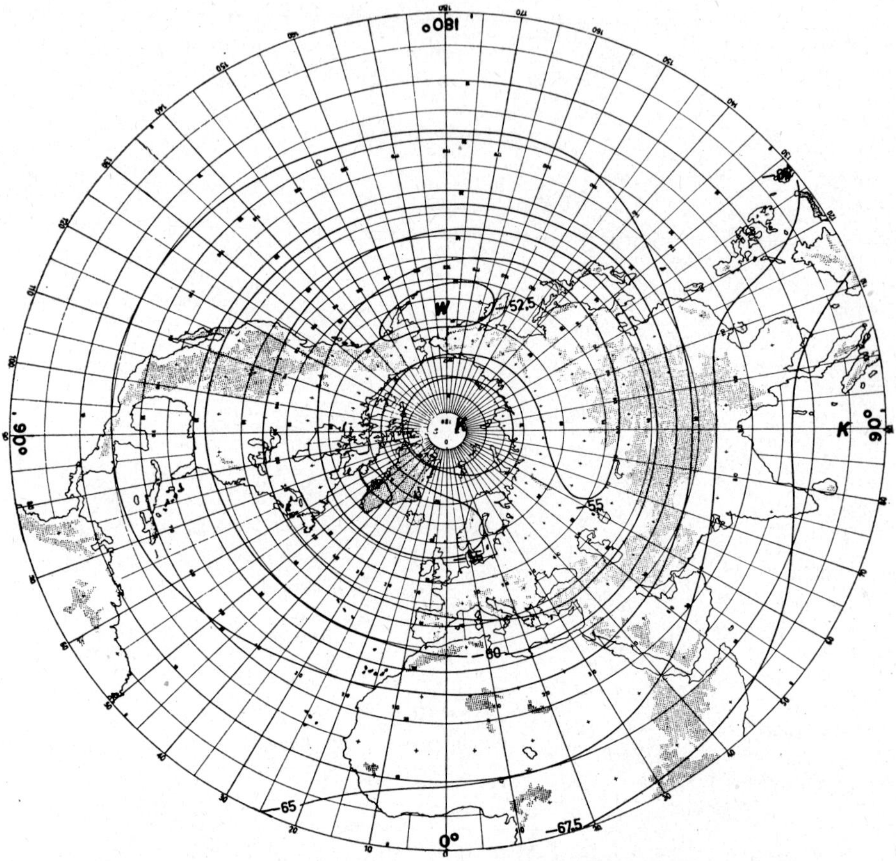

FIG. 10. 50-mb isotherms (°C), October (1949–1953), from Wege [37].

2.1.2. April. The surface pattern shows a belt of high pressure extending from Siberia across the central Arctic to America, separating the Aleutian and Icelandic lows. At 850 mb the polar ridge is reduced and the Icelandic low has moved north near Spitsbergen. At 500 and 300 mb a strong cold (— 35°C, — 55°C) polar low is centered over the polar basin with slight axes extending toward Greenland and Japan.

At 200 mb the low is still centered at the pole but there is a most interesting temperature distribution. The American sector is relatively warm (— 50°C at about 70°N) and the Siberian side cold (— 58°C at about 50°N). Admittedly this range over a hemisphere is not very large but here it represents almost the extremes on the map.

At 100 mb and at 50 mb, as a consequence of the asymmetrical temperatures, the low moves toward the colder Eurasian side of the pole with increasing height. The warm center moves westward from northern Canada at 200 mb to the Bering Straits at 50 mb, and the general pattern of isotherms becomes more concentric and centered more closely about the pole than at lower levels. The low can be said to be a warm one from 100 mb and higher, and it always continues to be warmest toward the western Pacific even at 20 mb. The subtropical Pacific high becomes more intense and moves northward with height from 100 mb where it is south of 25°N, so that it reaches 40°N at 50 mb, and into the Bering Sea and Alaska at 30 and 20 mb.

These patterns are so different in the Pacific and Atlantic sectors that one is forced to wonder if the unusual warming toward Japan at all levels from 100 mb upward may not be due in part to relatively large radiation errors of the radiosondes used during the period (1949–1953). If the effects are real, however, then here again, Wege's explanation invoking large-scale downward motions in a diffluent area, appears a most reasonable one.

2.1.3. July. In July, a weak cold surface high serves as a bridge across the Arctic between the strong highs over the oceans, while weak thermal lows dominate the continents. At 850 mb the surface Arctic high already becomes a low which now bridges the two lows over the continents. At 500 and 300 mb a cold (— 25°C, — 45°C) central polar vortex is established which has a trough extending in the direction of Baffin Island. Between 300 and 200 mb the temperature field reverses so the low becomes warm at 200 and 100 mb, but still maintains its Baffin trough. Finally, at 50 mb, the warm low becomes a warm (— 45°C) high. This temperature reversal near 200 mb is due to the decrease of temperature with height at low and temperate latitudes, while high latitudes remain near isothermal. This surface is near the relatively isopycnic layer which occurs about 50 mb above the tropopause at temperate and high latitudes as pointed out by Kochanski [42] and Belmont [42a]. The Asian subtropical high increases its intensity and also moves northward from 200 mb upward into the stratosphere. At 200 mb it extends to 30°N and at 100 mb it is at 40°N. At 50 mb it merges with the polar high and the resultant high is centered near 70°N on the Siberian coast. The two thermal warm centers at 200 mb, at the pole and over the Himalayas, merge to a single deep warm polar center at 100 mb (— 43°C).

For "summer," the warm polar high remains on the Siberian side of the pole at 30 and 20 mb, but at 15 mb is shown on the American side of the pole again due to a warm center at 20 mb extending from that direction. No information is available for the Siberian area, so the existence of a similar warm center there is unknown.

2.1.4. October. The surface pressure pattern resembles April with the Arctic high connecting continental highs, but the gradients are stronger. At 850 mb the Arctic again has an elongated low between the oceanic lows, and at 500 and 300 mb the low becomes a cold Arctic low, with a strong trough over eastern America and eastern Siberia.

At 200, 100, and 50 mb the polar vortex axis shifts slightly over to central Siberia but the American trough is still directed toward Baffin Island. The asymmetrical temperature pattern as in April exists again with a slight difference: the Siberian cold center extends closer to the pole at 100 and 50 mb. This stronger cooling in the fall at high latitudes increases at 30 and 20 mb. The cold center (— 60°C) is over Spitsbergen and Greenland at these levels which again brings the polar low to that side of the pole.

2.2. Mean Cross Sections

From the excellent source of maps described above, Wege and colleagues in addition prepared cross sections for several longitudes: 90°W, 160°W, 140°E, 40°E, and 10°E. For each of these are shown temperature, dew point (to 6 km), and pressure-height, all for each of mid-seasonal months. From these cross sections showing the large-scale vertical thermal gradients, the integrated effect of variable thicknesses from equator to pole can be compared easily. It is not intended to describe the cross sections here since they reflect the same features as the maps. Their main use will be to make available a source of quantitative material for detailed comparative calculations. If the section could be prepared for all months, it would help to pin down the months of greatest change in the seasonal reversal of circulation in the stratosphere, assuming a longer and more representative series of data for the stratosphere becomes available. Analyses of periods shorter than a month during the reversal months would be interesting but can hardly be expected with present monthly data summarization procedures, except for a daily series of individual sections during a reversal period. Further reference to recent cross sections will be given in Section 4.

To summarize in a single sentence, the main feature shown by these new maps is that the circulation pole is displaced toward the Pacific in the troposphere and low stratosphere but toward the Atlantic in the middle stratosphere.

3. Variability in the Stratosphere

The stratosphere has generally been considered a region of little change, dominated by slow radiative effects rather than rapid dynamic influences. However, in the past few years, especially in high latitudes, large variations in temperature have been observed from day to day, month to month, and year to year. No doubt many of the first observations of extreme changes were discarded or treated as *prima facie* evidence of instrumental errors, as some originally probably were, especially when made with earlier models of radiosondes. But as improved balloons and instruments permitted higher ascents with greater frequency and accuracy, it became apparent that unusual and spectacular changes were indeed real. Owing to the importance of these recent discoveries, some attention to their beginnings may be of interest.

3.1. "Sudden Warmings"

It was Scherhag who first observed that the radiosonde ascents taken at Berlin from January to March, 1952 showed remarkable changes of temperature at highest levels. The observations were taken with the newly inaugurated USAF radiosonde with an exposed thermistor, thus reducing radiation errors from previous instruments. Changes of about 30 to 50°C were noticed over relatively small numbers of days. Confirmation of these patterns was found in independent values at Copenhagen and Thule. The first notice of these changes appeared as a note by Scherhag in the *Tägliche Wetterbericht* for March 14, 1952. Another announcement was issued by Schweitzer [43] a month later.

From these first two preliminary notices Willett [44] prepared an article in which he attempted to show the warming data were the first "direct measurement of thermal effects produced in the higher atmosphere by a sudden solar disturbance." He linked the solar disturbance with sea level pressure increase in the North American Arctic, through the location there of the north magnetic axis. He felt the anticyclogenesis of eastern Alaska was related, not to cyclonic activity in the Pacific as Bodurtha postulated, but to solar influences. A few months later, Scherhag's detailed discussion appeared [45] in which he pointed out that the strong southwest wind flow pattern at the warming levels was in no way changed during the period; that the warming appeared first at highest levels (10 mb) and only gradually warmed later at lower levels down to 200 mb, over a period of a week. The first Berlin warming appeared between January 26 and 28, 1952 (there was no observation on January 27), reached its warmest value January 30, then cooled again approximately to its previous values. A similar second warming began on February 22, 1952, just 26 days later, when values at 15 mb rose from — 55 to — 13°C, a most improbable figure, and doubtless many corrected it by 50°C as Scherhag pointed out. The stratospheric winds during this period were steadily from the east, and so continued far into March. It must be admitted today that despite present awareness of instrumental errors in aneroids and thermistors, of lag and radiation and other errors, that the temperature patterns of the Berlin data, however incorrect in absolute value, must indeed reflect sudden disturbances at stratospheric levels.

Scherhag also noted that the temperature at Thule rose from January 12 to February 22, 1952 by over 30°C at levels above 14 km.

To explain these phenomena he examined solar disturbances as evidenced by daily magnetic character. A magnetic storm occurred from February 23 to 24, and a solar eruption was observed on February 22, a few hours prior to the first observations of high stratospheric temperatures. Another increase in magnetic character on February 5 and 6 was said to coincide with the beginning of a slow warming at Thule, and a third increase with the Berlin warming of January 27. In a later paper [46] he realized that the radiation from the solar disturbance would not reach the earth until 31 hr after the warming began. A statistical comparison of extraterrestrial conditions with warmings led him to conclude that solar effects also influence surface pressure changes. Scherhag explained the main problem of why all solar eruptions do not result in sudden warmings by the requirement that they occur only in a high latitude stratospheric cold center which has moved south at the end of the winter night into the influence of the sun. Since they are initially cold and contain low ozone amounts, it might be possible, he suggested, that solar radiation could penetrate lower down into the ozone layer and influence these stratospheric levels. In other seasons the solar radiations are absorbed at much higher levels owing to higher ozone content.

Scherhag assumed that cold Arctic night air would be poor in ozone because of the lack of sunlight. However, it has been demonstrated [47] in a single test, that even in mid-winter, about as normal ozone values can be found at the North Pole as at latitude 50°. This experiment used a Dobson instrument in an airplane, during a full moon. Further, newer impressions of flow patterns in the stratosphere now lead one to expect sufficient latitudinal mixing across the boundary of the polar night to prevent any "ozone gap," as Normand termed it [48], from developing.

Wexler, in a 1956 lecture published in 1958 [49], pointed out the improbability of solar influences and explained the warmings by normal processes of adiabatic subsidence and advection, utilizing daily 50-mb charts for February, 1952. These hemispheric charts, prepared prior to March, 1953 are among the first to demonstrate the slowly moving highly contrasting, large-scale contour patterns which apparently are typical of this level. A − 75°C cyclone over Baffin Bay and a − 50°C ridge over the Gulf of Alaska moved southeastwards over Europe. The ridge warmed as it moved and the resulting tight thermal gradient could have caused the observed warming at that level (8°C/24 hr). Wexler pointed out that an adiabatic sinking of 1 cm/sec over 48 hr would alone account for the 40°C rise at higher levels. He concluded that a combination of the advected pressure patterns and subsidence are most reasonable explanations for the Berlin warmings. Large differences in the temperatures of stratospheric lows and highs were earlier reported by Riehl, 1950 [50].

After Scherhag's paper appeared, Scrase in 1953 [51] mentioned that his 1951 data [52] for Lerwick (60°N) and Downham Market (53°N) also showed abnormally higher temperatures, from February 5 to 20, 1951 by about 20°, 25°, and 40°F at 18, 24, and 30 km, respectively. Easterly flow existed from February 1 to 12 above 18 km. He found no solar connection.

McIntosh, 1953 [53], also discounted the solar activity relationship by showing that the February 23, 1952 storm was not a very strong disturbance compared with others in the same month, and that it was unlikely a solar flare occurred. He points out that since total ozone does not follow a solar cycle, or increase at a time of flare, probably no effective variation in solar radiation reaches the ozone layer. He agrees that ozone offers the only possible link of solar variations with surface weather reactions.

In 1954 attention was called [42a] to the unusually warm mean stratosphere at Ice Island T-3 in January, 1953 as compared with random published climatic data then available for the Arctic stratosphere. This was attributed to strong anticyclonic development from the Alaskan–Pacific sector into the central Arctic. In a later paper [13] when more appropriate climatic data became available, it was found that not only T-3 but most stations in northern Canada were equally warmer that month than appeared to be normal.

In 1956, Warnecke [54] analyzed data for Alert (83°N) for the winters of 1951 to 1954, and found sudden high stratospheric temperatures during January and February, 1952 and March, 1954. He found no evidence of geomagnetic relations, and generally explained stratospheric temperature rises by advection of very extreme gradients of temperature and height found between contour centers at 50 and 100 mb. Alert warmed 24°C at 100 mb from February 3 to 10, 1952 as the pressure pattern slowly moved, and similarly in March, 1954.

Wexler and Moreland [55], in agreement with Warnecke, showed that 50-mb cold lows and warm highs may be independent of lower levels, but emphasized that the large thermal contrasts between these centers appeared to be due mainly to vertical motions rather than advection. The warm and cold centers move slowly and appear to have no seasonally fixed position from year to year.

Further examples of sudden warmings in the Arctic using 25-, 50-, and 100-mb charts were reported by Craig and Hering [56] for the January–February, 1957 season. Craig points out that the warming occurred despite strong cold advection by 100-knot winds from an area where temperatures were − 70°C. Vertical descent of about 5 cm/sec sustained for one week over a 3000-mile trajectory were thus required to create the observed warming. The stratospheric maps for this winter were studied in more detail by Hering and Salmela [23] who also noted the great persistence of pattern from day to day yet marked changes from year to year at high latitudes.

The most extensive case history so far appears to be that for the 1957 warming over North America which has been widely reported by Teweles [57] and others [40, 56, 58, 59]. Teweles points out that for western North American stations there are very strong warmings above 25 mb, but none below that level. He attributes the wave motion of the Arctic stratospheric jet, not to variable radiation effects of the underlying surface, but, like Wege, to the Japanese jet. This he finds causes, in the Aleutian area, rising motions below the jet axis level, and sinking motions above that level, which produce a semipermanent cold low in the troposphere, and a warm ridge in the stratosphere. The ridge may retrograde as in 1957 over Siberia, or progress over Canada as in 1958.

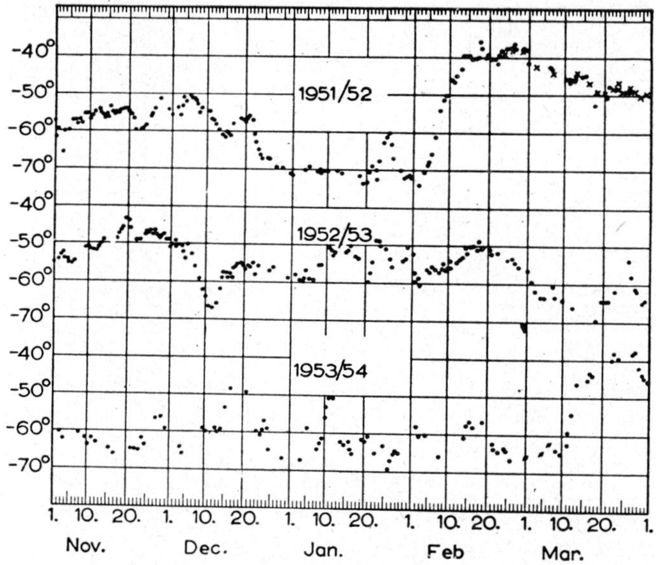

FIG. 11. Temperatures at the 100-mb level over Alert, from Warnecke [54].

The sudden warmings may or may not mark the end of the polar winter. Many of the authors already cited have found temperature increases followed by decreases. By plotting the temperature for the entire winter (November–April), a definite time of final warming, if any, and the associated seasonal reversal of circulation can be determined. The solar contribution to the seasonal reversal of temperature gradient is a function of solar altitude and may or may not coincide with a sudden warming over a given area which is due to shifting pressure patterns and to vertical motion systems. Most likely, the separate effects are masked by phase differences among themselves.

Table III presents warming dates since 1951 for the North American Arctic. The 1952–1953 season appears to have been so warm that no further relative warming was noticeable to March 20. This is the same winter for which the T-3 stratosphere was unusually warm. It is apparent from the table that "sudden warmings" occur probably each year and are a normal feature of the stratosphere at high latitudes, although most instances occur above the levels presently reached regularly by radiosonde, especially during the Arctic winter. Figures 11 and 12, from references [54, 58] show the abrupt character of these warmings for the years 1952–1957, as noticed at 100 mb over northern Canada.

TABLE III. Examples of seasonal warmings.

Date	Temperature change	Pressure level (mb)	Stations	Source
Feb. 5–20, 1951	22°C	10	British Isles	[51]
Feb. 2–9, 1952	25°C	50	N. Canada, N. W. Greenland	[61]
	35°C	100	Alert	[54, 62]
1952–1953	—	100	Alert	[54, 62]
	[Warm Nov. 1 to Feb. 15, cooled to March 20, then irregular. No sudden warming noted.]			
Mar. 10–20, 1954	25°C	100	Alert	[54, 62]
Jan 1–10, 1955	17°C	100	Alert, Resolute, Eureka, Coral Harbor	[60, 63]
Mar. 20–Apr. 1, 1956	23°C	100	Alert, Resolute, Eureka, Coral Harbor	[60, 63]
Jan. 23–30, 1957	40°C	25	S. Greenland, Iceland, S. E. Canada	[56]
Feb. 1–15, 1957	25°C	100	Alert, Eureka, Mould Bay, Norman Wells	[58]
Jan. 31–Feb. 4, 1957	50°C	20	Thule (N. America, N. Pacific)	[57]
Jan. 29–Feb. 7, 1957	45°C	50	Frobischer (+ 34°C/24hr)	[40]
Jan. 24–30, 1957	35°C	30	Ship B	[40]
Feb., 1957	50°C	10	Churchill (much warmer than normal)	[59]
Jan. 24–28, 1958	31°C	25	S. of Greenland	[64]
Jan. 14–17, 1959	22°C	25	St. Cloud, Minn.	General Mills Data

Little is yet known concerning the hemispheric-wide pattern of these phenomena. All studies so far have been confined to the American Arctic.

It will be most interesting to see how the Eurasian Arctic warming patterns coincide with those of the West, and whether similar effects are found in the Antarctic. Godson and Lee [58] warn that the structure of the warming

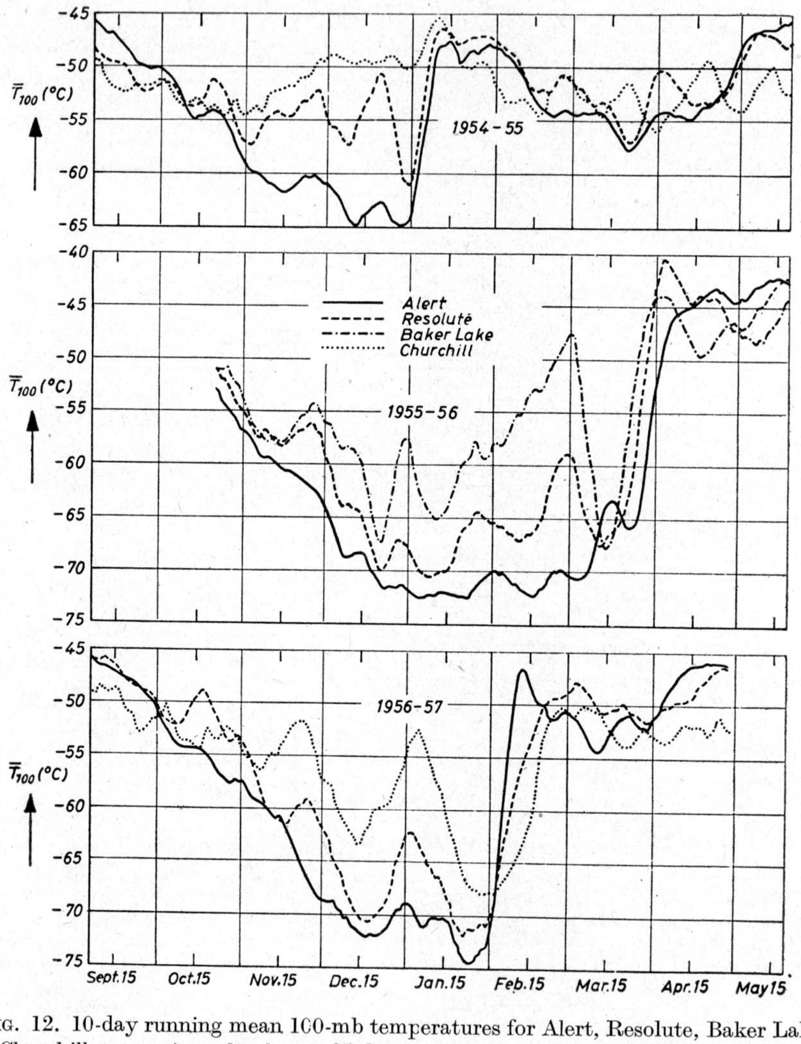

FIG. 12. 10-day running mean 100-mb temperatures for Alert, Resolute, Baker Lake and Churchill, approximately along a N–S line, from Godson and Lee [58].

process is exceedingly complex and its full explanation will probably be delayed to the day when better data for a wide ring of the circumpolar area become accessible.

A recent advance in stratospheric analysis, which will help clarify these sudden changes, and which will be of great value for demonstrating the gross patterns at higher stratospheric levels, is the series of 10-mb charts by

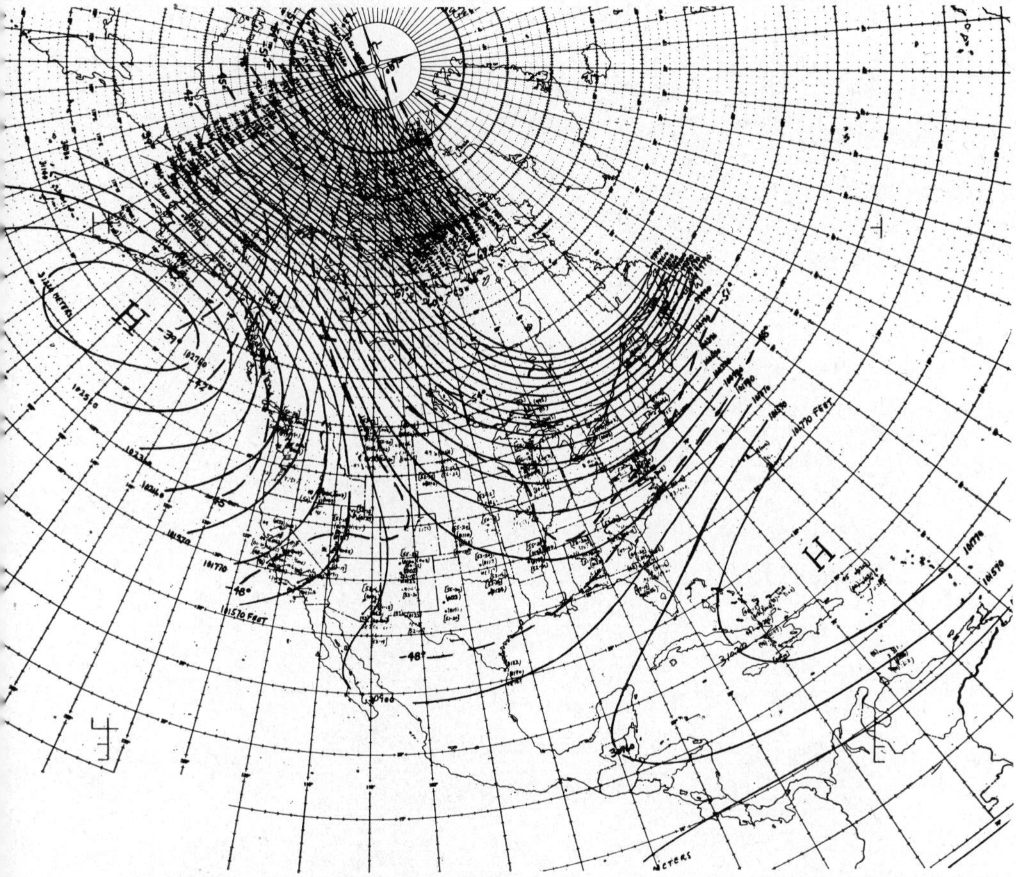

IG. 13. 10-mb chart, 15 January 1958, Contours at 60 meter intervals. Pacific High = 31,320 ters, Caribbean High = 31,020 meters, Arctic Low = 28,560 meters. Isotherms are dashed lines 3°C intervals, ranging from − 39°C at Fairbanks to − 69°C at Eureka. From Teweles and ger [65].

Teweles and Finger [65] who prepared a monthly series for the period July, 1957 to June, 1958. Each map represented a three-day period centered on the fifteenth of each month. Sample maps are shown in Figs. 13 and 14. Since patterns change slowly, this technique permitted a better chance to get sufficient coverage for this level which represents our highest frontier in map analysis to date. The charts were prepared for North America and showed systems of continental dimensions which may be at least partly the result of data unreliability. Observational errors at these levels are at present too great to reveal detail with any assurance.

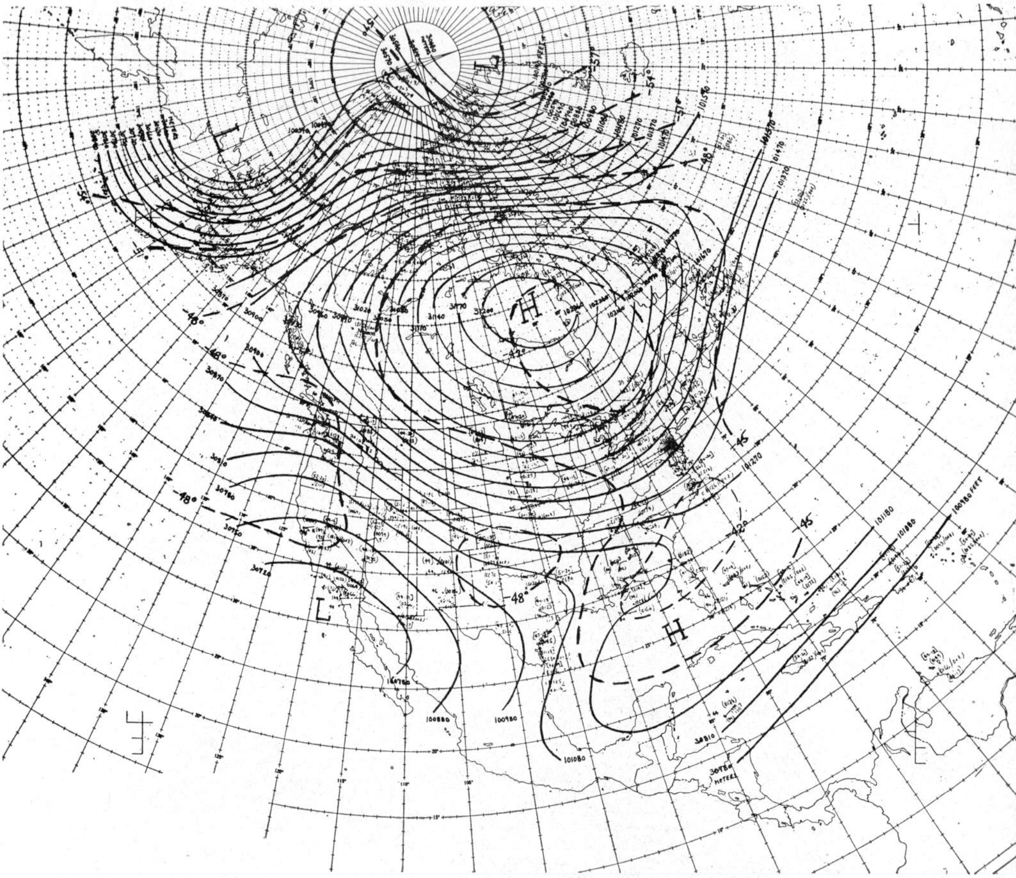

FIG. 14. 10-mb chart, 15 February 1958. Contours at 30 meter intervals. Canadian High = 31, meters, Bering Strait Low = 30,600 meters. Temperatures vary from − 42°C over James Bay a Florida to − 54°C over Western Alaska. From Teweles and Finger [65].

It is to be hoped more frequent analyses of this type can be continued on a regular basis. It is interesting to note again, despite the great reduction in density at this level and the previously assumed tranquility, what strong gradients of heights and of temperature exist, especially at highest latitudes, and how extreme are the changes which result from their movement. Surely, the dynamics of the polar stratosphere cannot be ignored longer,

3.2. *The Arctic Jet*

The next step toward explaining these temperature anomalies came through an investigation of the Arctic stratospheric wind field, and to a first synoptic study of the "Arctic jet." Lee and Godson [60] presented a paper at

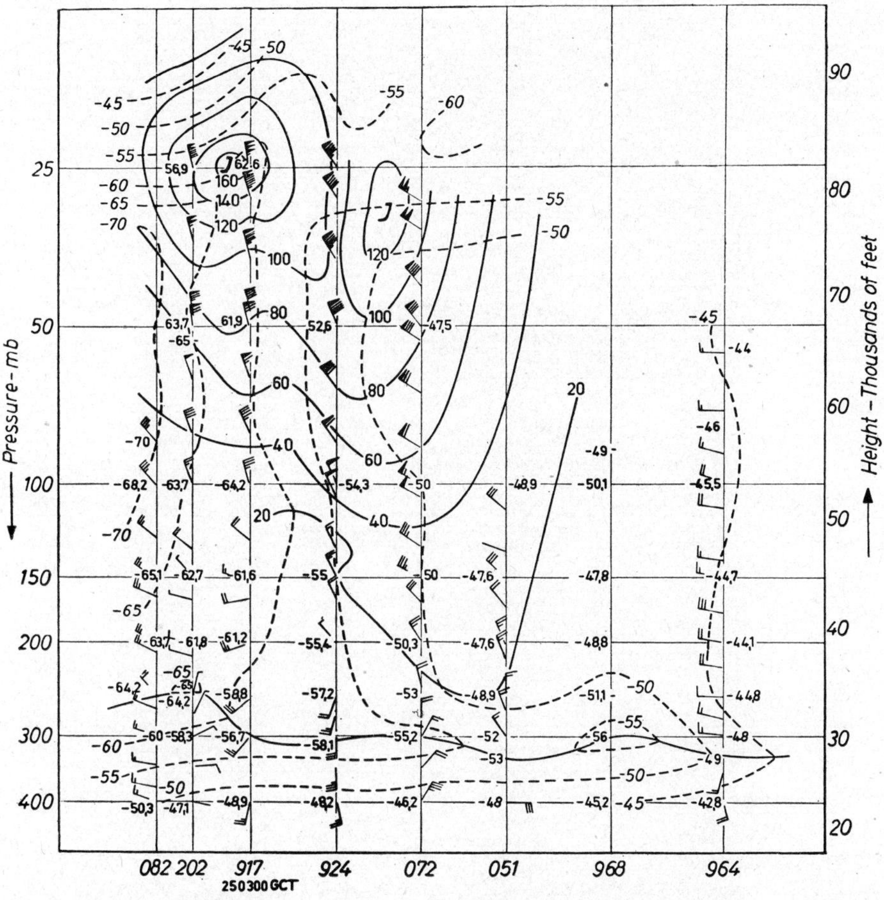

Fig. 15. Vertical cross-section oriented approximately NE–SW for 1500 CGT February 26, 1956. Tropopause is indicated by thick solid line, isotherms (°C) by broken lines, and isotachs of observed northwesterly winds in the stratosphere by thin solid lines (kt). From Lee and Godson [60].

the Oslo Arctic Symposium in 1956 which revealed again a sudden warming, this time in March, 1956 at 100 mb at Canadian Arctic stations. Above this intensely baroclinic field was found a jet at 25 mb (83,000 ft) with winds of 150 knots or more, as, for example, in Fig. 15 for February 26, 1956. The existence of such a jet had been inferred previously from mean meridional cross sections (Flohn [66] and Kochanski [67], for example), but this was the first synoptic demonstration we have seen. The maintenance of this jet is attributed to the differential heating in the ozone layer across the boundary of the polar night. Hence, it moves south in fall and north again in spring when it apparently dissolves as the Arctic thermal field is reversed. However, Wexler and Moreland [55] believed that the polar night jet may be found along the southern edge of cold lows at 20 km, rather than as a regular circumpolar feature of the Arctic night. Murgatroyd [68] shows a winter jet from 20 to 80 km centered at 60°N and perhaps the 25 km jet is but a lower portion of Murgatroyd's deep, high stratospheric westerly current.

In the past year a more detailed analysis of the wind and temperature fields in the Canadian Arctic was made by Godson and Lee [58] for the three winters, 1954/1955–1956/1957. The three years were very different with regard to the 100-mb temperature patterns. Temperature waves with periods of 20 to 30 days were noted which may terminate in a rapid cooling and rapid warming bringing an end to the Arctic stratospheric winter. Further cross sections showed the Arctic jet stream, usually near 70–75°N, with winds which were observed once at 230 knots.

3.3. The Seasonal Reversal of Flow

One of the basic properties of the polar atmosphere is the complete seasonal reversal of the upper stratospheric circulation, a process whose effects extend southward into temperature latitudes. Whipple in 1935 [69] suggested summer easterlies and winter westerlies to explain sound propagation measurements. This seasonal regime has been well established since then for many latitudes from direct balloon data. However, the details on a monthly basis are just beginning to become accessible.

Darling [61] presented monthly direction frequency data at 50 mb over the U.S. which showed that the frequency of easterlies became predominant in May and the reverse in October. Scrase [52] in 1951 presented some first British high-level rawinsonde results, and demonstrated the sudden onset of easterly winds above 50,000 ft at the vernal equinox and their gradual transition to westerlies by the autumnal equinox. Burley *et al.* [70] used a five-year series of seasonal data which showed the apparent northward migration of the tropical easterlies. In July, all levels were easterly at and above 200 mb at 25°N and 30°N, 100 mb at 35°N, 80 mb at 40°N, and 50 mb at 45°N and 50°N. In April, easterlies prevailed north to 30°N and above 30 mb, in October to 30°N and 50 mb, along 80°W. Along 120°W these limits differed slightly.

Kochanski's [67] cross sections along 80°W from equator to pole and up to 30 km, showed similar features. The subtropical easterlies in January merge with easterlies as far north as 60° in April at levels above 20 km. merge with the polar easterly jet which is centered at 75°N in July and which extends down to 12 km, and then in October retreat to 25°N.

The Wege map series shows that a mean high first appears at 50 mb in May and lasts through August, but over Baffin Island a weak remnant of the winter trough prevails throughout June and July, and even becomes a closed low in August, which is inserted as a bay into the otherwise circumpolar warm high. And since these are mean maps, one may expect that many individual lows at 50 mb, especially over eastern Canada, contribute to the mean summer trough. Hence, we should expect that over the Baffin Bay region there will be considerable variation from a purely easterly flow, as well as lesser interruptions of the resultant flow pattern elsewhere over the hemisphere.

Both Kochanski [71] and Lowenthal and Arnold [29] demonstrate the daily variability of winds measured with special equipment to 120,000 ft in January and August, 1955. Goose Bay (53°N) and Narsarssuak (61°N), both of which are near the east Canadian trough at 50 mb, show large frequencies of easterly wind in January at all levels from 700 to 10 mb. In August when the mean map of 50 mb still shows a warm high over the Arctic, save for the Baffin Low, westerlies were found in the stratosphere with respectable frequency.

Lowenthal and Arnold [29] point out that the winds for January, 1955 consist of two opposite flow patterns above 50 mb: mainly easterly the first half of the month and westerly the other half. The change from easterly to westerly was a large-scale phenomenon since it was noticed at several stations over a large area on the same day. They emphasize that the accepted zonal flow pattern of the stratosphere is an exception rather than a rule, due to prolonged interruptions of as long as two weeks. In this instance winds at Belmar, New Jersey (40°N) had easterlies above 60,000 ft consistently from October to January 20th; thus the equinox need not always serve as a reversal date over a given area. Comparative observations were not taken at stations other than Belmar before January first; hence, the area of stratospheric easterlies during the fall of 1954 is unknown.

Occasionally these "strange" easterlies, which are not to be expected in a model showing only westerly flow, are found only in shallow layers of 50 mb. Some of these instances which have come to our attention are given in Table IV. It is to be anticipated that increasingly frequent reports will appear as rawinsondes or balloons begin to penetrate the upper stratosphere. The explanation for them is partly evident from the foregoing account of sudden warmings caused by the advection of slowly moving, closed contour patterns.

To generalize the reversal pattern, we will draw upon some preliminary findings from work in progress on the monthly variation of winds over the Atlantic and Pacific portions of the Northern Hemisphere. It is still too early

TABLE IV. "Anomalous" winds

Date	Height	Place	Direction*	Source
Winter, 1947–1948	"Lower stratosphere"	New Mexico	East	[72]
Feb. 1–12, 1951	21 km	British Isles	East	[51]
Jan., Feb., 1952	"Lower stratosphere"	Tateno, Japan	East	[72a]
Feb., 1952	70,000 ft.	N. Canada, N.W. Greenland	East	[61]
Feb. 23–March, 1952	20 mb	Berlin	East	[45]
Jan. 18–30, 1953 (approx.)	80,000 ft	Pierre, South Dakota	East	[73]
Aug. 29–Sept. 2, 1954	50,000–90,000 ft	Narsarssuak	West	[29, 71]
Dec. 15, 1954	80,000 ft	31°N, Texas	East	[73]
Jan. 5–12, 1955 Jan. 3–17, 1955 Jan. 1–17, 1955	70,000 to 100,000 ft	Narsarssuak Goose Bay Belmar, New Jersey	East	[29, 71]
Feb. 9, 11, 1958	100,000 ft	Minneapolis	East	[73]
Dec. 1–8, 1958	100,000 ft	Minneapolis	East	[73]
Jan. 14–17, 1959	100,000 ft	Minneapolis	East	General Mills data

*"East" (or "West") includes all directions with easterly (or westerly) component.

to give details, but it appears that the normal seasonal change of westerlies to easterlies and back again should proceed about as follows: in the spring as the thermal gradient toward the poles weakens, the strong winter westerlies start decreasing in force at the highest levels first (above 10 mb) and from mid-latitudes to high latitudes. This process, including the northward migration of any polar night boundary jet is relatively rapid as the sun returns to the Arctic. The cold polar low becomes a warm low and finally a warm high in response probably to direct heating of ozone in the polar atmosphere. The reversed thermal wind first produces weakened polar westerlies which become weak easterlies and gradually extend farther southward and downward to 50 mb, merging in mid-latitudes with the tropical easterlies which propagate downward and northward. By July, the polar easterlies are strongest at highest levels and at low latitudes. As autumnal cooling commences, the warm high becomes a cold low first at 50 mb and then upward. Thus, the westerlies replace easterlies at 50 mb and progress upward, and then southward increasing in strength as winter approaches. Borden [24] illustrated the fall reversal process at 100,000 ft for Alaska and the U.S.

In using upper wind data it must be recognized that there is a strong bias toward light or calm winds at highest levels, or in favor of easterlies, due to the termination of flights with strong westerlies at lower levels. This bias is probably least at Tokyo, where owing to extremely strong westerlies, the precaution is taken, it is understood, of launching balloons upstream of the observing site to keep the balloon in sight longer. There also appear to be longitudinal differences which are natural consequences of the asymmetry of the circulation, and hence, several meridional profiles of the wind are required.

3.4. Summary

Summing up, it appears that despite the limitations of present radiosondes above 100 mb in the Arctic winter, the large fluctuations of temperature and wind on a daily, monthly, or annual basis which have been reported are representative of real phenomena. The fundamental characteristics which are now evident are:

1. There are relatively few contour centers at high levels, and they are generally cold lows and warm highs. The systems are slow moving and may maintain strong thermal gradient for long periods. The movement of these gradients over a station may cause rapid warmings or coolings. At times, these extreme gradients appear to be due to persistent vertical motions, although the large-scale organization of these vertical circulations has not yet been discussed.

2. The seasonal warming of the stratosphere depends essentially upon the seasonal cycle of insolation, but its timing at a given place may vary considerably from year to year. Alternating pulses of warming and cooling, some

of which may begin suddenly and last for many days or weeks, demonstrate the presence of other factors. Ozone distribution must be of great importance as it is a strong absorber of solar ultraviolet radiation and since the levels at which the large thermal contrasts appear are in the ozone layer. The final seasonal warming may occur at almost any time during the winter.

3. The seasonal reversal of circulation appears in the mean to start early in the spring at highest levels and in the polar regions, and progress downward rather rapidly to 50 mb, and southward. In the fall, westerlies reappear first at 50 mb in the polar regions and progress upward and southward.

4. An Arctic jet appears established with a center near 60 km and at 60°N in winter. An easterly jet at 50 km in mid-latitudes is a feature of the summer stratosphere.

5. Among the factors responsible for Arctic–Antarctic differences in stratosphere temperatures, two that have been suggested recently are the influence of radiation from the earth's surface as a function of surface temperature, and of reflected radiation as a function of the albedo of the surface or cloud cover. The other causes may be traced to land–sea distribution, both in the polar regions and in the hemisphere as a whole, and topography.

In conclusion, the more one studies temperatures and winds at 100 mb or above, especially in the polar winter, the more the need for suitably designed equipment becomes apparent. The rapidly growing literature on "sudden warmings," coupled with the recent 10-mb maps, reveal a new kind of meteorological domain which is perhaps best studied on a case history basis owing to the great extent of the systems and the intense gradients between them. Ordinary statistics for these heights at fixed stations are not very useful until a large enough record is available—perhaps 20 years or more from now. Meanwhile, intensive studies of the dynamics of individual cases may reveal the best clues to the processes such as large-scale vertical motions which are to be expected at these remarkable levels and latitudes.

4. General Survey of Recent Advances

Because of the hundreds of research papers written in the past eight years since the subject was last summarized in the *Compendium of Meteorology* [1], it seems appropriate to survey this recent literature now for the help of those wishing a "fresh start" in the field of Arctic meteorology. It is the purpose here to mention papers only briefly, as a general guide to recent developments. Together with the sources mentioned in Section 1, this should provide a summary of available data and recent research which can be applied to the problems of the Arctic atmosphere. Since many topics are interrelated there can be no strict distinction by subject, but rather only loose grouping to assist in finding those of interest.

4.1. General Circulation

The early controversy concerning the mean polar surface pressure, begun some hundred years ago and involving Maury, Ferrel, Dove, Helmholtz, and Hann among many others, and culminating with the glacial anticyclone theory of Hobbs, has been settled. The latter theory and its history were reviewed by Hare in 1950 [74], and a few additional final papers appeared at about that time and since then [3, 4, 5, 17, 75, 76]. The new ideas of fairly normal mean Arctic basin pressure were adopted in the 1952 edition of the U.S. Weather Bureau Normal maps [32] which corrected the previous 1946 exaggerated high pressure [76a]. The 1952 patterns are generally confirmed by Namias [77] in the light of more recent data.

The Arctic is a key area in any model of the large-scale vertical meridional circulation due to the probability that the springtime breakdown of the polar vortex is accompanied by downward motions causing large temperature rises and increases in ozone and radioactivity. Theoretical models have been proposed at least as early as Maury [78] and Ferrel [79], but as the polar winter stratosphere is still mostly unobserved present meridional circulation hypotheses must rely on indirect evidence from ozone, water vapor and radioactivity. As in all such idealized atmospheric models the variability in longitude and season and from year to year is so large that only many years of data can show which portions of recent suggestions are typical (*e.g.* Martell [80] and his references).

With regard to horizontal flow patterns, it is now recognized, as on the Wege maps [37, 38, 40], that there is a "bipole" pattern of pressure and temperature during the cold half of the year. This was pointed out in 1949 by Flohn [81] for upper levels, although as long ago as 1855, Dove called attention to the same feature for the surface, by the isotherms on his circumpolar mean maps [82]. Perhaps Dove should be given the credit for first demonstrating 100 years ago the asymmetry of the surface thermal field due to land-sea distribution. The problem is still of importance in examining tropospheric and even stratospheric distributions as it appears that the surface land-sea distribution with its different heat sources, together with other factors, do influence the stratosphere [83, 41]. Measurements of the asymmetry of the general circulation have been outlined by LaSeur, 1954 [84], Frenzen [85], and Kobayashi [86], who have found that the pole of the circulation is displaced toward the Pacific Ocean from the geographic pole. According to a report published in 1955 [87], the influence of the Gulf Stream heat source usually causes the Icelandic low to be farther north than the Aleutian low and this favors more frequent development of meridional ridges into the polar regions on the Atlantic side of the pole.

One of the factors causing the asymmetry which has not previously been appreciated is that of surface albedo. Sverdrup [88] and Schumacher [89] have recently pointed out how important albedo can be in polar regions where snow cover can reflect most of the incoming radiation.

The effect of the intense radiational cooling of the polar regions in winter upon lower latitude circulation is not well understood although a relationship appears reasonable. Austin and Krawitz [90] suggest that there is a resulting baroclinic instability which leads to meridional horizontal flow to restore the heat loss. Petterssen [91] had earlier advanced a theory of the general circulation in which radiative processes create dynamic instability and result in meridional circulations to re-establish radiative equilibrium. The polar regions were found to be a major source of vorticity which is fed into the subpolar vorticity sinks (Aleutian and Icelandic lows). Flohn and Seidel [92] compared Arctic tropospheric temperatures with European circulation indices, but could not find a direct relationship. It was noted that anomalies in mid-latitude circulations, such as blocking highs, occurred with asymmetric shifting of Arctic cold air to one side of the polar region. Namias [36] approaches the problem by using anomalies from "normal," but to date there is too little reliable data for any general conclusions. The variations of the "index cycle," the periodical, irregular expansion and contraction of the circumpolar upper level vortex, appears related to the growth of surface polar anticyclones and subtropical highs, but the underlying physics remains obscure.

4.2. Upper-Air Climatology

4.2.1. Upper-Air Temperature. Investigations of the Arctic upper-air temperature field have been numerous. Flohn has contributed many comprehensive papers dealing with various aspects of the Arctic upper atmosphere during the past ten years [17, 93–96]. His most complete paper is probably "Zur Aerologie der Polargebiete" [17] which summarizes all available data to 1952 and is still current. Other studies have been made by Kochanski [42, 67, 71, 97, 98] who presented some of the first upper stratospheric material. Although his data are mainly from the U.S., his area extends frequently over the North American Arctic. He pointed out the two isopycnic layers just above the tropopause and near 30 mb, demonstrated an early 10-mb isotherm chart, presented vertical motion patterns for the stratosphere, described the thermal structure of troughs and ridges in the lower stratosphere, prepared mean 25- and 50-mb charts, showed that layers above 50 mb have no relation to lower level wind patterns, and prepared a mean cross section to 30 km along 80°W which was a common standard until recently supplanted by Wege's [40].

Murgatroyd [68] reviews temperatures from 20 to 100 km, for summer and winter for all latitudes. He shows that the stratospheric inversion lies over the poles in winter at from 30 to 65 km while it is at about 20 to 50 km at lower latitudes. Gaigerov [9] doubts that the 50-km maximum exists at all in winter since temperatures at North Pole 4 decreased continually to 24 km, and he stressed the need for thorough investigation of the winter thermal field with proper equipment.

The apparent "disappearance" of the winter tropopause due to the continued decrease of temperature with height is attributed by Wexler and Moreland [55] to the ascent being taken in a cold low whose temperature is maintained by vertical motions. Individual extreme temperatures have been observed as low as − 81°C (Thule) [22], − 85°C (Jan Mayen) [89], − 81°C (Thule) [45], and Gaigerov [9] reported − 81°C from North Pole 4 on January 4, 1956, which he claimed to be the lowest Arctic temperature observed at that time. It is interesting that the Jan Mayen values were reported during the same winter in February, 1956. Such values make doubtful the opinion of Wexler and Moreland that even individual extreme minimum temperatures in the Arctic are 5 to 10°C warmer than mean monthly soundings at equivalent latitudes in the Antarctic. That the mean winter Arctic is warmer than the mean Antarctic ascent would seem more likely accepted.

Other presentations of the Arctic tropopause are contained in papers by Defant [99] who classifies a spectrum of ascent types based on a six-day mean by latitude (without regard for pressure pattern), and by Moore [100] who gives its mean pressure and temperature and shows it is higher in summer in the Arctic than the Antarctic, and higher in winter in the Antarctic than Arctic, which are in agreement with Soviet results relayed by Taubert [16]. There have been numerous mean meridional temperature cross sections based upon observations, which extend to polar latitudes [66, 67, 68, 40], but that by Poage, 1954 [101] is of special interest for the Polar Basin at levels below 500 mb since it is based upon a two-year series of Ptarmigan reconnaissance flight dropsondes released between Fairbanks and the Pole. Other cross sections founded mainly upon theoretical radiational estimates have also been prepared [102, 103].

The course of the seasonal variation of upper-air temperature does not yet seem to be clear. Murgatroyd [68] finds autumnal stratospheric temperatures warmer than in spring, perhaps owing to a thermal lag from lower levels, but Dobson [104] expected the maximum temperature at high levels to be advanced toward spring owing to larger ozone concentrations then. Probably there is so much real variability from year to year, month to month, and place to place in stratospheric temperature, that a definite answer to this must await more reliable measurements gathered over many years for each place. Salmela [105] studied the monthly variation of temperature at 100 and at 50 mb along a north-south section from Alert to the Caribbean, based on 1953–1957 data, and found that distortion of the five-year mean was caused by a single unusual season. Namias [77] points out additional examples. Diurnal variations (i.e., the difference between day and night) is the basis at the present of determining radiation correction to radiosonde readings. However, it is quite possible that such diurnal changes are caused by other instrumental errors [29], or perhaps are even real [106].

The characteristic thermal feature of the lower troposphere in the polar regions is the prevalent inversion. Since Sverdrup's pioneer description, few quantitative studies have appeared. Poage [101] analysed dropsonde records in a parallel study to Sverdrup's and, in general, confirmed the earlier conclusions. Flohn [93], using mean monthly ascents for Siberia, and Belmont [42a] doing the same for North America and Antarctic stations, presented some preliminary features. When the radiosonde observations for Ice Island T-3 became available, a detailed study based upon significant levels was possible [15]. One result of the latter study and of Flohn's was that the inversion magnitude increased as the surface temperature decreased. The time changes of inversion structure are too complex to be explained without complete measurements of radiation balance, vertical motion, and advection. Possibly an approach to this can be made with IGY data as now being collected in the Antarctic, and as already analyzed by Liljequist [107] for the lowest 10 meters. A Soviet work on the subject [108] has also been reported but was not available for review.

The opposite type of thermal gradient, superadiabatic instability during Arctic winter, has been reported by Robinson [109].

4.2.2. Upper-Air Wind. The Arctic wind field has been discussed in Sections 3.2 and 3.3; however, a few additional observations may be made. There are several meridional cross sections from pole to equator showing the zonal component of the resultant vector, such as Mintz and Dean (to 20 km) [110], Mironovitch (to 22 km) [111], Kochanski (to 26 km) [67], Murgatroyd (20–100 km) [68], Flohn (to 30 km) [66], Moore (to 15 km) [112], and Koslowski (to 20 km) [113]. Koslowski's sections are geostrophic winds from Wege's 1949–1953 maps, for summer and winter. Estimated values of the vector mean wind are tabulated for levels from 20,000 to 100,000 ft for January and July, 30°–70°N, and for 20° intervals of longitude in [114]. Murgatroyd points out there are westerlies from 10 to 90 km in winter at mid- and high latitudes and apparently they remain as weak westerlies from 40 to 60 km at high latitudes even in summer, although in mid-latitudes they become easterly.

Wexler and Moreland [55] point out that there may be polar westerlies at 20 km in summer since at that level pressure decreases from 55° to 90°. They feel that rather than a circumpolar high, there is a belt of high pressure between 70 and 75°N. However, they agree that these westerlies weaken with height and become easterlies near 30 km. In support of this, Salmela's time sections [105] showed mean westerlies at 50 mb (20 km) all year for latitudes 65–90°N.

Campbell [115] summarized the position and number of jets at 500 mb from two years of *Täglicher Wetterbericht* maps. He finds a two-jet condition most common along 80°W and 20°E (near 68° and 40°N) but occasionally there are three jets (near 80°, 57°, 36°N).

4.3. Synoptic Studies

To investigate frontal structure, Reed [116, 117] prepared a series of 63 cross sections from Kodiak to Alert from February to June, 1955. He found a shallow "Arctic front" is intense near the surface but disappears at 700 mb. Pacific fronts moving inland are diffuse near the ground and are most intense at about 500 mb. A review of various other interpretations of the Arctic front is given by [118].

Kunkel [27] made frequency counts of cyclones for the summers of 1952–56 and verified Dzerdzeevski's 1945 observation that the central Arctic has a high frequency of cyclonic activity north of the anticyclonic belt which lies near 75°N. Keegan [119] in a parallel study for the winter months, also confirmed from recent, more complete data, that cyclones were frequent in the Arctic again as maintained by [4] and Flohn [17]. He found most were warm and entered from the Atlantic, while those which originated in the Arctic had cold cores. An example is given of the rapidity with which pressure patterns can change (1050 mb high to a 1000 mb low in three days) and the consequent need for reports from the central Arctic. He included anticyclones in his study and was able to verify Bodurtha's [120] results on the location of a maximum frequency of anticyclonic activity in the Yukon. Trajectories of pressure centers have also been compiled by Berry *et al.* [121].

Anticyclogenesis in the polar regions remains one of the most baffling problems. Bodurtha's statistical study found that eastern Alaska is one of the most frequent locations for anticyclogenesis (pressure rise of at least 7mb/24 hr). His model has the following features: warm air advection in the troposphere west of Alaska, with ascending motions causing cold 200-mb temperatures; cold air in low levels over the area of anticyclogenesis; strong cold air advection at 200 mb.

An examination from a theoretical standpoint of the problem of maintaining the anticyclonic circulation in the Arctic was made by Fjørtoft [122]. He found that a net warming surrounding the anticyclonic area was associated with subsidence and horizontal divergence at the ground. The surface land-sea characteristics are basically responsible for the lack of a purely zonal circulation.

The counterpart to Arctic anticyclones are the typical cold lows of high latitudes. Klug [123] has found that these lows move in the direction of the mean resultant wind vector 300 to 500 km around the low at 500 mb. The Arctic cold poles, centers of coldest air at 500 mb, have also been extensively investigated by Scherhag [58] who has mapped their trajectories which may be followed for many weeks or months as they are steered around the polar latitudes. Reed and Tank [27] examined the changes in the thermal field of a cold low at 700 mb as it intensified. They found rising motion was mainly responsible for the net cooling although infrared radiation also produced a slight cooling which equaled the advective warming.

The McGill Group [26, 124–126] has attempted to specify the pressure pattern of daily maps using a statistical concept known as orthogonal polynomials. Such an array can represent surfaces with which a second (pressure or contour) field may be correlated. This permits the many individual grid values to be described with relatively few (15 or more as desired to improve accuracy) correlation coefficients. With such a set of values further correlations of the original field become possible, either with itself in time, or with individual fields within the larger field. This technique was applied to a study of persistence and to kinetic energy.

Wilson [127] has attempted the difficult task of typing Arctic circulation patterns at sea level, using the polynomial specifications, and examining their persistence. Requiring a duration of four days at or above a specified persistence level, she found 45 % of all days (in 1955) were in persistent periods. Persistence averaged six days. Changes in the large scale pattern occurred abruptly. Following Shapiro's examination of geomagnetic activity with circulation [128], Wilson found a minimum persistence 14 days after 14 key geomagnetic days, in agreement with Shapiro, but owing to the very limited sample, the conclusions are indefinite.

The contour variance was also related to kinetic energy and the relations used to interpret the kinetic energy variations of a 500-mb map series by Godson and MacFarlane [129]. However the relationship turned out to be too loose to make the contour variance a useful parameter in this application.

4.4. Surface Climatology

Most of the literature of the past ten years described some phase of the surface climate or weather. Nevertheless, the classic examples of Sverdrup *et al.* [130] on the Canadian Arctic and Greenland, and of Georgi [131] on Eismitte must still be referred to. Some of the recent descriptions are: Dzerdzeevskii [5], Becker [132], Somov [6], and Meyer [133] on the central Arctic; Fristrup [134] and Hamilton [135–137] on North Greenland; Georgi [138] and Expeditions Polaires Françaises [139] on Eismitte; Rae [140] on the Canadian Arctic; and a general description of the entire Arctic, but based on mainly pre-war data, by Petterssen *et al.* [141]. A more recent discussion of the Eurasian sectors is available now by Putnins and Stepanova [20] which includes some aerological data also.

A study of strong surface winds at seven North American Arctic stations over a two-year period was made [27] which showed strong winds always occurred with above normal pressure gradients. Alert appears to have more frequent strong winds than Thule (based on only two reports each day). Orographic control was evident since most winds were downslope or parallel to coastlines.

Visibility peculiarities in the Arctic have been treated by Fritz [142] and Mitchell [143], and ice fog has been carefully investigated by Bell *et al.* [144].

Solar radiation measurements made at Ice Island T-3 are described by Fritz [145], and general problems of observation are discussed by Schumacher [146]. An exhaustive study of radiation and heat transfer by Liljequist [147], despite the fact that it refers primarily to the Antarctic, will be as fundamental to the Arctic as Sverdrup's work. A study of the summer heat budget at Barrow is available by Bryson [148]. The influence of underlying heat sources upon polar air was discussed by Burbidge [149], Craddock [150], and Johnson [151]. Miller [152] explained the influence of snow cover on local climate in Greenland. A detailed study of ice-cap meteorology is given by Orvig [153], and he reviews the general subject of snow and ice climate [28]. An ice atlas has been recently issued [154]. The influence of land and sea distribution on the free atmosphere is discussed by Flohn [83].

4.5. Arctic–Antarctic Comparisons

As has been seen in the past paragraph, it is difficult to separate Antarctic from Arctic physical processes. Although there is no intent to enter into a review of the southern polar region, it is instructive to compare the two polar regions. Their differences in circulation enable the basic causes to be identified a little easier. We will mention several papers which have presented comparative data; with the great activity in the Antarctic during the IGY, it is to be expected that many more will soon appear.

Gibbs [155] found that Northern Hemisphere circulation is slower than that in the Southern Hemisphere owing to the greater mixing of anticyclones and cyclones in the north which is caused by land-sea distribution. Moore [100], Rubin [156], Schumacher [89], Court [157], Flohn [17, 96], Hofmeyer [158], and Belmont [13, 72a] discussed tropopause differences and compared mean radiosonde ascents. The contrasts in the marine climates are treated by Holcombe [159] and Krauss [160]. The latter believes that the heat of freezing given to the atmosphere by the ocean is returned to the surface in the Arctic but not in the Antarctic. Pogosian [161] compares the annual ranges of temperature at North Pole Stations 4 and 7 with those at Amundson–Scott (South Pole) station for the same period. Bannon [162] finds a steeper spring warming in the Antarctic stratosphere than in the Arctic. Sverdrup [88] and

Schumacher [89] point out the importance of albedo differences on the radiation balance for both troposphere and stratosphere. Willett [44] attempted to account for the difference in polar stratospheric circulations by solar-terrestrial effects. It will be of great interest to compare occurrences of sudden warmings in the Antarctic with those reported in Section 3 when the newly observed data are published.

Wexler and Moreland [55] point out that the flow pattern must be considered in comparing the vertical temperature profiles in the Arctic and Antarctic, or even along the same latitude due to differences in vertical motions.

4.6. Forecasting Studies

About the only efforts toward practical studies in Arctic forecasting by synoptic methods have been in [163] which presented preliminary climatological contingency tables for specific weather situations, map series showing "abnormal" developments, tracks of pressure centers, and some empirical forecast rules on the movement of lows, and fog forecasting. These studies, for the Navy, are concerned with the Aleutian–Bering and Baffin Bay areas. For aids to local forecasting at given stations, the Technical Circulars of the Canadian Meteorological Office and the local forecast studies of the U.S. Air Weather Service are valuable. A series of synoptic maps to illustrate typical development of cyclones and anticyclones in the Arctic has also been compiled by the U.S. Navy [164]. They are selected from the Weather Bureau Historical Map Series.

Estoque [165] has developed a dynamic prediction model for use in numerical weather prediction which he applied graphically to a month of 500- and 1000-mb maps of the Edmonton Arctic Forecast Group. His results indicate that the method can be applied to the Arctic as successfully as to mid-latitudes. Reed [166] has also developed a prediction model which incorporates a term for the heating of Arctic air by water. This heating term was an improvement over an earlier model, but from a test forecast series it was found to be useful only after a cyclone had developed.

4.7. Vertical Motions and Ozone

Vertical motions have been seen above to be of prime importance in explaining the maintenance of intense gradients at stratospheric levels, in accounting for the temperature field in cold lows, and in helping to produce anticyclogenesis. Kochanski [42] explained the lower stratospheric (up to 100 mb) thermal field which has cold ridges and warm troughs, as being associated with vertical motions which are organized in such fashion that air descends from ridge to trough and ascends from trough to ridge. Above 100 mb the contour pattern returns to the same thermal pattern as in the troposphere and the vertical motions are correspondingly reversed.

Wexler [49, 55] points out that these stratospheric vertical motions probably extend well into the ozonosphere, and suggests that independent ozone measurements could provide estimates of vertical motion at 50 mb. This is because increases in ozone are associated with descending currents and decreases with ascending motions. He hopes that 50-mb analyses over the west European network will provide a test of these relations.

The reasons for the late winter sudden increase in ozone or for the high latitude maximum in total ozone, are still matters of conjecture. The successful spectroscopic measurement of ozone in the polar night near the North Pole [47] confirms Normand's expectation that the "ozone gap" does not exist [48]. A commonly held theory is that in the late winter transport from middle to high latitudes is increased by deep cyclonic systems or horizontal meridional circulations, thus advecting ozone-rich air from sunlit lower latitudes. Wexler and Moreland [55] favor the sequence of radiational cooling which leads to instability of zonal flow, and to ozone transport along the resulting meridional blocking flow. This implies that the ozone distribution along a latitude should vary in accordance with the pressure pattern. Ozone data from the Canadian Arctic during the IGY may help verify these ideas.

Godson and Lee in their fundamental exposition of these problems [58] state that "the importance of ozone observations in the Arctic during polar darkness, and of the determination of the vertical distribution of ozone on all possible occasions cannot be overemphasized."

Gaigerov [9] also emphasizes the need for investigations of the winter stratosphere thermal field and especially of the vertical ozone distribution. He calls for special high-level balloons and meteorological rockets. He also mentions a special characteristic of ozone in polar regions: it has a longer half-life due to low temperature there, and to the reduced amount of organic or dust particles for the ozone to oxidize.

One further comment may be offered on this topic. It is evident that vertical motion is a critical factor in explaining both ozone and temperature variations in the stratosphere. Eventually there will be a network of ozonsonde stations to provide vertical measurements of ozone and temperature, but we will still lack satisfactory means to estimate vertical motions independently of the other two variables. In order to obtain some experimental values from which to establish general relations among these parameters, constant level balloons which have been properly instrumented could be utilized to great advantage. They could best approximate the trajectory of an air parcel, which would eliminate the great source of error of the adiabatic method when applied to usual weather charts. A single balloon could cover wide expanses of the polar regions over areas not likely to be sampled from the ground and do so more economically than by aircraft. A few such flights over areas for which good synoptic charts are available could give independent evidence of these theories relating vertical motion, ozone and temperature changes in the stratosphere, especially at the end of the polar night.

4.8. *Solar-Terrestrial Relations*

Murgatroyd has pointed out that the lower stratosphere has some coupling to the troposphere since ozone, temperature, and wind can be correlated with the tropopause [68]. For at least a hundred years reputable scientists have tried to find a similar coupling between solar activity and the troposphere or even stratosphere. The nearest established connection still seems to be the effect of solar magnetic storms upon the ionosphere, and the production of visible aurorae. The subject of solar-terrestrial relations is a large one and of concern here only to the extent that in the polar regions these possible relations appear to have the best chance for discovery through the presence there of geomagnetic poles and aurorae. The similarity of the earth's magnetic field to the mean contours of the upper troposphere has been frequently noted. The apparently bipole magnetic field is in reality a somewhat deformed single-pole field. Flohn [167] and Mironovitch [168] point out that the bipole contour pattern is probably caused by land-sea distribution and orography rather than magnetic field influences.

From Maury's time, 1851, when he speculated on atmospheric circulation and magnetism and indeed invoked "terrestrial electricity" to guide his vertical meridional circulation [169], to today, no reliable relationships seem to have been found; but the search continues. A few may be mentioned: Wexler [170] checked possible effects of ozonosphere heating on sea level pressure, and Riehl [171] compared the 500-mb circulation in the auroral belt with solar flares; London *et al.* [103] concluded that no relations of temperatures and heights at three stations in the auroral zone could be found with geomagnetic activity, and Wexler [172] showed that the downward propagation of disturbances from the upper ozonosphere or higher, is considerably damped, and that it is more likely that tropospheric disturbances are propagated upward to affect the upper atmosphere significantly. A detailed bibliography of the 350 or so attempts since 1826 is given in [C].

5. Conclusion

The unsolved problems of Arctic meteorology are largely those of meteorology in general now that observing facilities in the Arctic are able to describe the atmosphere almost as well as in mid-latitudes. Each of the topics mentioned in the last section is thus a problem requiring research. However, of greatest interest are probably those questions which always lie just above the levels where one can observe regularly. In this instance, these are the large-scale features of the middle stratosphere and above, through which there is the hope of further explaining the "unusual" features found at high latitudes.

Arctic Meteorology Bibliographies

A. Sutcliffe, R. C., ed. (1958). "Polar Atmosphere Symposium, Oslo, 1956," Pt. 1 (Meteorology Section), 341 pp. Pergamon, New York.

B. Bonnlander, B. H., and Belmont, A. D. (1955). Bibliography on polar atmospheric circulation. Appendix 1 to Status Report No. 5, Contract AF 19(604)-1141, Arctic Meteorology Research Group, McGill University, Montreal.

C. Meteorological Abstracts and Bibliography. Bibliographies: Climate of the Arctic, Vol. 5, No. 9 (Sept. 1954); Synoptic Meteorology of the Arctic, Vol. 7, No. 1 (Jan. 1956); Solar Weather Relations, Vol. 8, No. 1 (Jan. 1957). Am. Meteorol. Soc., Boston.

D. Arctic Institute of North America (1953–1957). "Arctic Bibliography," Vols. 1–7. U.S. Department of Defense, Washington, D.C.

E. U. S. Weather Bureau, Foreign Area Section, Office of Climatology. (1957). "Bibliography on the Climate of Greenland," 473 pp. Washington, D.C.

References

1. Dorsey, H. G. (1951). Arctic meteorology. *In* "Compendium of Meteorology" (T. F. Malone, ed.), pp. 942–951. Am. Meteorol. Soc., Boston, Massachusetts.
2. Van Rooy, M. P., ed. (1957). "Meteorology of the Antarctic." Weather Bureau, Pretoria, South Africa.
3. Dzerdzeevskii, B. L. (1945). Circulation models in the troposphere of the central Arctic. *Akad. Nauk S.S.S.R.*, 40 pp. Translated (1954) in Scientific Report No. 3, Part 2, Contract AF 19(122)-228, Department of Meteorology, University of California, Los Angeles.
4. Dzerdzeevskii, B. L. (1945). The circulation of the atmosphere in the central polar basin. Vol. II, Part 3 of: Reports of the Drifting Station "North Pole." Northern Sea Route Admin. 1941–1945, pp. 64–200. Translated (1954) in full as Scientific Report No. 6, Contract AF 19(122)–228, Department of Meteorology, University of California, Los Angeles.
5. Dzerdzeevskii, B. L. (1946). The distribution of atmospheric pressure over the Central Arctic. *Meteorol i Gidrol.* **1**, 33–38. Translated (1954) in Scientific Report No. 3, Part 1, Contract AF 19(122)–228, Department of Meteorology, University of California, Los Angeles.
6. Somov, M. M. (1954–1955). Observational data of the scientific research drifting station of 1950–1951, Vols. 1–3 (in Russian). Translated by Am. Meteorol. Soc., Boston.
7. U.S. Office of Technical Services. (1958). Soviet Bloc International Geophysical Year Information (Feb. 14, 1958–weekly). OTS Report No. PB 131,632. Washington, D.C.
8. Crary, A. P. (1956). Arctic ice islands research. *Advances in Geophys.* **3**, 3, 7, 8.
9. Gaigerov, S. S. (1957). Atmospheric processes in the central Arctic. *Priroda* **12,** 27–34. Translated (1958) by Canada, Defense Research Board, Translation No. T-282-R.

10. Gaigerov, S. S. (1956). Some results of aerological observations on the drifting station "Severnyi Polius-4" Summer, 1955 (in Russian). *Meterol. i Gidrol.* **6**, 3–13.
11. Laktionov, A. F. (1955). "The North Pole," pp. 347–426. Morskoi Transport, Moscow. Translated (1956) by K. R. Whiting as "Recent Soviet Scientific Investigations in the North Polar Regions." Documentation Research Div., Research Studies Inst., Air University, Maxwell Air Force Base, Alabama. 104 pp.
12. Burkhanov, V. F. (1957). Achievements of Soviet geographic exploration and research in the Arctic. *In* "Across the Ocean on Drifting Ice," published by National Publishing House for Geographic Literature, Moscow. Canada (1957). Defense Research Board, Translation T-253-R, 26 pp.
13. Belmont, A. D. (1959). Upper air temperatures at Ice Island T-3, 1952–1954. *Geophys. Res. Paper No.* 63, vol. II, *U.S.A.F. Cambridge Res. Center*, Bedford, Mass.
14. Belmont, A. D. (1954). Final report on Arctic meteorological research, Contract AF 19(122)-228, Department of Meteorology, University of California, Los Angeles.
15. Belmont, A. D. (1958). Lower tropospheric inversions at Ice Island T-3. *In* "Polar Atmosphere Symposium, Oslo, 1956" (R. C. Sutcliffe, ed.), Pt. 1, pp. 215–284. Pergamon, New York.
16. Anonymous. (1954). Über neue sowjetische Forschungen und Entdeckungen in der Zentral-Arktis. *Izves. Akad. Nauk S.S.S.R. Ser. Geog. No. 5* (in Russian). Translated (1955) into German by H. Taubert *Petermann's Geograph. Mitt.* **99**, 70–77.
17. Flohn, H. (1952). Zur Aerologie der Polargebiete. *Meteorol. Runds.* **5**, 81–87, 121–128. Translated (1955) as "The aerology of the polar regions," Appendix to Status Report No. 4, Contract AF 19(604)-1141, Arctic Meteorology Research Group, McGill University, Montreal.
18. Ratner, B. (1957–1958). Upper-air climatology of the U.S. Part 1: Averages for isobaric surfaces. Part 2: Extremes and standard deviations of average heights and temperatures. *U.S. Weather Bur. Tech. Pap.* 32.
19. Bedel, B. (1954). Les observations météorologiques de la Station Française du Groenland; Conditions atmosphériques en altitude du 17 September 1949 au 10 Août 1951. 119 pp. Expéditions Polaires Françaises, Missions P.-É Victor, La Météorologie Nationale, Paris.
20. Putnins, P., and Stepanova, N. (1956). Climate of the Eurasian Northlands. *In.* "Dynamic North." Washington, D.C., Technical Assistant to Chief of Naval Operations for Polar Projects. OPNAV PO3-30. 104 pp.
21. Henry, T. J. G., and Armstrong, G. R. (1949). "Aerological Data for Northern Canada," 271 pp. Meteorol. Branch, Dept. of Transport, Toronto, Canada.
22. Tolefson, H. B. (1957). Atmospheric temperature observations to 100,000 feet for several climatological regions of the Northern Hemisphere. *Natl. Adv. Com. Aeronaut., Tech. Notes* 4169.
23. Hering, W. S., and Salmela, H. A. (1957). A climatological study of wind and temperature at 100, 50 and 25 mb over North America during the 1956–57 winter season. *USAF Cambridge Res. Center*, GRD TM-57-27; Also in *GRD Research Notes* **1**, 19–50.
24. Borden, T. R. Jr. (1958). The wind field at 100,000 feet during the first six months of the IGY (July 1957–December 1957). *USAF Cambridge Res. Center, GRD Research Notes* **1**, 51–70.
25. Jacobs, I. (1957–1958). 5-bzw. 40 jährige Monatsmittel der absoluten Topographien der 1,000 mb-, 850 mb-, 500 mb-, und 300 mb- Flächen sowie der relativen Topographien 500/1,000 mb und 300/500 mb über der Nordhemisphäre und ihre monatlichen Änderungen. *Meteorol. Abh.* **4**(1), 325 pp.; *Meteorol. Abh.* **4**(2), 168 pp.

26. Hare, F. K. (1958). Studies in arctic meteorology. Final Report, Contract AF 19(604)-1141, Arctic Meteorology Research Group, McGill University, Montreal.
27. Reed, R. J. (1958). Synoptic studies in arctic meteorology. Final Report, Contract AF 19(604)-1298, Meteorology Dept., University of Washington, Seattle.
28. Hare, F. K., and Orvig, S. (1958). The arctic circulation, a preliminary review. Suppl. to Final Report, Contract AF 19(604)-1141, Arctic Meteorology Research Group, McGill University, Montreal.
29. Lowenthal, M., and Arnold, A. (1957–1958). High altitude radiosonde flights, January 1955, Part 2. *U.S. Army Signal Res. and Development Lab. Tech. Rept.* 1962; Part 3. *U.S. Army Signal Res. Tech. Memo.* No. M–1933.
30. U.S. Air Weather Service (1955). Accuracies of radiosonde data. *Tech. Rep.* 105–133.
31. Leviton, R. (1954). Height errors in a rawin system. *USAF Cambridge Res. Center, AF Surveys in Geophys. No.* 60, 17 pp.
32. U.S. Weather Bureau. (1952). Normal weather charts for the Northern Hemisphere. *Tech. Pap. No.* 21.
33. Anonymous. (1953). Normalwerte des Luftdruckes auf der Nordhemisphäre für die Periode 1900–1939. *Meteorol. Abh.* **2** (1), 126 pp.
34. Kochanski, A., and Wasko, P. E. (1953). Analysis and wind flow at the 50 and 25 mb levels. *U.S. Air Weather Service Tech. Rept.* 105–96, 76 pp.
35. Heastie, H. (1955–1956). Average height of the standard isobaric surfaces over the area from the North Pole to 55°N in January. *Great Britain, Meteorol. Research Com.* MRP 918. Average height of the standard isobaric surfaces over the area from the North Pole to 55°N in July. *Ibid* MRP 981: Average height of the standard isobaric surfaces over the area from the North Pole to 55°N in April and October. *Ibid* MRP 1020.
36. Namias, J. (1956). The general circulation of the lower troposphere over arctic regions and its relation to the circulation elsewhere. *In* "Polar Atmosphere Symposium, Oslo, 1956" (R. C. Sutcliffe, ed.), Pt. 1, pp. 45–61. Pergamon, New York.
37. Wege, K. (1957). Druck-, Temperatur- und Strömungsverhältnisse in der Stratosphäre über der Nordhalbkugel. *Meteorol. Abh.* **5** (4), 55 + 90 pp.
38. Wege, K. (1958). Druck- und Temperaturverhältnisse der Stratosphäre über der Nordhalbkugel in den Monaten März, Mai, September, November und Dezember. *Meteorol. Abh.* **7** (1), 17 + 49 pp.
39. Goldie, N., Moore, J. G., and Austin, E. E. (1958). Upper air temperatures over the world. *Great Britain, Meteorol. Office Geophys. Mem. No.* 101.
40. Wege, K., Leese, H., Groening, H. V., and Hoffmann, G. (1958). Mean seasonal conditions of the atmosphere at altitudes of 20 to 30 km and cross sections along selected meridians in the Northern Hemisphere. *Meteorol. Abh.* **6** (4).
41. Wahl, E. W. (1958). Mean monthly 300 and 200-mb contours and 500, 300 and 200-mb temperatures for the northern hemisphere. *USAF Cambridge Res. Center, Geophys. Res. Pap. No.* 57, 80 pp.
42. Kochanski, A. (1954). Thermal structure and vertical motion in the lower stratosphere. *U.S. Air Weather Service Tech. Rept.* 105–129, 36 pp.
42a. Belmont, A. (1954). Ice Island T-3 temperatures and comparison with other polar stations. Report No. 4, Contract AF 19(122)-228, Department of Meteorology, University of California, Los Angeles.
43. Schweitzer, H. (1952). Haben die Eruptionen auf der Sonne Einfluss auf das Wetter der Erde. *Umschau Forschr. Wiss. U. Tech.* **52**, 247.
44. Willett, H. C. (1952). Atmospheric reactions to solar corpuscular emissions. *Bull. Am. Meteorol. Soc.* **33**, 255–258.

45. Scherhag, R. (1952). Die explosionsartigen Stratosphärenerwärmungen des Spätwinters 1951/1952. *Ber. Deut. Wetterd. U.S. Zone* **38**, 51–63.
46. Scherhag, R. (1953). Rätselhafte Temperatursprünge in der Stratosphäre. *Umschau Forschr. Wiss. U. Tech.* **53**, 646–648.
47. Douglas, A. E., Herzberg, G., and Rose, D. C. (1956). Atmospheric ozone near the North Pole. *In* "Vistas in Astronomy" (A. Beers, ed.), Vol. 2, pp. 874–877. Pergamon, New York.
48. Normand, C. (1953). Atmospheric ozone and the upper-air conditions. *Quart. J. Roy. Meteorol. Soc.* **79**, 39–50.
49. Wexler, H. (1958). A meteorologist looks at the upper atmosphere. *In* "Atmospheric Explorations" (H. G. Houghton, ed.), pp. 79–100. Wiley, New York.
50. Riehl, H. (1950). Variations in the structure of high-level cyclones. *Bull. Am. Meteorol. Soc.* **31**, 291–294.
51. Scrase, F. J. (1953). Relatively high stratospheric temperatures of February 1951. *Meteorol. Mag.* **82**, 15–18.
52. Scrase, F. J. (1951). Radiosonde and radar wind measurements in the stratosphere over the British Isles. *Quart. J. Roy. Meteorol. Soc.* **77**, 483–488.
53. McIntosh, D. H. (1953). Solar and terrestrial relationships. *Meteorol. Mag.* **82**, 11–15.
54. Warnecke, G. (1956). Ein Beitrag zur Aerologie der arktischen Stratosphäre. *Meteorol. Abh.* **3**(3), 60 pp.
55. Wexler, H., and Moreland, W. B. (1958). Winds and temperatures in the Arctic stratosphere. *In* "Polar Atmosphere Symposium, Oslo, 1956" (R. C. Sutcliffe, ed.), Pt. 1, pp. 71–84. Pergamon, New York.
56. Craig, R. A., and Hering, W. S. (1957). An "explosive warming" at 25 mb in January 1957. *USAF Cambridge Res. Center,* GRD TM-57-17; Also in *GRD Research Notes* **1**, 81–102.
57. Teweles, S. (1958). Anomalous warming of the stratosphere over North America in early 1957. *Monthly Weather Rev.* **86**, 377–396.
58. Godson, W. L., and Lee, R. (1958). High-level fields of wind and temperature over the Canadian Arctic. *Beitr. Phys. Atmos.* **31**, 40–68.
59. Lowenthal, M. (1957). Abnormal mid-stratospheric temperatures. *J. Meteorol.* **14,** 476.
60. Lee, R., and Godson, W. L. (1957). The arctic stratospheric jet stream during the winter of 1955–1956. *J. Meteorol.* **14**, 126–135.
61. Darling, E. M. (1953). Winds at 100 mb and 50 mb over the U.S. in 1952. *Bull. Am. Meteorol. Soc.* **34**, 458–461.
62. Scherhag, R. (1958). The role of tropospheric cold-air poles and of stratospheric high-pressure centers in the arctic weather. *In* "Polar Atmosphere Symposium, Oslo, 1956" (R. C. Sutcliffe, ed.), Pt. 1, pp. 101–117. Pergamon, New York.
63. Godson, W. L. (1958). Canadian meteorological program for the IGY. *Roy. Meteorol. Soc. Can. Branch Pub.* **8**(2).
64. Teweles, S. and Finger, F. G. (1958). An abrupt change in stratospheric circulation beginning in mid-January 1958. *Monthly Weather Rev.* **86**, 23–28.
65. Teweles, S., and Finger, F. G. (1958). Contours and isotherms at 10 mb, July 1957 to June 1958. U.S. Weather Bureau Report for U.S. National Committee for the IGY. Unpublished manuscript.
66. Flohn, H. (1950). Die planetarische Zirkulation der Atmosphäre bis 30 km Höhe. *Ber. Deut. Wetterd. U.S. Zone* **12**, 156–161.
67. Kochanski, A. (1955). Cross sections of the mean zonal flow and temperature along 80°W. *J. Meteorol.* **12**, 95–106.

68. Murgatroyd, R. J. (1957). Winds and temperatures between 20 km and 100 km—A review. *Quart. J. Roy. Meteorol. Soc.* **83**, 417–458.
69. Whipple, F. J. W. (1935). The propagation of sound to great distances. *Quart. J. Roy. Meteorol. Soc.* **61**, 285–308.
70. Burley, M. W., Ritchie, E. M., and Gray, C. R. (1957). Winds of the upper troposphere and lower stratosphere over the U.S. *Monthly Weather Rev.* **85**, 11–17.
71. Kochanski, A. B. (1956). Wind, temperature and their variabilities to 120,000 feet. *U.S. Weather Service Tech. Rep.* 105–142.
72. Gutenberg, B. (1949). New data on the lower stratosphere. *Bull. Am. Meteorol. Soc.* **30**, 62–64.
72a. Nojima, H. (1954). The easterlies in the lower stratosphere observed over Honjo and Tateno in January and February 1952 (in Japanese). *J. Aerol. Obs. Tateno, Japan* **5**, 233–244.
73. Mantis, H. (1958). Personal communication on results of University of Minnesota constant level balloon flights, 1953–1958.
74. Hare, F. K. (1951). Some climatological problems of the Arctic and Sub-Arctic. *In* "Compendium of Meteorology" (T. F. Malone, ed.), pp. 952–964. Am. Meteorol. Soc., Boston.
75. Georgi, J. (1948). Bemerkungen zur glazialen Antizyklone. *Ann. Meteorol.* **1**, 277–283.
76. Matthes, F. E., and Belmont, A. D. (1950). The glacial anticyclone theory examined in the light of recent meteorological data from Greenland, Part 2. *Trans. Am. Geophys. Un.* **31**, 174–182.
76a. Rodewald, M. (1950). Zur Frage der Luftdruckverhältnisse in der Arktis. *Ann. Meteorol.* **3**, 284–290.
77. Namias, J. (1958). Synoptic and climatological problems associated with the general circulation of the Arctic. *Trans. Am. Geophys. Un.* **39**, 40–51.
78. Maury, M. F. (1850). On the general circulation of the atmosphere. *Proc. Am. Assoc. Advance. Sci.* **3**, 126–147.
79. Ferrel, W. (1898). "A Popular Treatise on the Winds", 2nd ed., p. 139. Wiley, New York.
80. Martell, E. A. (1959). Atmospheric aspects of Strontium 90 fallout. *Science,* **129**, 1197–1206.
81. Flohn, H. (1949). Mittlere Druck- und Strömungsverhältnisse in der Troposphäre (Winter). *Deut. Wetterd. U.S. Zone Grosswetterlagen Mitteleuropas* **2**, 105–107.
82. Dove, H. W. (1855). "Die Verbreitung der Wärme in der nördlichen Hemisphäre" (with 2 maps: 1. Karte. . . 2. Karte der Nordpolarländer). D. Reimer, Berlin.
83. Flohn, H. (1943). Kontinentalität und Ozeanität in der freien Atmosphäre. *Meteorol. Z.* **60**, 325–331.
84. LaSeur, N. F. (1954). On the asymmetry of the middle-latitude circumpolar current. *J. Meteorol.* **11**, 43–57.
85. Frenzen, P. (1956). Criteria for the occurrence of an eccentric circumpolar vertex in the atmosphere inferred from model experiments. *J. Meteorol.* **13**, 124–125.
86. Kobayashi, S. (1956). On the asymmetry of circumpolar currents (in Japanese). *J. Meteorol. Res.*, Tokyo. **8**, 503–507.
87. U.S. Air Weather Service. (1955). Ultrapolar circulation patterns. *Tech. Rept.* 105–143.
88. Sverdrup, H. U. (1954). Some problems of arctic meteorology. *Proc. Toronto Meteorol. Conf. 1953*, pp. 69–73. Royal Meteorol. Soc., London.

89. Schumacher, N. J. (1958). Temperature, height and humidity (Maudheim). "Norwegian-British-Swedish Antarctic Expedition, 1949–1952, Scientific Results," Vol. 1, Pt. 1, Norsk Polarinstitutt, Oslo.
90. Austin, J. M., and Krawitz, L. (1956). 50-millibar patterns and their relationship to tropospheric changes. *J. Meteorol.* **13,** 152–159.
91. Petterssen, S. (1950). Some aspects of the general circulation of the atmosphere. *Centenary Proc. Roy. Meteorol. Soc.* pp. 120–155. Royal Meteorol. Soc., London.
92. Flohn, H., and Seidel, G. (1958). Recent studies on the arctic troposphere and its teleconnections. *In* "Polar Atmosphere Symposium, Oslo, 1956" (R. C. Sutcliffe, ed.), Pt. 1, pp. 62–70. Pergamon, New York.
93. Flohn, H. (1944). Zum Klima der freien Atmosphäre über Sibirien. I. Temperatur und Luftdruck in der Troposphäre über Jakutsk. *Meteorol. Z.* **61,** 50–57; (1947). II. Die regionale winterliche Inversion. *Meteorol. Runds.* **1,** 1.
94. Flohn, H. (1947). Zum Klima der Kältepole der Erde. *Meteorol. Runds.* **1**(1/2), 25–26.
95. Flohn, H. (1947). Grundzüge der atmosphärischen Zirkulation über Sibirien und dem angrenzenden Polarmeer. *Polarf.* **2,** 143–149.
96. Flohn, H. (1951). Die Zirkulation der Atmosphäre in den Polargebieten. *Polarf.* **3**(1), 58–64. Translated (1954) by Am. Meteorol. Soc., Boston.
97. Kochanski, A., and Wasko, P. E. (1956). Daily wind flow at the 50- and 25-mb levels. *Bull. Am. Meteorol. Soc.* **37,** 8–13.
98. Kochanski, A., and Wasko, P. E. (1956). Mean wind flow at the 50- and 25-mb levels. *Bull. Am. Meteorol. Soc.* **37,** 61–69.
99. Defant, F., and Taba, H. (1958). The breakdown of zonal circulation. *Tellus* **10,** 430–450.
100. Moore, J. G. (1956). Average pressure and temperature of the tropopause. *Meteorol. Mag.* **85,** 362–368.
101. Poage, W. C. (1954). The dropsonde record from Alaska to the North Pole, April 1950–April 1952. Scientific Report No. 2, Contract AF 19(122)-228, Department of Meteorology, University of California, Los Angeles.
102. Ohring, G. (1957). The radiation budget of the stratosphere. Scientific Report No. 1, Contract AF 19(604)-1738, Research Division, College of Engineering, New York University, New York, 42 pp.
103. London, J., Ohring, G., and Ruff, I. (1956). Radiative properties of the stratosphere. Final report, Contract AF 19(604)-1285, Department of Meteorology, New York University, New York.
104. Dobson, G. M. B., Brewer, A. W., and Cwilong, B. M. (1946). Meteorology of the lower stratosphere. *Proc. Roy. Soc.* (*London*) **A185,** 144–175.
105. Salmela, H. A. (1958). A preliminary study of the annual variation of the height and temperature at 100 and 50 mb along a North–South cross section from Alert to Guantanamo. *USAF Cambridge Res. Center, GRD Research Notes* **1,** 71–85.
106. Belmont, A. D. (1957). Apparent diurnal variations of Arctic stratospheric temperatures. *Trans. Am. Geophys. Un.* **38,** 462–468.
107. Liljequist, G. H. (1956). Energy exchange of an antarctic snow-field. *In* "Norwegian-British-Swedish Antartic Expedition, 1949–1952, Scientific Results," Vol. 2, Pt. 1D, pp. 237–298. Norsk Polarinstitutt, Oslo.
108. Vorontsov, P. A. (1956). Aerological investigations of surface inversions. *Glavnaia Geofiz. Obs. Trudy, Vypusk* **63**(125), 77–102.
109. Robinson, E. (1955). Some instances of unstable surface temperature conditions during an Arctic winter. *Arctic* **8,** 148–157.
110. Mintz, Y., and Dean, G. (1952). The observed mean field of motion of the atmosphere. *USAF Cambridge Res. Center, Geophys. Res. Pap. No.* 17, 65 pp.

111. Mironovitch, V. (1953). Répresentation de la circulation atmosphérique générale pour une coupe aérologique méridienne à travers les deux hemisphères. *Compt. rend.* **236,** 404–406.
112. Moore, J. G. (1956). Cross-section of the mean zonal component of geostrophic wind. *Meteorol. Mag.* **85,** 167–171.
113. Koslowski, G. (1958). Überlegungen zu einem Geschwindigkeitsdiagramm in der Atmosphäre. *Meteorol. Abh.* **6**(2), 32 pp.
114. USAF Cambridge Research Center. Geophysics Research Directorate (1957). "Handbook of Geophysics for Air Force Designers." Cambridge, Massachusetts.
115. Campbell, W. J. (1958). The arctic jet stream. Scientific Report No. 1, Contract AF 19(604)-3063, Meteorology Dept., University of Washington, Seattle.
116. Reed, R. J. (1958). Arctic weather analysis. *In* "Polar Atmosphere Symposium, Oslo, 1956" (R. C. Sutcliffe, ed.), Pt. 1, pp. 124–136. Pergamon, New York.
117. Reed, R. J. (1958). Arctic synoptic analysis. *In* "Contributions to the Study of the Arctic Circulation", Scientific Report No. 7, Contract AF 19(604)-1141, Arctic Meteorology Research Group, McGill University, Montreal, pp. 48–59.
118. U.S. Air Weather Service. 7th Weather Group (1956). The "arctic front." *Tech. Memo. No.* **15.**
119. Keegan, T. J. (1958). Arctic synoptic activity in winter. *J. Meteorol.* **15,** 513–521.
120. Bodurtha, F. T. (1952). An investigation of anticyclogenesis in Alaska. *J. Meteorol.* **9,** 118–125.
121. Berry, F. A., Owens, G. V., and Wilson, H. P. (1954). Arctic track charts. *Proc. Toronto Meteorol. Conf.* 1953 pp. 91–102. Royal Meteorological Society. London.
122. Fjørtoft, R. (1958). The problem of the maintenance of the low-level anticyclonic circulation in the Arctic. *In* "Polar Atmosphere Symposium, Oslo, 1956" (R. C. Sutcliffe, ed.), Pt. 1, pp. 87–92. Pergamon, New York.
123. Klug, W. (1956). Der Mechanismus der Steuerung von hochreichenden Zyklonen. *Meteorol. Abh.* **3**(2), 42 pp.
124. Hare, F. K., Godson, W. L., MacFarlane, M. A., and Wilson, C. V. (1958). Specification of pressure fields and flow patterns in polar regions. Scientific Report No. 3, Contract AF 19(604)-1141. Arctic Meteorology Research Group, McGill University, Montreal.
125. MacFarlane, M. A. (1958). The Wadsworth orthogonal polynomial technique over the north polar area during 1955. Scientific Report No. 3, Suppl. 1, Contract AF 19(604)-1141. Arctic Meteorology Research Group, McGill University, Montreal.
126. MacFarlane, M. A. (1958). The Godson orthogonal harmonic technique over the north polar area during 1955. Scientific Report No. 3, Suppl. 2, Contract AF 19(604)-1141, Arctic Meteorology Research Group, McGill University, Montreal.
127. Wilson, C. V. (1958). Synoptic regimes in the lower arctic troposphere during 1955. Scientific Report No. 6, Contract AF 19(604)-1141, Arctic Meteorology Research Group, McGill University, Montreal, 100 pp.
128. Shapiro, R. (1956). Further evidence of a solar-weather effect. *J. Meteorol.* **13,** 335–340.
129. Godson, W. L., and MacFarlane, M. A. (1958). Pressure-contour variance and kinetic energy over the Arctic. Scientific Report No. 5, Contract AF 19(604)-1141, Arctic Meteorology Research Group, McGill University, Montreal.
130. Sverdrup, H. U., Petersen, H., and Loewe, F. (1935). Klima des kanadischen Archipels und Grönlands. *In* "Handbuch der Klimatologie" (W. Köppen and R. Geiger, eds.), Vol. 2/K. Borntraeger, Berlin.

131. Georgi, J. (1935). Die Eismittestation. *In* "Deutsche Grönland-Expedition Alfred Wegener, 1929 und 1930/31" (K. Wegner, ed.), Vol. 4 (2), pp. 191–386. Brockhaus, Leipzig.
132. Becker, R. (1953–1954). Das Sommerklima am Nordpol. *Ann. Meteorol.* **6**, 251–252.
133. Meyer, H. K. (1955). Das Wetter in der Nähe des Nordpols. *Meteorol. Runds.* **8**, 35–39, 117–122.
134. Fristrup, B. (1952). Climate and glaciology of Peary Land, North Greenland. *Intern. Geodetic Geophys. Un. Assoc. Sci. Hydrol. Brussels 1952* **1**, 185–193.
135. Hamilton, R. A., and Rollitt, G. (1957). British North Greenland Expedition, 1952–54. Climatological tables for the site of the Expedition's base at Brittannia Sound and the station on the Inland-Ice "Northice." *Medd. Grønland* **158**(2), 83 pp.
136. Hamilton, R. A., and Rollitt, G. (1957). British North Greenland Expedition, 1952–54. Meteorological Observations at "Northice," Greenland. *Medd. Grønland* **158**(3), 45 pp.
137. Hamilton, R. A. (1958). The meteorology of north Greenland during the midsummer period. *Quart. J. Roy. Meteorol. Soc.* **84**, 142–158.
138. Georgi, J. (1953–1954). Bemerkungen zum Klima von "Eismitte", Grönland. *Ann. Meteorol.* **6**, 283–295.
139. Expéditions Polaires Françaises, Missions P.-É. Victor (1954). Les Observations Météorologiques de la Station Française au Groenland. Conditions atmosphérique en surface, 5 Sept. 1949 au Juin 1950. Fasc I & II. La Météorologie Nationale, Paris.
140. Rae, R. W. (1951). "Climate of the Canadian Arctic Archipelago" 90 pp. Meteorol. Branch, Dept. of Transport, Toronto, Canada.
141. Petterssen, S., Jacobs, W. C., and Haynes, B. C. (1956). Meteorology of the Arctic. *In* "Dynamic North." Washington D.C., Technical Assistant to Chief of Naval Operations for Polar Projects. OPNAV P03-3.
142. Fritz, S. (1958). On the "arctic whiteout". *In* "Polar Atmosphere Symposium, Oslo, 1956" (R. C. Sutcliffe, ed.), Pt. 1, pp. 182–186. Pergamon, New York.
143. Mitchell, J. M. (1958). Visual range in the polar regions with particular reference to the Alaskan Arctic. *In* "Polar Atmosphere Symposium, Oslo, 1956" (R. C. Sutcliffe, ed.), Pt. 1, pp. 195–211. Pergamon, New York.
144. Bell, G. B., Thuman, W. C., St. John, G. A., Wiggins, E. J. (1954). Investigation of ice fog phenomena in the Alaskan area. Final Report, Contract AF 19(122)-634, Stanford Research Institute, Stanford, Calif.
145. Fritz, S. (1958). Solar radiation measurements in the Arctic Ocean. *In* "Polar Atmosphere Symposium, Oslo, 1956" (R. C. Sutcliffe, ed.), Pt. 1, pp. 159–166. Pergamon, New York.
146. Schumacher, N. J. (1958). Some problems of meteorological observing in polar regions. *In* "Polar Atmosphere Symposium, Oslo, 1956" (R. C. Sutcliffe, ed.), Pt. 1, pp. 212-214. Pergamon, New York.
147. Liljequist, G. H. (1958). Long-wave radiation and turbulent heat transfer in the antarctic winter and the development of surface inversions. *In* "Polar Atmosphere Symposium, Oslo, 1956" (R. C. Sutcliffe, ed.), Pt. 1, pp. 167–181. Pergamon, New York.
148. Bryson, R. A. (1956). Preliminary estimates of the surface heat budget, summer, clear days at Point Barrow, Alaska. Department of Meteorology, University of Wisconsin, Madison.
149. Burbidge, F. E. (1951). The modification of continental polar air over Hudson Bay. *Quart. J. Roy. Meteorol. Soc.* **77**, 365–374.

150. Craddock, J. M. (1951). The warming of Arctic air masses over the Eastern North Atlantic. *Quart. J. Roy. Meteorol. Soc.* **77**, 355–364.
151. Johnson, C. B. (1948). Anticyclogenesis in eastern Canada during the spring. *Bull. Am. Meteorol. Soc.* **29**, 47–55.
152. Miller, D. H. (1956). The influence of snow cover on local climate in Greenland. *J. Meteorol.* **13**, 112–120.
153. Orvig, S. (1954). Glacial-meteorological observations on icecaps in Baffin Island. *Geografis. Ann.* **36**, 193–318.
154. U.S. Hydrographic Office (1955). "Ice Atlas of the Northern Hemisphere," 106 pp. Washington, D.C.
155. Gibbs, W. J. (1953). Comparison of hemispheric circulations with particular reference to the western Pacific. *Quart. J. Roy. Meteorol. Soc.* **79**, 121–136.
156. Rubin, M. J. (1953). Seasonal variations of the antarctic tropopause. *J. Meteorol.* **10**, 127–134.
157. Court, A. (1951). Antarctic atmospheric circulation. *In* "Compendium of Meteorology" (T. F. Malone, ed.), pp. 917–941. Am. Meteorol. Soc., Boston, Massachusetts.
158. Hofmeyr, W. L. (1957). Upper air over the Antarctic. *In* "Meteorology of the Antarctic" (M. P. Van Rooy, ed.), pp. 173–199. Weather Bureau, Pretoria, South Africa.
159. Holcombe, R. M. (1958). Similarities and contrasts between the arctic and antarctic marine climates. *In* "Polar Atmosphere Symposium, Oslo, 1956" (R. C. Sutcliffe, ed.), Pt. 1, pp. 9–17. Pergamon, New York.
160. Krauss, W. (1955). Zum System der Meeresströmungen in den höheren Breiten. *Deut. Hydrog. Z.* **8**, 102–111.
161. Pogosian, C. P. (1958). The singular circulation of the atmosphere in Antarctica (in Russian). *Meteorol. i Gidrol.* **8**, 1–10.
162. Bannon, J. K. (1958). Stratospheric temperatures over the Antarctic. *Quart. J. Roy. Meteorol. Soc.* **84**, 434–436.
163. U.S. Navy Bureau of Aeronautics. (1953). Guide to Arctic Forecasting. Part 1. Aleutian Islands and the Bering Sea area. Part 2. The Canadian Archipelago and the Davis Straits area. (AROWA). NAVAER 50–1P–505, Norfolk, Virginia.
164. U.S. Navy Bureau of Aeronautics. (1953). Arctic Weather Maps (AROWA). Norfolk, Virginia.
165. Estoque, M. A. (1957). Dynamical prediction of the arctic circulation. Scientific Report No. 4, Contract AF 19(604)-1141, Arctic Meteorology Research Group, McGill University, Montreal, 10 pp.
166. Reed, R. J. (1958). A graphical prediction model incorporating a form of non-adiabatic heating. *J. Meteorol.* **15**, 1–8.
167. Flohn, H. (1952). Atmosphärische Zirkulation und erdsmagnetisches Feld. *Ber. Deut. Wetterd. U.S. Zone* **38**, 46–51.
168. Mironovitch, V. (1956). Les pôles de la circulation atmosphérique générale et les pôles magnétiques terrestres. *Beitr. Phys. Atm. S.* **29**, 123–128.
169. Maury, M. F. (1851). On the probable relation between magnetism and the circular tion of the atmosphere. Appendix, 17 pp. Washington Astronomical Observations for 1846.
170. Wexler, H. (1950). Possible effects of ozonosphere heating on sea-level pressure. *J. Meteorol.* **7**, 370–381.
171. Riehl, H. (1956). On the atmospheric circulation at 500 mb in the auroral belt. *J. Geophys. Res.* **61**, 525–534.
172. Wexler, H. (1956). A look at some suggested solar-weather relationships. Tech. Report No. 2. Inst. for Solar-Terrestrial Research, High Altitude Observatory, University of Colorado, pp. 21–30.

PHASE RELATIONS OF SOME ROCKS AND MINERALS AT HIGH TEMPERATURES AND HIGH PRESSURES*

George C. Kennedy

Institute of Geophysics, University of California, Los Angeles, California

1. Introduction

Perhaps no field of research in modern science has been so completely dominated by a single individual as has the field of research in high pressures. Prior to the last war only two laboratories in the United States were actively engaged in research in this field. These were the laboratories of P. W. Bridgman and his colleagues at Harvard University and the laboratory of L. H. Adams and colleagues at the Carnegie Institution of Washington, Washington, D.C. In post-war times a dozen other laboratories have become actively involved in high-pressure research. This review article will deal largely with the post-war high-pressure work of geophysical interest. Bridgman's extensive contributions are summarized in his volume [1]. Almost all of the post-war high-pressure work has been applied research with results primarily of interest to mineralogists and geophysicists, while Bridgman's early research was in the physics of matter at high pressure.

Only work extending to at least 10,000 bars† will be considered in this

*Publication No. 138, Institute of Geophysics, University of California, Los Angeles.

†The results of various laboratories are normally reported in pressure units according to the peculiar prejudice of the investigator. Pressure units in use are kilogram per centimeter, atmosphere, bar, and pounds per square inch. Fortunately a kilogram per square centimeter, an atmosphere, and a bar are all numerically almost the same, particularly when account is taken of the fact that errors in estimating friction and in estimating dilation of parts of apparatus may amount to 5 to 10% of the total pressure value reported. These various units are related exactly as given in Table I.

paper. A great amount of research has been conducted in geochemical laboratories as well as in industrial and physical laboratories in the low-pressure region of a few hundred or a few thousand atmospheres. Pressures of 10,000 bars and beyond require special materials and special techniques which more or less serve as a sharp cutoff between the high-pressure field and the field, now commonplace in industry and university laboratories, of medium to low pressures.

TABLE I. Comparison of various pressure units.

	Bars	Normal atmosphere	kg/cm²	1 lb/in²
1 bar (10^6 dyne/cm²)	1	0.986924	1.019716	14.5038
1 normal atmosphere	1.013249	1	1.033226	14.6960
1 kg/cm²	0.980665	0.967842	1	14.2234
10 lb/in²	0.689474	0.68046	0.70307	10

A surprisingly small number of laboratories, even today, are actively engaged in research at high pressures, in spite of the greatly expanded interest in high pressure and in its application to mineralogical and geophysical problems. Laboratories currently engaged in research are those of Francis Birch at Harvard University, the laboratories of F. R. Boyd and H. Yoder at the Geophysical Laboratory, Carnegie Institution of Washington, the laboratory of Rustum Roy and his colleagues at the Pennsylvania State University, the laboratory of Julian Goldsmith at the University of Chicago. the laborarory of D. T. Griggs and G. C. Kennedy at the Institute of Geophysics, University of California, and the laboratory of Tracy Hall at Brigham Young University, Utah. In addition to these research laboratories, advances have been made in the industrial laboratories of the General Electric Company and L. Coes has stimulated an enormous amount of work from his early results on mineralogical transitions at the laboratories of the Norton Company in Worcester, Massachusetts.

In addition to these groups, using static high pressures, new techniques have recently been developed in which very high transient pressures are obtained by shock waves. The data from these shock wave experiments are becoming of increasing importance to geophysicists and are likely to increase in importance in the future.

Pressures obtainable by shock waves, even though transient, are an order of magnitude higher than those obtained to date by static methods. Most of he shock wave data have stemmed from work sponsored by the U.S. Atomic

Energy Commission. Very important and interesting work using shock waves with exceedingly high transient pressures has been reported recently from the U.S.S.R. Most of the results from shock wave experiments have been reviewed lately by Rice *et al.* [2].

2. Apparatus

2.1. Early Equipment

Three principal types of apparatus are being used in the various high-pressure laboratories. Most of these designs stem back to early designs by Bridgman, although they have been modified and tinkered with almost beyond recognition in some cases.

Much of Bridgman's apparatus cannot be directly converted to the study of geophysical problems. Most of Bridgman's research in the field of high-pressure work was carried out essentially at room temperature or over a small temperature interval. Within this temperature range, steels remain strong and modern heat-treated steels do not loose their temper or properties. Unfortunately, the numbers of most interest to geophysicists and geochemists involve the study of phenomena at temperatures up to 1000°C and more. This means that external heating can no longer be applied to the apparatus and methods must be found of inserting furnaces within the high-pressure chamber. The only exception to this general rule is that of Bridgman's piston-anvil device of Carboloy [3]. Carboloy may be heated to as much as 500°C and still retain the large portion of its compressive strength and hardness.

Two general types of internally heated apparatus are in current use. One type involves gas as a pressure medium, useful up to approximately 25,000 bars where most gases freeze in the pressure line; the other type involves solid pressure media. With the latter, carbon resistance heaters are normally used and the pressure-transmitting media are either pyrophyllite, boron nitride, talc, or some other inorganic substance which retains its plasticity to high temperatures.

Perhaps the earliest internally heated vessel was that designed and built by Smyth and Adams [4] at the Carnegie Instution of Washington. A furnace was placed in a rather large and cumbersome pressure container, the lid of which was held in place by a hundred ton oil ram. The pressure medium used was CO_2 and pressures up to 4 kb were readily obtainable. With this rather cumbersome apparatus, Goranson made the exceedingly important early measurements on the solubility of water in granite melts and in albite melts [5, 6]. A small sample of the mineral under investigation was sealed in platinum or gold envelopes along with a small amount of water. The pressure of the gaseous medium was transmitted to this sample by collapse of the walls of the small envelope until pressure on water and the sample inside the sealed-off envelope was the same as outside gas pressure. The sample was placed in the hot spot of the furnace in the pressure container. Insulated leads carried power and thermocouples through the wall of the bomb to outside measuring devices. This particular apparatus, although not "high pressure" in the sense of the present article, was the forerunner of various internally heated pressure vessels used today.

2.2. High-Pressure Gas Equipment

The first high-pressure design, capable of 10,000 bars with an internal furnace, is that of Birch and Law [7] who modified Bridgman's apparatus. Birch's apparatus was used in determining the elastic constants of rocks at high temperature.

A permutation of Birch's gear was constructed at the Carnegie Institution of Washington by Yoder [8] where a small, platinum wound furnace, capable of higher temperatures, was substituted for the larger furnace used by Birch in heating his rock specimens. Yoder has used this apparatus very extensively in determining phase relations of minerals in equilibrium with water and in melting points of minerals. His pressure medium is argon gas, and the sample, with or without water, is sealed in platinum tubes, the technique developed by Goranson. Pressure is transmitted through the platinum tubing to the water vapor inside.

The major difficulty with both Goranson's, Birch's, and Yoder's apparatus is that the entire furnace assembly must be extracted after each run in order to fetch out the sample and replace it with a new run. This is a very laborious and time-consuming procedure, and pressure packings, frequently disturbed, are troublesome.

Two permutations of this general design, involving a system where little other than the sample is extracted from the furnace with all major packing left in place, have been built by Kennedy, and Griggs. Griggs' design involves a Carboloy rod sealed into the opening by a controlled-clearance packing of the type designed by Newhall [9].

In addition to the intensifier needed to bring the cylinder to pressure, a separate intensifier applies almost an equal pressure to the controlled-clearance packing thus sealing off any leak around the Carboloy rod. When the pressure on the controlled-clearance packing is released, the Carboloy rod may be freely extracted, the sample withdrawn, and a fresh sample inserted. This apparatus is extremely rapid in operation, as there is practically no lost time in interchanging the samples. As many as 15 runs in a single day have been made with this apparatus and scores of runs in sequence have been made without lost time for overhauling packings and repairs of parts (see Fig. 1).

The apparatus designed by Kennedy involves a somewhat different feature. A furnace inside the pressure container is wound on a thick-walled platinum tube, and the sample is carried on the end of a platinum rod into the thick-walled tube, much as in Tuttle's cold-seal apparatus [10]. In this apparatus the sample is not placed in a separate sealed capsule but is acted upon directly by water or any desired mixture of gases. The pressure inside the platinum

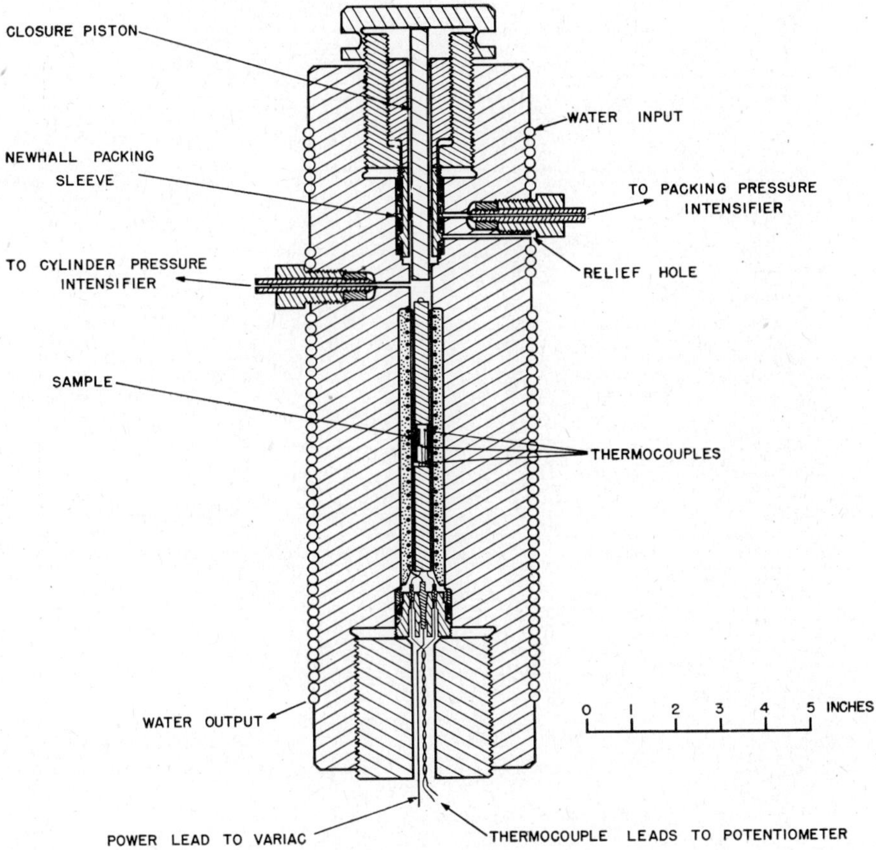

FIG. 1. Controlled-clearance apparatus.

tube is exactly balanced, via a mercury-filled U tube or silicon-filled U tube, with the pressure outside the thick-walled platinum tubing. The sample may be readily extracted by backing off the packing holding the platinum rod in place. This apparatus has the advantage that equilibrium between sample and two or more gases may be studied, such as equilibrium between a silicate and H_2O plus CO_2 in known ratios. However, the major disadvantage of this apparatus is the necessity for rather cumbersome exterior plumbing, a balancing block, and the necessity of keeping track of the position of the meniscus in the balancing block, normally tracked by means of a sensing electrical needle, as illustrated in Fig. 2.

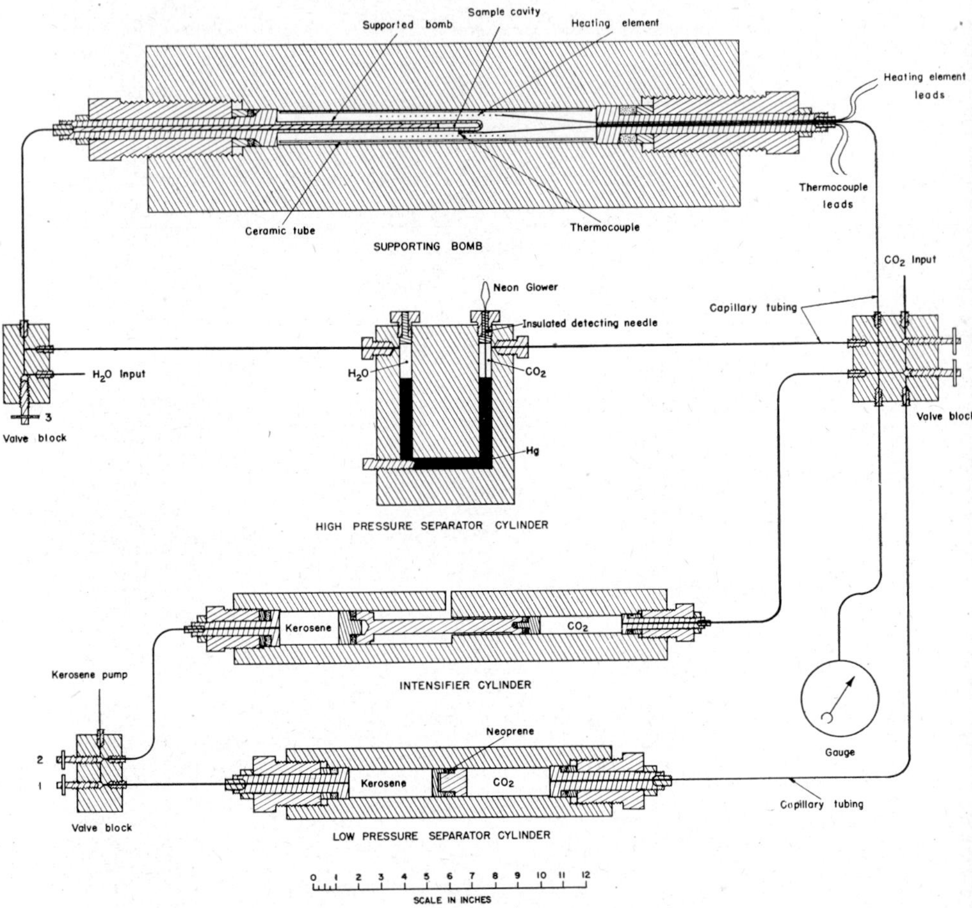

FIG. 2. Diagram of high-temperature–high-pressure apparatus involving external balancing U tube.

These last four related kinds of gas apparatus have a maximum capability of approximately 10,000 bars. Robertson *et al.* [11] have described a multiple-piece gas apparatus, modified from Bridgman, capable of pressures up to 25,000 bars. This apparatus is shown in Fig. 3 and is, in principle, similar to apparatus described by Bridgman [12]. Pressures very much beyond 25,000 bars are almost impossible to obtain in gas apparatus as gases which can be readily utilized as pressure media freeze at pressures much greater than this, and cannot be pumped into pressure cylinders. For higher pressures, solid pressure-transmitting media with radically different designs are required.

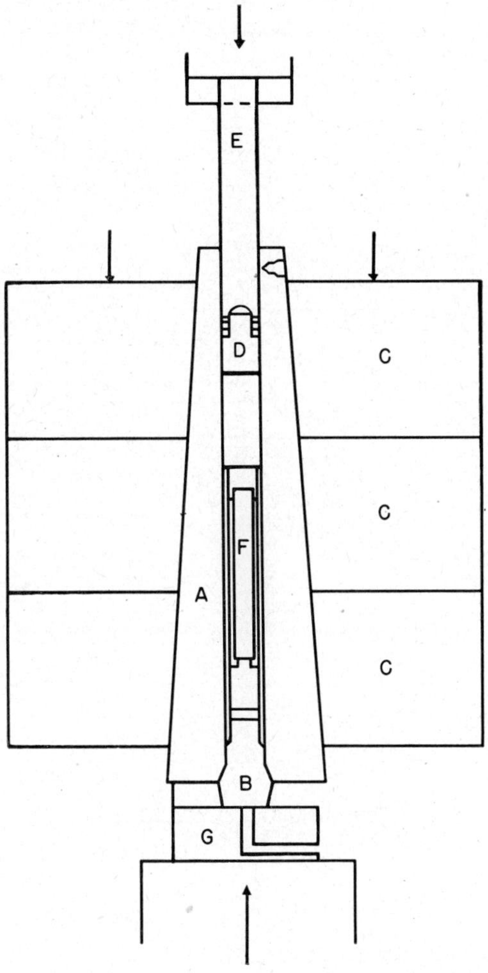

FIG. 3. Internal arrangements of high-pressure system. *A*, high-pressure cylinder; *C*, supporting rings; *B*, lower closure, with electrical leads; *D*, piston mushroom, with packing; *E*, piston; *F*, furnace.

2.3. *Solid High-Pressure Equipment*

The earliest solid-pressure apparatus, capable of high temperatures and high pressures, known to the writer, was that built by Coes [13] at Norton Company. Coes, for the first time, succeeded in synthesizing many of the high pressure phases known to occur in nature. Among the phases made in

Coes' gear were jadeite, kyanite, sillimanite, dense polymorphs of kyanite, a dense polymorph of quartz known as coesite, many of the garnets including the magnesium garnet, pyrope, and diamond (S. S. Kistler, oral communication). Coes was able to use this apparatus successfully at 45 kb and, by virtue of a graphite heating element, to temperatures of more than 1000°C. Coes' synthesis of various high-pressure mineral phases, known in natural deposits, served as a great stimulus to high-pressure investigations. Much of

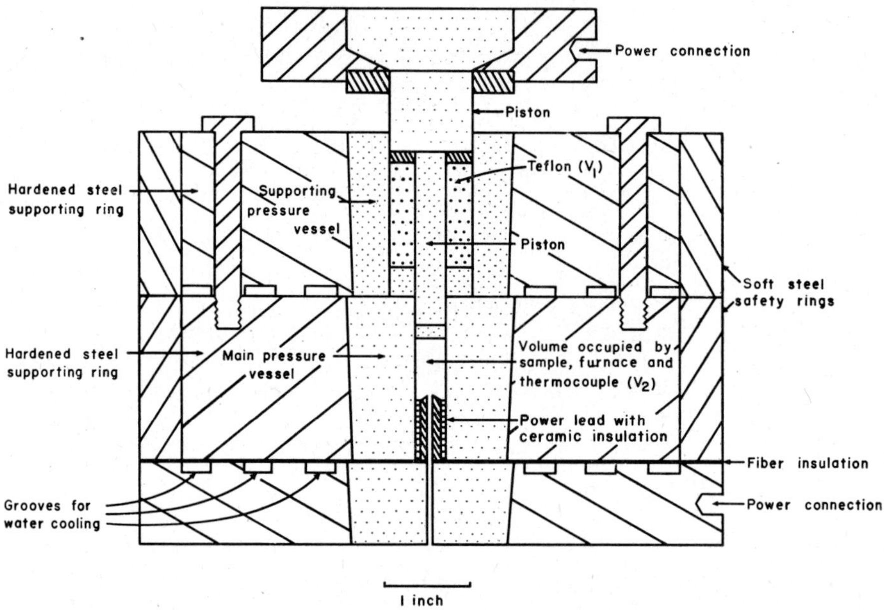

FIG. 4. Apparatus for use in the pressure range over 50 kb. Steel parts are ruled, carbide parts stippled.

the current work in the field of mineral relations at high pressures stems directly from the impetus given to this field by Coes' work at Norton Company. The pressure he achieved was limited only by the strength of the carbide piston used to raise the pressure in the cylinder.

A most spectacular advance has been made in designs for solid-pressure apparatus by Tracy Hall. This investigator was responsible for construction at the General Electric Company of apparatus capable of achieving pressures up to 180,000 bars and of attaining temperatures as high as 2000°C. Hall's apparatus has recently been described [14]. This apparatus involves tapered pistons, supported by pyrophyllite, which thrust into a hole in a carbide block. Successive shrunk-on rings of steel prevent fracture of the carbide. Hall, now at Brigham Young University, has, with great ingenuity, developed another high-pressure device which achieves pressures and temperatures almost comparable to that of the device he developed for diamond synthesis at the General Electric Company. He has utilized the major features of Bridgman's piston-anvil device [15, 16] in his design, but has converted Bridgman's essentially two-dimensional apparatus into a three-dimensional apparatus. Four tetrahedral pistons are simultaneously advanced on a tetrahedron of pyrophyllite containing the sample. Provision is made for an internal furnace. Pressures of 100,000 bars at temperatures of 2000°C can be readily obtained with this gear [17].

Boyd [18] has developed some effective high-pressure apparatus with certain novel design features (see Fig. 4). In Boyd's equipment the carbide piston, which is advanced into the high-pressure chamber, is stiffened and

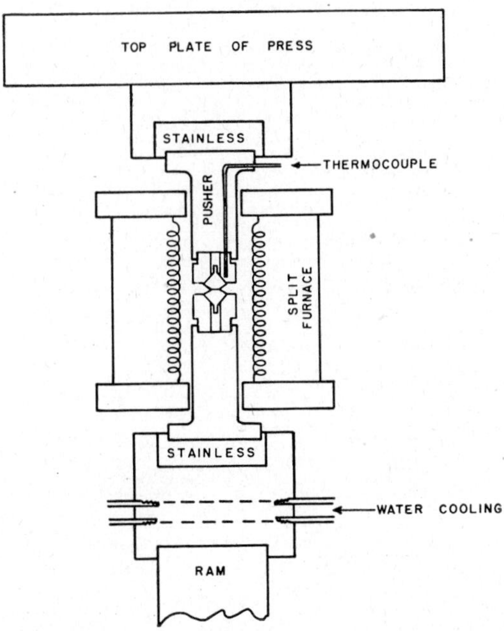

FIG. 5. Bridgman's modified piston-anvil device.

strengthened by being immersed in Teflon at high pressures. This process has been used successfully to 100,000 bars and 1500°C and the design is probably capable of even higher pressures and temperatures.

Griggs and Kennedy [3] describe a simple two-piston high-pressure device. This was a modification of earlier Bridgman apparatus equipped with an external heater. This device, shown in Fig. 5, is simple in construction and operation and has proved useful in mineralogical phase work. Temperatures and pressures obtainable are shown in Fig. 6.

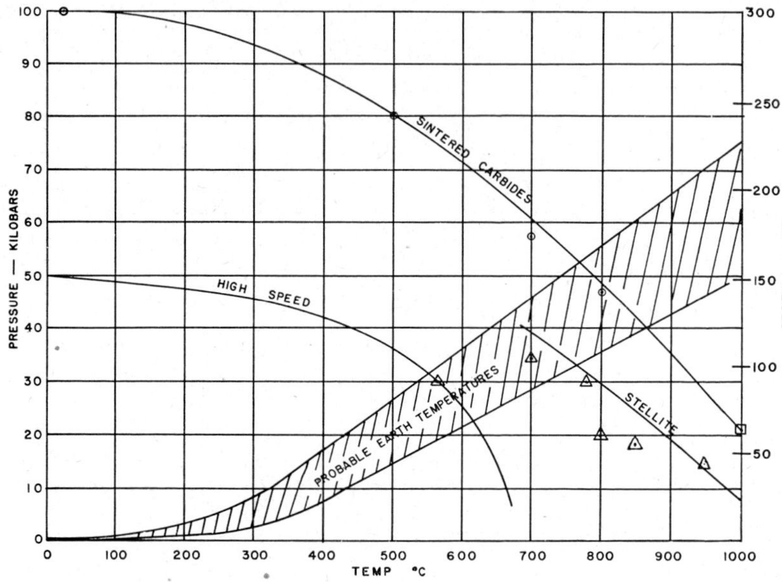

FIG. 6. Pressures that have been attained with various pistons.

3. Results

Four general categories of results of work at high pressures have recently appeared. These all deal with the stability of the mineralogical phases which make up the earth's crust. These results are: melting points of minerals under high confining pressures, melting curves for minerals under high water pressure, vapor pressures of hydrous-anhydrous mineral assemblages at high water pressures, and equilibrium curves for solid-solid phase changes at high temperatures and pressures.

3.1. Melting Points of Minerals under High Confining Pressure

Surprisingly, melting points as a function of pressure have been determined for only a very few minerals. Data are available only for albite, diopside, and iron.

The albite data are from the work of Paul Le Comte in the laboratory of Francis Birch (written communication). These results have been recently confirmed and extended by LaMori and Newton working in the writer's laboratory. Yoder [19] has measured the melting point of diopside to 10 kb. Boyd and England [20] have extended the data to 35 kb.

H. M. Strong of the General Electric laboratories has recently determined the melting point of iron to 96 kb (written communication). These three melting point curves are shown in Fig. 7.

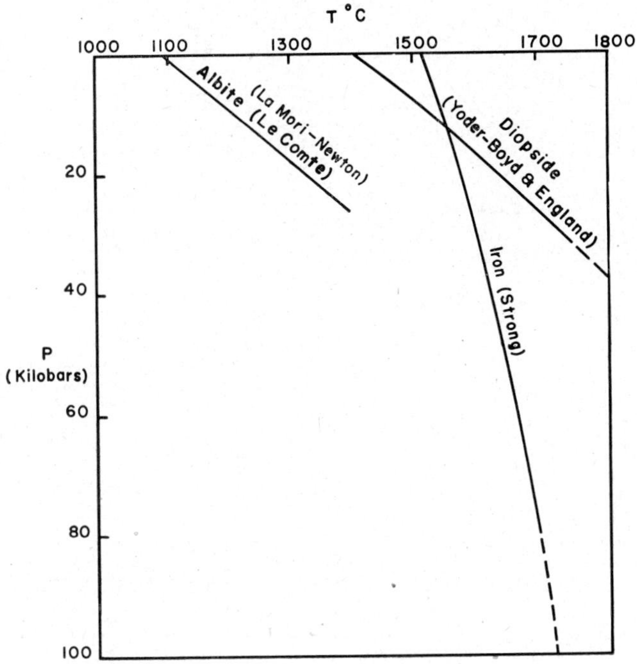

FIG. 7. Melting point curves at high pressure for three minerals.

3.2. Melting Points of Minerals under High Water Pressures

The melting curves of a variety of minerals have been determined under high water pressures. Confining pressure increases the melting temperatures of minerals as seen in Fig. 7 but high water pressure decreases the melting temperature of minerals by virtue of the solubility of water in the melt.

A summary of the available information on melting curves under high water pressure has been published recently by Yoder [21] (see Fig. 8) showing melting curves of (1) SiO_2—H_2O: Tuttle and England [22]; Yoder (unpublished, 1955); Stewart (unpublished, 1957). (2) $CaAl_2Si_2O_8$—H_2O: Yoder (unpublished, 1953); Stewart (unpublished, 1957). (3) $NaAlSiO_4$—H_2O: Yoder (unpublished, 1958). (4) $KAlSi_3O_8$—H_2O: Goranson [6]; Bowen and Tuttle [23]; Yoder, Stewart, and J. R. Smith (unpublished, 1957). (5) $NaAlSi_3O_8$—H_2O: Goranson [6]; Bowen and Tuttle [23]; Yoder (unpublished, 1953); Yoder, Stewart, and J. R. Smith (unpublished, 1957). (6) $CaMgSi_2O_6$—H_2O: Yoder (unpublished, 1953).

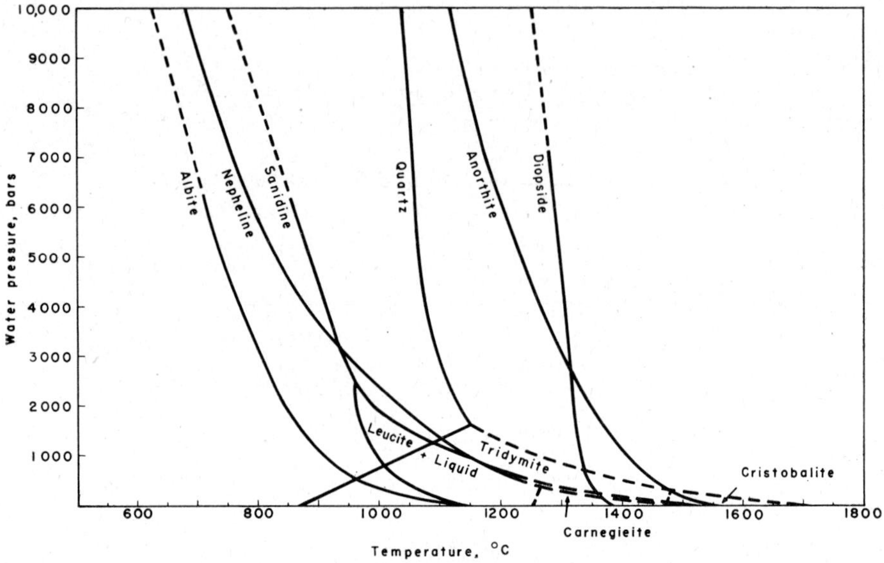

FIG. 8. Melting curves of minerals under high water pressure.

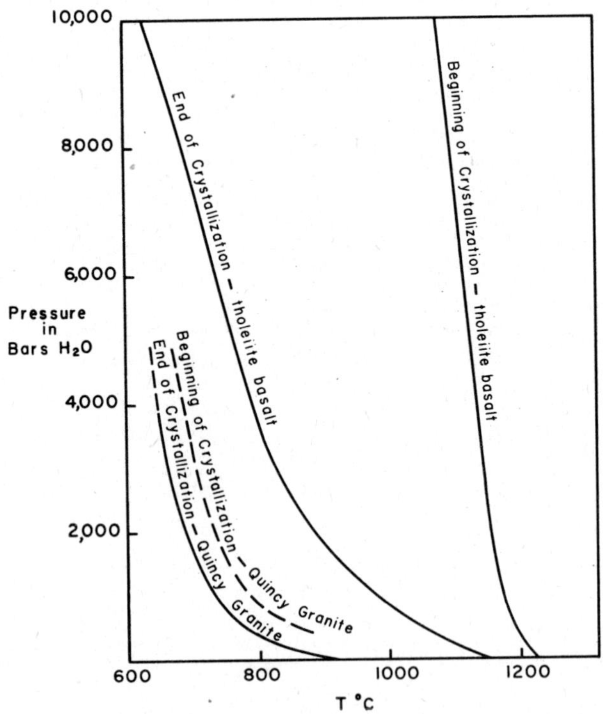

FIG. 9. Melting curves of basalts and granite under high water pressure.

In addition to these melting-point curves for individual minerals the melting of basalts and granite under high water pressure has been investigated. Yoder and Tilley [24] present data on the solidus-liquidus curves for a natural tholeiite basalt from Kilauea, Hawaii (see Fig. 9). Data to 4-kb H_2O pressure are available for granite [25], and the solidus curve is shown in Fig. 9. Unfortunately, detailed data on the liquidus curve for granite are not available. However, some results from the Tuttle–Bowen work indicate the liquidus for granite lies only 30–40° above the solidus and it has been so indicated in Fig. 9. The difference in separation between the solidus-liquidus curves for granite and for basalt emphasize the difference in origin of these two contrasting rock types.

3.3. *Vapor Pressures of Hydrous Phases*

The vapor pressures of a number of hydrous-anhydrous mineral equilibrium assemblages have been determined. These are shown in Fig. 10. Data are

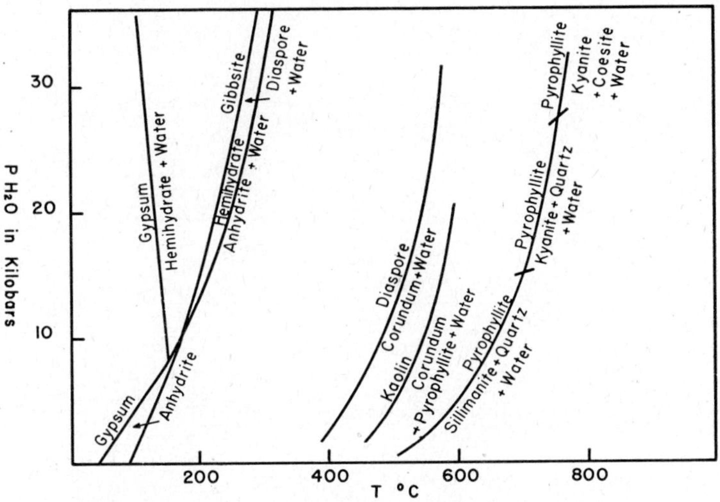

FIG. 10. H_2O vapor pressures of some hydrous minerals.

available for gypsum = anhydrite plus water, gypsum = hemihydrate plus water, and hemihydrate = anhydrite plus water (Kennedy and Lyon, unpublished). Data for aluminum oxide and water are available. The curves shown are those for gibbsite = diaspore plus water and diaspore = corundum plus water (Kennedy, in press, *American Journal of Science*). Composite curves showing the boundary of kaolin = sillimanite plus pyrophyllite plus water, kaolin = kyanite plus pyrophyllite plus water, pyrophyllite = sillimanite plus quartz plus water, pyrophyllite = kyanite plus quartz, water, and pyrophyllite = kyanite plus coesite plus water are shown (Kennedy, unpublished).

3.4. *Solid-Solid Phase Changes*

A large number of solid-solid phase transitions have been investigated at high temperatures and pressures. The calcite-aragonite equilibrium boundary was first determined by Jamieson [26] from measurements of conductivity of saturated solutions of calcite and aragonite. MacDonald's [27] experimental curve (Fig. 11) closely agrees with Jamieson's curve. Work by Clark [28] in a hydrostatic pressure system, contrasting markedly with MacDonald's piston-anvil apparatus, is in essential agreement.

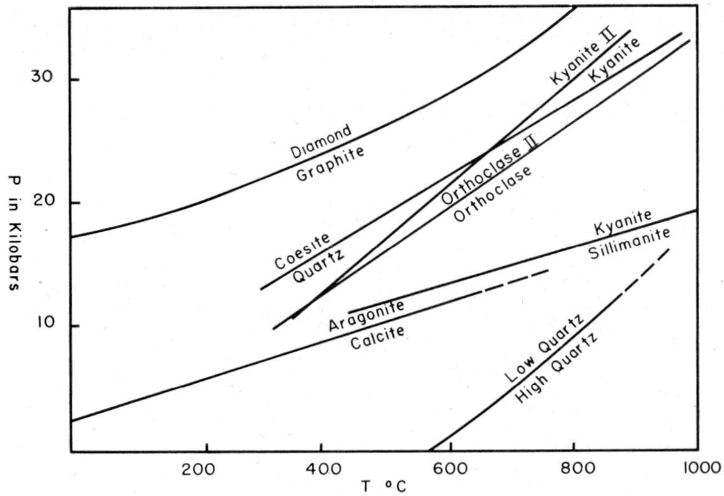

FIG. 11. Some solid-solid phase changes.

The quartz-coesite transition has been determined by MacDonald [29] over the temperature interval 400 to 600°C. MacDonald's curve has been slightly modified and extended by Kennedy (unpublished) and is shown in Fig. 11. Also shown in this figure are orthoclase to orthoclase II and kyanite to kyanite II (Kennedy, unpublished). The high quartz to low quartz boundary is from Yoder [30]. The kyanite-sillimanite boundary has been studied experimentally at high temperatures by Clark *et al.* [31] and by Kennedy [3]. The agreement between the two sets of data is excellent. The graphite-diamond curve is drawn from the various theoretical studies from thermochemical data by Goranson, Liljeblad, Berman, Simon, and MacDonald and has been confirmed by the experimental work of Bovenkerk, *et al.* [14] and Bridgman [32].

Several other important phase changes in common minerals are known to exist but have not yet been studied in sufficient detail. Ringwood [33] reports a transition in fayalite from orthorhombic to spinel structure taking place at 400°C and 45 kb. Unfortunately the slope of the phase boundary is not known.

Phase changes have been sought to 50 kb and 500°C but not found in dolomite and enstatite.

In addition to simple solid-solid phase transitions much more complex changes at high pressures take place in certain systems. Figure 12 shows the relationships between grossularite and hydrogrossularite, with the assemblage anorthite and wollastonite (Pistorius and Kennedy, unpublished). The amount of quartz involved in the reaction changes regularly as temperature and pressure change.

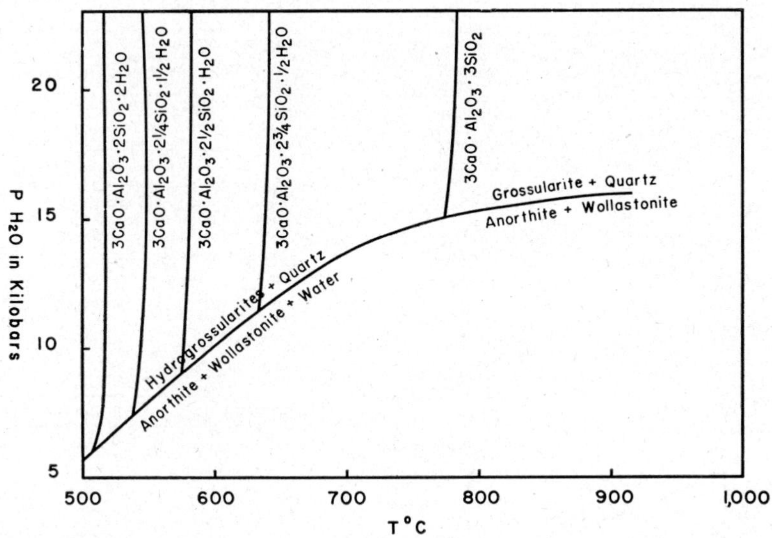

FIG. 12. Grossularite-hydrogrossularite relations at high temperatures and water pressures.

A related diagram from the system $CaO—Al_2O_3—SiO_2$ is shown in Fig. 13. The relations between the phases anorthite, zoisite, and lawsonite are shown as functions of temperature and H_2O pressure. The hydrates, zoisite and lawsonite, are of particular interest as their stability fields are dominantly controlled by H_2O pressure, not by temperature, thus these phases serve as sensitive indicators of water pressure in the earth's crust (Pistorius and Kennedy, preliminary diagram, unpublished).

In several systems, composition of the solid phase changes as pressure on the system changes. Boyd and England [34] have published a phase diagram for nepheline ($Na_2O . Al_2O_3. 2SiO_2$) (see Fig. 14). At pressures above 9 kb a jadeite-like phase begins to exsolve from the nepheline and increases in amount as pressure is raised. This suggests the nepheline breaks into two phases one richer in silica than nepheline and the other poorer in silica. In addition, nepheline undergoes a solid-solid transition in part of the P–T field to nepheline II.

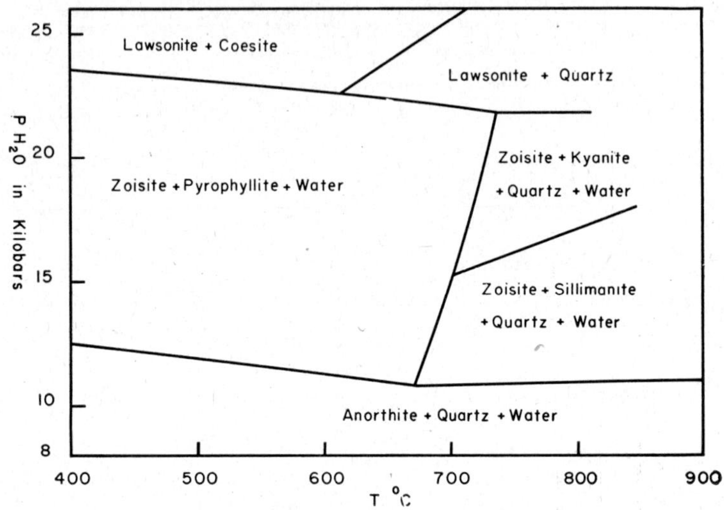

FIG. 13. Phase relationships between lawsonite, zoisite, and anorthite.

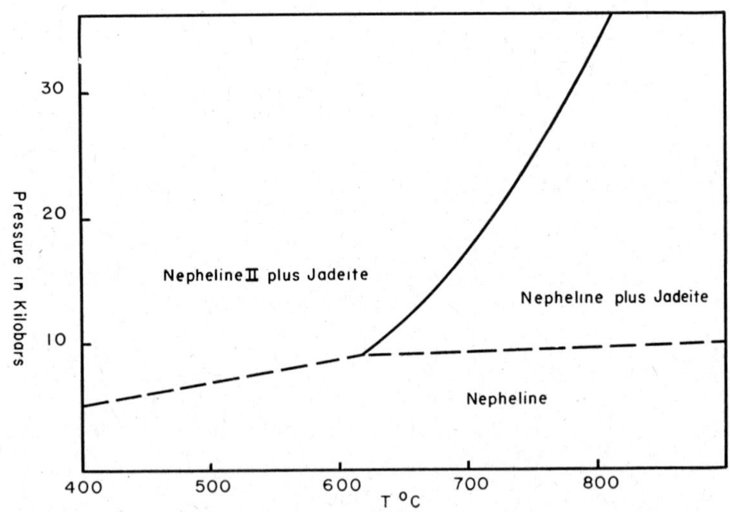

FIG. 14. Pressure-temperature diagram for the composition $Na_2O \cdot Al_2O_3 \cdot 2SiO_2$

Further interesting examples of pressure exsolution of a dense phase are found in the system involving nepheline ($Na_2O \cdot Al_2O_3 \cdot 2SiO_2$), albite ($Na_2O \cdot Al_2O_3 \cdot 6SiO_2$), and jadeite ($Na_2O \cdot Al_2O_3 \cdot 4SiO_2$). In the "wet" system where pressure is pressure H_2O (Kennedy, unpublished), melting relations mask the reaction nepheline plus albite equal 2 jade (see Fig. 15).

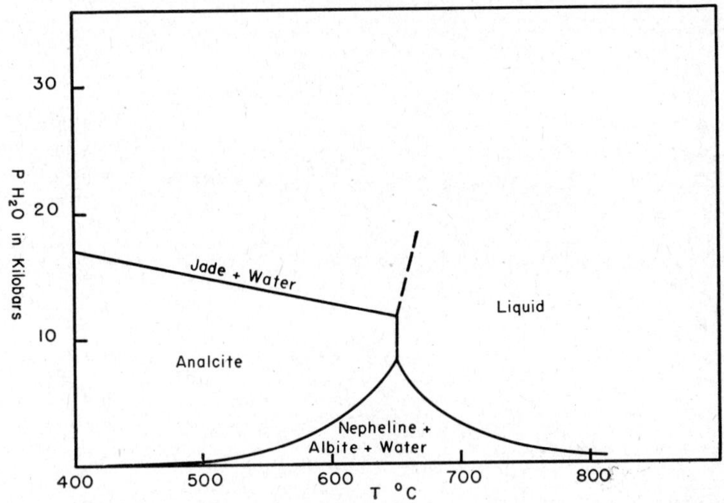

FIG. 15. H_2O pressure-temperature diagram for the composition $Na_2O \cdot Al_2O_3 \cdot 4SiO_2$.

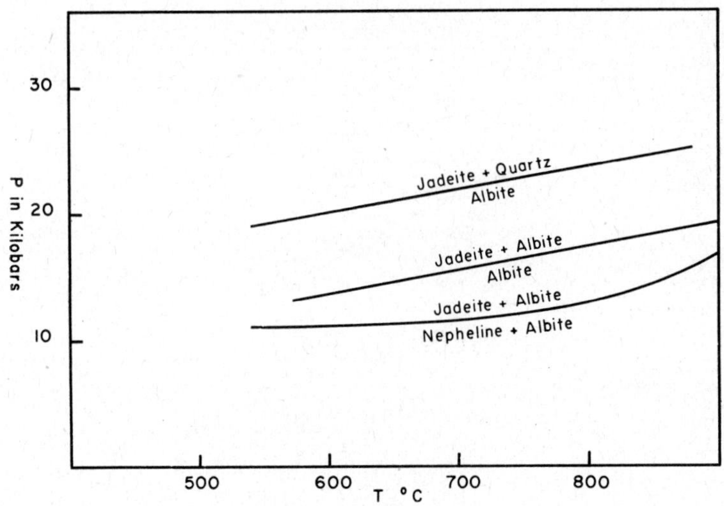

FIG. 16. Phase relations between nepheline, albite, and jadeite.

In the dry system, melting relations do not mask the phase changes involving nepheline, albite, and jadeite. Relations between these phases, as worked out in the piston-anvil device, are shown in Fig. 16 (Kennedy, unpublished).

Nepheline and albite are mutually soluble at high temperatures and high pressures. Nepheline, saturated with albite and thus containing more silica than stoichiometric nepheline, undergoes a phase change to silica-deficient jadeite with albite remaining unchanged. Albite dissolves in the jadeite as pressure is raised. Pure albite exsolves jade along the boundary as indicated in Fig. 16 and at a somewhat higher pressure, breaks down completely to jadeite plus quartz. The relations between jadeite and nepheline plus albite as shown in this figure contrast markedly with the phase diagram as published by Robertson *et al.* [11], but the actual experimentally determined points as presented by Robertson *et al.* disagree only slightly with the diagram presented here.

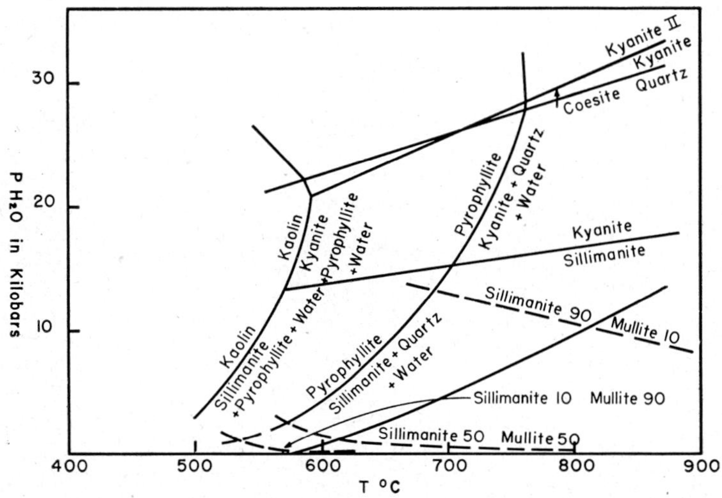

FIG. 17. The system $Al_2O_3—SiO_2—H_2O$ at high pressures and temperatures.

Pressure solid-solution relations also shown in the system $Al_2O_3 . SiO_2 . H_2O$ (Fig. 17) (Kennedy, unpublished), established by the breakdown of kaolinite at high temperature and pressures, show stability fields of various phases in the system. The field of andalusite was not determined. Of particular interest is the solid-solution relationship shown between mullite ($3Al_2O_3 . 2SiO_2$) and sillimanite ($Al_2O_3 . SiO_2$). Pure mullite was synthesized by breaking down kaolin at very low pressures. At 1-kb H_2O pressure, half-breed compounds form of approximate composition sillimanite 50 %-mullite 50 %. Only at pressures of approximately 13 kb, along the pyrophyllite boundary, were compositions of 90 % sillimanite-10 % mullite formed, and pure 100 % sillimanites were formed only at still higher pressures. These relationships are most puzzling and seem to be at odds with the compositions of naturally occurring minerals. Several dozen naturally occurring sillimanites all showed

100 % sillimanite composition, no half-breed compositions are known from natural occurrences. It is clear that the last word has not yet been said about the relationship of sillimanite to mullite and further investigations are needed.

4. Conclusion—with Words of Warning

It is very difficult to achieve equilibrium between condensed silicate phases, particularly in the low-temperature and pressure region. Many of the diagrams presented in the figures associated with this paper are "preliminary" in the sense that conclusive evidences of equilibrium have not been obtained in many cases, i.e., the reverse reactions at the field boundaries have not always been demonstrated. The general relationships are believed to be correct, but it is certain that refined work by different experimental techniques may change the field boundaries somewhat.

The most startling thing from the geological point of view is that high-pressure polymorphism of common silicate phases known at the surface of the earth appears to be the rule rather than the exception. In general, however, the depth at which common polymorphs are formed appears to be greater than current geological thinking would admit. Of particular pertinence are the considerable depths in the earth's crust needed to form the dense phases grossularite, zoisite, lawsonite, kyanite, and jadeite. It must be remembered that "impurities" in the crystal such as iron in grossularite, zoisite, and jadeite, drastically modify the stability fields of these minerals so that they are stable at much lower pressures than those indicated here.

References

1. Bridgman, P. W. (1952). "The Physics of High Pressures," 445 pp. G. Bell and Sons, London.
2. Rice, M. H., McQueen, R. G., and Walsh, J. M. (1958). Compression of solids by strong shock waves *Solid State Phys.* **6,** 1–63.
3. Griggs, D. T., and Kennedy, G. C. (1956). A simple apparatus for high temperatures and pressure. *Am. J. Sci.* **254,** 722–735.
4. Smyth, F. H., and Adams, L. H. (1923). The system calcium oxide-carbon dioxide. *J. Am. Chem. Soc.* **45,** 1167–1184.
5. Goranson, R. W. (1931). The solubility of water in granite magmas, *Am. J. Sci.* **22,** 481–502.
6. Goranson, R. W. (1936). The solubility of water in albite-melt. *Trans. Am. Geophys. Un. 17th Ann. Meeting Rep. and Pap. Volcanology.*
7. Birch, F., and Law, R. (1935). Measurement of compressability at high pressures and high temperatures. *Bull. Geol. Soc. Am.* **46,** 1219–1250.
8. Yoder, H. S., Jr. (1950). High-low quartz inversion up to 10,000 bars. *Trans. Am. Geophys. Un.* **31,** 827–835.
9. Johnson, D. P., and Newhall, D. H. (1953). The piston gage as a precise pressure measuring instrument. *Trans. Am. Soc. Mech. Engrs.* **75,** 301–310.
10. Tuttle, O. F. (1949). Two pressure vessels for silicate-water studies. *Bull. Geol. Soc. Am.* **60,** 1727–1729.

11. Robertson, E. C., Birch, F., and MacDonald, G. J. F. (1957). Experimental determination of jadeite stability relations to 25,000 bars. *Am. J. Sci.* **255,** 115–137.
12. Bridgman, P. W. (1930). The resistance of nineteen metals to 30,000 Kg. cm². *Proc. Am. Acad. Arts Sci.* **72,** 157–202.
13. Coes, L., Jr. (1955). High pressure minerals. *J. Am. Ceram. Soc.* **38,** 298.
14. Bovenkerk, H. P., Bundy, F. P., Hall, H. T., Strong, H. M., and Wentorf, R. H. (1959). Preparation of diamond. *Nature* **184,** 1094–1098.
15. Bridgman, P. W. (1935). Effects of high shearing stress combined with high hydrostatic pressure. *Phys. Rev.* **48,** 825–847.
16. Bridgman, P. W. (1937). Shearing phenomena at high pressures, particularly in inorganic compounds. *Proc. Am. Acad. Arts Sci.* **71,** 387–460.
17. Hall, H. T. (1958). Some high pressure, high temperatures design considerations: Equipment for use at 100,000 atmospheres and 3,000°C. *Rev. Sci. Instrum.* **29,** 267–275.
18. Boyd, F. R., and England, J. L. (1958). Development of high pressure apparatus. *Carnegie Inst. Wash. Year Book* **57,** 170–173.
19. Yoder, H. S., Jr. (1952). Change of melting point of diopside with pressure. *J. Geol.* **60,** 364–374.
20. Boyd, F. R., and England, J. L. (1958). Melting point of diopside under high pressure. *Carnegie Inst. Wash. Year Book* **57,** 173–174.
21. Yoder, H. S. (1958). Effect of water on the melting of silicates. *Carnegie Inst. Wash. Year Book* **57,** 189–191.
22. Tuttle, O. F., and England, J. L. (1955). Preliminary report on the system $SiO_2—H_2O$ *Bull. Geol. Soc. Am.* **66,** 149–152.
23. Bowen, N. L., and Tuttle, O. F., (1950). The system $NaAlSi_3O_8—KalSi_3O_8—H_2O$ *J. Geol.* **58,** 489–511.
24. Yoder, H. S., and Tilley, C. E. (1956), Natural tholeiite basalt—water system *Carnegie Inst. Wash. Year Book* **55,** 169–170.
25. Tuttle, O. F., and Bowen, N. L. (1958). The origin of granite in the light of experimental studies in the system $NaAlSi_3O_8—KAlSi_3O_8—SiO_2—H_2O$. *Geol. Soc. Am. Mem.* **74,** 83.
26. Jamieson, J. C. (1953). Phase equilibrium in the system calcite-aragonite. *J. Chem. Phys.* **21,** 1385–1390.
27. MacDonald, G. J. F. (1956). Experimental determination of calcite-aragonite equilibrium relations at elevated temperatures and pressures. *Am. Mineral.* **41,** 744–756.
28. Clark, S. P., Jr. (1957). A note on calcite-aragonite equilibrium. *Am. Mineral.* **42,** 564–566.
29. MacDonald, G. J. F. (1956). Quartz-coesite stability relations at high temperatures and pressures. *Am. J. Sci.* **254,** 713–721.
30. Yoder, H. S., Jr. (1950). High-low quartz inversion up to 10,000 bars. *Trans. Am. Geophys. Un.* **31,** 827–835.
31. Clark, S. P., Robertson, E. C., and Birch, F. (1957). Experimental determination of kyanite-sillimanite equilibrium relations at high temperatures and pressures. *Am. J. Sci.* **255,** 628–640.
32. Bridgman, P. W. (1947). An experimental contribution to the diamond synthesis. *J. Chem. Phys.* **15,** 92–98.
33. Ringwood, A. E. (1958). Olivine-spinel transition in fayalite. *Bull. Geol. Soc. Am.* **69,** 129–130.
34. Boyd, F. R., and England, J. L. (1956). Phase equilibrium at high pressures. *Carnegie Inst. Wash. Year Book* **55,** 154–157.

AUTHOR INDEX

Numbers in parentheses are reference numbers and are included to assist in locating references in which the authors' names are not mentioned in the text. Numbers in italics indicate the page on which the reference is listed.

SUBJECT INDEX

R

S

T

W